THE SQUARE KILOMETRE ARRAY:
AN ENGINEERING PERSPECTIVE

Edited by:

PETER J. HALL
*International SKA Project Office,
Dwingeloo, The Netherlands*

Reprinted from *Experimental Astronomy*
Volume 17, Nos. 1–3, 2004

Library of Congress Cataloging-in-Publication Data is available

ISBN 1-4020-3797-x
1-4020-3798-8

Published by Springer,
P.O. Box 17, 3300 AA Dordrecht, The Netherlands.

The selection of an SKA site requires comprehensive characterization of the radio-frequency environment at candidate locations. While the SKA will feature advanced radio-frequency interference (RFI) mitigation, the science goals also demand a radio-quiet setting. As part of an international engineering collaboration, a specialist team from The Netherlands Foundation for Research in Astronomy (ASTRON) is working with local engineers in Argentina, Australia, China and South Africa to investigate candidate sites. The picture on the cover shows the facility established by ASTRON and South African engineers at the remote K3 candidate site, in the Karoo wilderness of South Africa. Similar camps will be set-up for measurements in the other countries during 2005. Photo credit: Rob Millenaar, ASTRON

Printed on acid-free paper

All Rights Reserved
© Springer 2005
No part of the material protected by this copyright notice may be reproduced or utilized in any form or by any means, electronic or mechanical, including photocopying, recording or by any information storage and retrieval system, without written permission from the copyright owner.

Printed in the Netherlands

TABLE OF CONTENTS

Richard Schilizzi / Foreword   1

Peter Hall / Introduction   3

P.J. Hall / The Square Kilometre Array: An International Engineering Perspective   5–16

SKA DEMONSTRATORS AND RESULTS

David Deboer, Rob Ackermann, Leo Blitz, Douglas Bock, Geoffrey Bower, Michael Davis, John Dreher, Greg Engargiola, Matt Fleming, Girmay-Keleta, Gerry Harp, John Lugten, Jill Tarter, Doug Thornton, Niklas Wadefalk, Sander Weinreb and William J. Welch / The Allen Telescope Array   19–34

Stefan J. Wijnholds, Jaap D. Bregman and Albert-Jan Boonstra / Sky Noise Limited Snapshot Imaging in the Presence of RFI with LOFAR's Initial Test Station   35–42

Kjeld van der Schaaf, Chris Broekema, Ger Van Diepen and Ellen van Meijeren / The LOFAR Central Processing Facility Architecture   43–58

Stefan J. Wijnholds, A. ger de Bruyn, Jaap D. Bregman and Jan-Geralt bij de Vaate / Hemispheric Imaging of Galactic Neutral Hydrogen with a Phased Array Antenna System   59–64

A. van Ardenne, P.N. Wilkinson, P.D. Patel and J.G. bij de Vaate / Electronic Multi-Beam Radio Astronomy Concept: Embrace a Demonstrator for the European SKA Program   65–77

Ray Norris / The Australian SKA New Technology Demonstrator Program   79–85

ANTENNAS

E.E.M. Woestenburg and J.C. Kuenen / Low Noise Performance Perspectives of Wideband Aperture Phased Arrays   89–99

W.A. van Cappellen, J.D. Bregman and M.J. Arts / Effective Sensitivity of a Non-Uniform Phased Array of Short Dipoles   101–109

Germán Cortés-Medellín / Low Frequency End Performance of a Symmetrical Configuration Antenna for the Square Kilometre Array (SKA)   111–118

Roger Schultz / Radio Astronomy Antennas by the Thousands   119–139

John S. Kot, Richard Donelson, Nasiha Nikolic, Doug Hayman, Mike O'Shea and Gary Peeters / A Spherical Lens for the SKA   141–148

Marianna V. Ivashina, Jan Simons and Jan Geralt bij de Vaate / Efficiency Analysis of Focal Plane Arrays in Deep Dishes   149–162

Meyer Nahon, Casey Lambert, Dean Chalmers and Wen Bo / Model Validation and Performance Evaluation for the Multi-Tethered Aerostat Subsystem of the Large Adaptive Reflector   163–175

Wenbai Zhu, Rendong Nan and Gexue Ren / Modeling of a Feed Support System for Fast   177-184

John D. Bunton / Cylindrical Reflectors   185–189

RF SYSTEMS

J. Bardin, S. Weinreb and D. Bagri / Local Oscillator Distribution Using a Geostationary Satellite   193-199

Suzy A. Jackson / RF Design of a Wideband CMOS Integrated Receiver for Phased Array Applications   201–210

DATA TRANSPORT

D.H.P. Maat and G.W. Kant / Fiber Optic Network Technology for Distributed Long Baseline Radio Telescopes   213–220

Ralph Spencer, Roshene McCool, Bryan Anderson, Dave Brown and Mike Bentley / ALMA and e-MERLIN Data Transmission Systems: Lessons for SKA   221–228

SIGNAL PROCESSING

Alle-Jan van der Veen, Amir Leshem and Albert-Jan Boonstra / Array Signal Processing for Radio Astronomy   231–249

John D. Bunton / SKA Correlator Advances   251-259

Steven W. Ellingson / RFI Mitigation and the SKA   261–267

R.H. Ferris and S.J. Saunders / A 256 MHz Bandwidth Baseband Receiver/Spectrometer — 269–277

Timothy J. Adams, John D. Bunton and Michael J. Kesteven / The Square Kilometre Array Molonglo Prototype (SKAMP) Correlator — 279–285

Kjeld van der Schaaf and Ruud Overeem / COTS Correlator Platform — 287–297

John D. Bunton and Robert Navarro / DSN Deep-Space Array-Based Network Beamformer — 299–305

DATA PROCESSING AND SOFTWARE

T.J. Cornwell and B.E. Glendenning / Software Development for the Square Kilometre Array — 309–315

A.J. Kemball and T.J. Cornwell / A Simple Model of Software Costs for the Square Kilometre Array — 317–327

T.J. Cornwell / SKA and EVLA Computing Costs for Wide Field Imaging — 329–343

Colin J. Lonsdale, Sheperd S. Doeleman and Divya Oberoi / Efficient Imaging Strategies for Next-Generation Radio Arrays — 345–362

SYSTEM ISSUES AND MISCELLANEOUS

Jaap D. Bregman / System Optimisation of Multi-Beam Aperture Synthesis Arrays for Survey Performance — 365–380

John D. Bunton and Stuart G. Hay / SKA Cost Model for Wide Field-of-View Options — 381–405

Jaap D. Bregman / Cost Effective Frequency Ranges for Multi-Beam Dishes, Cylinders, Aperture Arrays, and Hybrids — 407–416

John D. Bunton and T. Joseph W. Lazio / Cylinder – Small Dish Hybrid for the SKA — 417–422

B. Peng, J.M. Sun, H.Y. Zhang, T.Y. Piao, J.Q. Li, L. Lei, T. Luo, D.H. Li, Y.J. Zheng and R. Nan / RFI Test Observations at a Candidate SKA Site in China — 423–430

FOREWORD

The Square Kilometre Array (SKA) Project is a global project to design and construct a revolutionary new radio telescope with of order 1 million square meters of collecting area in the wavelength range from 3 m to 1 cm. It will have two orders of magnitude greater sensitivity than current telescopes and an unprecedented large instantaneous field-of-view. These capabilities will ensure the SKA will play a leading role in solving the major astrophysical and cosmological questions of the day (see the science case at www.skatelescope.org/pages/page_astronom.htm). The SKA will complement major ground- and space-based astronomical facilities under construction or planned in other parts of the electromagnetic spectrum (e.g. ALMA, JWST, ELT, XEUS, ...).

The current schedule for the SKA foresees a decision on the SKA site in 2006, a decision on the design concept in 2009, construction of the first phase (international pathfinder) from 2010 to 2013, and construction of the full array from 2014 to 2020. The cost is estimated to be about 1000 M€.

The SKA Project currently involves 45 institutes in 17 countries, many of which are involved in nationally- or regionally-funded state-of-the-art technical developments being pursued ahead of the 2009 selection of design concept. This Special Issue of Experimental Astronomy provides a snapshot of SKA engineering activity around the world, and is based on presentations made at the SKA meeting in Penticton, BC, Canada in July 2004. Topics covered include antenna concepts, software, signal transport and processing, radio frequency interference mitigation, and reports on related technologies in other radio telescopes now under construction.

Further information on the project can be found at www.skatelescope.org.

RICHARD SCHILIZZI
International SKA Project Director

INTRODUCTION

This SKA special issue of *Experimental Astronomy* is a snapshot of key engineering activities, ranging from antenna prototyping to software design. The emphasis which the SKA community places on prototyping and demonstration is apparent from the mix of included papers. Excellent early results from large endeavours such as the Allen Telescope Array and LOFAR underline the value of this emphasis but there are many additional encouraging reports from other projects. Most deal with crucial sub-systems and components, common to all SKA concepts. A series of whitepapers over recent years has illuminated important SKA design issues and, in this tradition, a feature of this volume is the inclusion of the first substantial papers dealing with SKA data processing and software.

I hope that you enjoy this summary of SKA engineering and related activities, and I thank authors, reviewers and *Experimental Astronomy* staff for their assistance in producing the volume. As Richard Schilizzi mentions in his Foreword, more information about the SKA is available at www.skatelescope.org. Readers interested in additional aspects of the project engineering, including costing exercises and industrial liaison, will find a growing list of material on the web site.

PETER HALL
International SKA Project Engineer and Guest Editor

THE SQUARE KILOMETRE ARRAY: AN INTERNATIONAL ENGINEERING PERSPECTIVE

P. J. HALL

International SKA Project Office, P.O. Box 2, 7990 AA Dwingeloo, The Netherlands
(E-mail: hall@skatelescope.org)

(Received 2 August 2004; accepted 4 October 2004)

Abstract. The pace of the international Square Kilometre Array (SKA) project is accelerating, with major concept reviews recently completed and a number of technology demonstrators well underway. First-round submissions to host the telescope were lodged by six countries. The SKA timeline currently shows a site decision in 2006, and one or more technology concepts chosen in 2008. The telescope is expected to be operational, in various phases, in the period 2015–2020. This paper gives a status review of the project, and outlines engineering concept development and demonstration projects.

Keywords: aperture synthesis, interferometry, international science projects, next-generation telescopes, novel antennas, radio telescopes

1. Introduction

The SKA radio telescope project is an international endeavour to build an aperture synthesis radio telescope with one million square metres of effective collecting area, operating the range 100 MHz to 25 GHz (Schilizzi, 2004). A major target of the project is to achieve a sensitivity gain of 100 relative to present-day radio interferometers. This is expressed in a key specification (Jones, 2004): over a wide frequency range the sensitivity metric, $A_{\text{eff}}/T_{\text{sys}}$, should be of order 20000. Here, A_{eff} is the effective collecting area (m^2) and T_{sys} is the system equivalent noise temperature (kelvin). With a canonical 50 K system temperature, the required A_{eff} is 10^6 m^2, or 1 km^2. While half the collecting area will be located in a central region of $\sim$5 km diameter, the full array will extend across trans-continental distances (Figure 1).

The SKA project rose to prominence in the late 1990s with the formation of the International SKA Steering Committee, an overseeing body currently consisting of representatives from 17 countries. The estimated construction budget is USD 1B (year 2004 dollars), a figure demanding many new technology developments in order to yield a cost per unit collecting area of one-tenth that of existing radio telescopes. An International SKA Project Office (ISPO) is now functional and the first International Director and Project Engineer commenced appointments in 2003 and 2004, respectively.

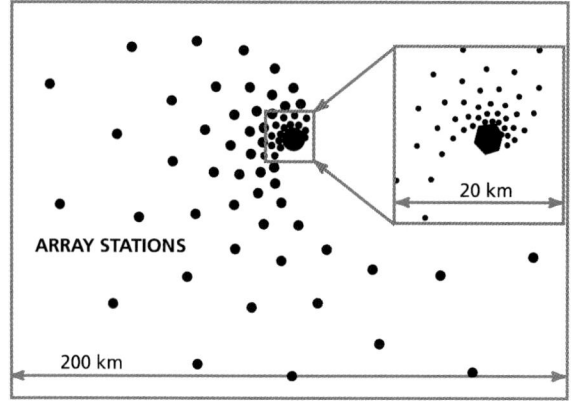

Figure 1. Example of a possible SKA configuration. In this model, patches of collecting area (stations) extend from a dense core in a log-spiral arrangement. The pattern continues to baselines of ~3000 km, with about 10% of the total collecting area being outside the scale depicted.

A recent series of whitepapers (ISPO Concept Whitepapers, 2003), or end-to-end descriptions of potential SKA designs, have proved invaluable in promoting science and engineering discussion, identifying areas in which there are deficiencies in knowledge or specification clarity, and stimulating new studies – including simulation of performance and cost tradeoffs. In effect, the whitepapers are slices through a complex problem and solution space, the sample solutions being used to illuminate critical issues and provoke still more imaginative designs. An updated science case for the SKA (Carilli and Rawlings, 2004) identifies five key areas of astronomy and cosmology, with the original driver – the evolution of structure in the primordial Universe – still figuring prominently.

As well as the many technology development projects underway, an important additional aspect of SKA engineering deals with site infrastructure design and costing. Initial siting proposals were received from Argentina, Australia, Brazil, China, South Africa and the USA (ISPO Site Whitepapers, 2003). A first costing study (Hall, 2003a) puts the infrastructure value of the project at around USD 250M, including a custom optical fibre communications network for at least the central array. Regardless of the site chosen there will be significant infrastructure challenges in areas such as remote power provision, active and passive environmental conditioning, and low-cost access road construction.

A decision on SKA siting is scheduled for 2006, with preceding measurements of radio-frequency (RF) interference at candidate locations being made both by site proponents and the ISPO, the latter via a contract with ASTRON, the Dutch national radio astronomy organization. These technical efforts provide a snapshot of the current RF environments and a parallel part of the site proposal process involves proponents examining the feasibility of establishing a radio-quiet zone for the central part of the SKA; such a zone would provide long-term interference protection, especially for the SKA's lower-frequency operations. It is expected that tropospheric attenuation and stability measurements will be made at sites short-listed on the basis of the RF environment data. This will ensure that acceptable performance is also possible at the highest operating frequencies of the SKA.

2. Whitepapers and convergence

One of the main outcomes of the whitepaper preparation and subsequent review processes has been the recognition that no one concept meets all the SKA performance goals. In particular, cost-effective designs meeting the high-frequency sensitivity requirements optimize T_{sys} at the expense of A_{eff}, giving inadequate sensitivity at low frequencies, where T_{sys} is dominated not by receiver contributions but by cosmic radiation from the Milky Way galaxy. Furthermore, the high frequency designs do not provide the independent multi-fielding (or area re-use) capability which is now established as an SKA target, at least at the low frequency end of the Telescope's operating band.

Despite the realization that different low and high frequency solutions may be required, the whitepapers did establish the commonality of a large amount of the SKA system design, independent of the selected antenna or associated RF technology. A design convergence process is therefore underway, involving astronomers and engineers working iteratively to explore the assumption that a hybrid – or composite – telescope will give a better match to the science goals. In this model, at least two antenna solutions might share common sites, signal transport and back-end infrastructure. In the end though, the costs incurred in designing, implementing and operating a composite instrument will need to be weighed against scientific gains. The choice of either a hybrid, or a "best fit" single concept, will involve intense astronomy and engineering interaction, and much debate centered on prioritizing key science goals.

While the hybrid model looks promising, its elements will most likely come from technologies described in the whitepapers. In parallel with the convergence work, a number of engineering groups are therefore working to validate pivotal technologies by means of demonstrators, of various scales. The intention is to have critical technology reviews of these demonstrators form the basis of a 2008 concept selection process.

The SKA concept whitepapers have been successful in setting out possible implementations of the Telescope but the diverse origins of the documents means that a variety of design and costing assumptions currently exist, making it difficult to directly compare designs. A cost/performance estimation project now underway builds on an emerging SKA system definition which recognizes the commonality in all concepts. The estimation tools will help compare concepts on the basis of agreed cost and performance assumptions, and via common assessment metrics (Horiuchi et al., 2004).

3. SKA concepts and demonstrators

Figure 2 is a diagrammatic view of the SKA, with typical data rates shown for the outer and inner parts of the array. It is obviously possible to divide the 10^6 m^2 of collecting area in many ways but the idea of a station, representing a geographically localized patch of collecting area, is basic to most designs. The number of stations, N, is a useful design classifier: large-N ($N > 300$) and small-N ($N < 50$) concepts may offer relatively high-fidelity imaging and low infrastructure costs, respectively. However, a series of simulations to investigate the trade-offs in large- and small-N designs is still in progress. With the SKA imaging specification set at 60 dB dynamic range over a wide field-of-view, the simulations themselves are non-trivial, requiring significant algorithm development and super-computing resources.

There are currently seven antenna concepts for the SKA being considered. Table I aggregates these by form and summarizes some key features, while Figure 3 is a visualization of various instruments. The descriptions of area division given in the table are representative of those so far investigated by proponents of the various concepts. It is important to realize that the SKA is much more than antennas, requiring as it does major developments in fields such as low-noise integrated RF systems, long distance data transmission at Tb/s rates, real-time signal processing at peta operation per second speeds, and highly complex computing systems spanning a number of software engineering fields. Still, apart from being the most visible part of the telescope, antennas will account for ~40% of the cost, despite intentions to make a "software telescope" by exploiting, as far as possible, the convergence of radio and computing technologies.

In the end, all ideas for realizing the SKA result in the transmission and processing of data from a given field-of-view, at given bandwidths and quantizing

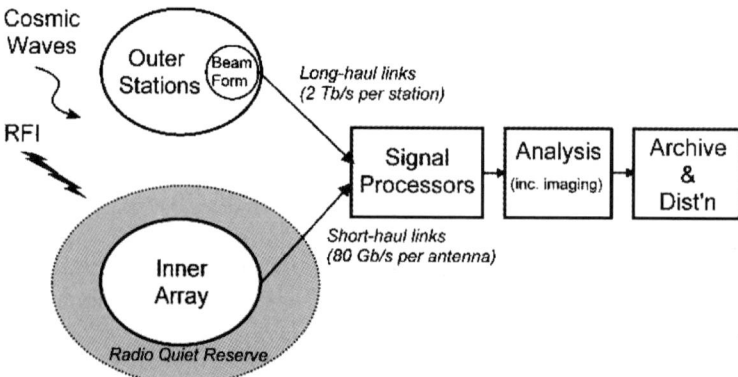

Figure 2. Simplified schematic view of the SKA. About half the total collecting area is contained within the inner array (<5 km diameter). Actual data rates from outer stations depend on available (or affordable) bandwidths on custom and/or commercial networks. The amount of signal aggregation (beam-forming) can be adjusted as transmission capacity evolves.

TABLE I
Summary of SKA concepts

Concept class	Concept name	Attributes	Challenges or issues	Initial demonstrators
Large diameter (>100 m) reflecting flux concentrators ($N = 30$–100 antennas; 1 antenna per station)	Large adaptive reflector (LAR) (Canada)	Filled station (high sensitivity to spatially extended radio emission) Small N may lower infrastructure costs Potentially wide frequency coverage	Large focal plane arrays for acceptable field-of-view High-fidelity imaging with small N	Scaled aerostat and control system (Canada)
	Kilometre-area radio synthesis telescope (KARST) (China)			Scaled version of five hundred metre aperture spherical telescope (FAST) (China)
	Cylindrical reflector (CR) (Australia)			SKA molonglo prototype (SKAMP) (Australia)
Small diameter (~12 m) reflecting flux concentrators (4500–8000 antennas, $N = 600$ stations)	Large N–small D array (LNSD) (USA)	Large N may yield very good imaging Good high-frequency sensitivity Highly versatile array Wide frequency coverage	No multi-fielding Poor low-frequency sensitivity	Allen Telescope Array (ATA) (USA)
	Pre-loaded parabolic dish (PPD) (India)			Raman Research Institute 12 m antenna (India)
Refracting flux concentrator (20000 antennas, $N = 300$ stations)	Luneburg lens (LL) (Australia)	Multi-fielding (quasi-optical beamforming)	10–15 GHz upper frequency limit with 3.5 m lenses New dielectric material required	1 m Luneburg lens with new artificial dielectric (Australia)
Aperture phased array (typically 200 000 elemental antennas for a constituent array operating at 500 MHz, $N = 100$ stations)	Aperture array tile (AAT) (Europe)	Multi-fielding (electronic beamforming) Highly agile field translation	1.5 GHz upper frequency limit <3:1 bandwidth; 3 constituent arrays needed to cover 100 MHz–1.5 GHz Low-cost realization needed	Thousand element array (THEA); (Europe)

accuracies, across various distance regimes. Not surprisingly, one arrives fairly naturally at a system definition which views a large part of the SKA design process as simply an exercise in optimizing the transport and processing of very large volumes of data: a perspective likely to appeal to industry and other associates operating outside the astronomy arena. While this view is useful, there is still a large, specialized, design task involving antennas, RF systems, analog-digital conversion and, in most concepts, station-level signal processing. Nevertheless, with the exception of some antenna-specific descriptors, these more specialized sections of the SKA are also amenable to definition in terms of standard sub-systems and components. Apart from the engineering imperatives in this area, a first exercise in standardized system design has underlined that a continuing definition process is essential to refining the science goals of the Telescope, and to ensuring that stakeholders have a clear idea of what the instrument will actually be able to do.

Turning to concept choice, practical technology demonstration is the basis of the SKA selection process. Aggressive project timescales mean that "blind alleys" need to be avoided and the general philosophy is to succeed, or fail, early in proof-of-concept demonstrators. SKA concept proponents have been encouraged to tackle critical issues early in their programs and to establish that the cost reduction factors necessary to make the SKA viable can, in fact, be achieved.

A practical example of this approach is the decision not to proceed further, at this point, with development of the Luneburg lens antenna in the SKA context. Despite notable achievements, including the patenting of a new dielectric material (Section 4), the Australian proponents used a highly-directed prototyping effort to conclude that the antenna is unlikely to be viable on SKA timescales. This decision was also bound to an evolving SKA science case favouring wide fields-of-view (FOVs) below 1 GHz over multiple, independent, FOVs at higher frequencies.

Two large-scale instruments currently being built will be especially important in the path to the SKA. The US Allen Telescope Array (ATA) will be central to verifying the feasibility of small reflector solutions incorporating new-generation cryogenic cooling techniques for broadband receivers. The LOFAR telescope, a geographically-distributed aperture array (30–240 MHz) being built in the Netherlands, will be the first software radio telescope and will address critical SKA concepts, including RF interference mitigation and area re-use via widely separated multi-fielding. Apart from the ATA and LOFAR, planned extensions to the US Very Large Array (VLA) and Deep Space Network (DSN), and to the European MERLIN and EVN telescopes (via the EVLA, initial DSAN, e-MERLIN and e-EVN projects) also have much to contribute to SKA demonstration, principally in the fields of data transport and software development. Similarly, many international project management and equipment production guidelines will undoubtedly be gleaned from the USD 0.5B Atacama Large Millimetre Array (ALMA) radio telescope now being constructed.

Large-scale SKA-specific technology proposals (USD 30M each) have recently been submitted to funding agencies in the USA and Europe. The US Technical Development Program seeks to demonstrate the viability of mass-production techniques for 12 m dish antennas and associated systems. In Europe, the Framework 6 proposal for SKA design studies aims to produce new insights into the economical manufacture of phased arrays, principally via the EMBRACE ("electronic multibeam radio astronomy concept") 500 m^2 aperture array demonstrator, but also via the development of focal plane arrays for large reflectors, an activity being pursued in collaboration with the Australian and Canadian SKA consortia. Regardless of the ultimate applicability to the SKA of particular antenna solutions, consortia in Australia, Canada, China, and India have framed further plans to construct large-scale pathfinder science instruments based on many of the concepts outlined in Table I. In the case of China, the hope is that a current bridging project to optimize engineering for the 500 m FAST demonstrator could lead into construction of FAST itself on a timescale of perhaps six years.

4. Pivotal technology and software

A recent review (Hall, 2003b) of updated SKA whitepapers produced a series of critiques for the various concepts and identified the major engineering challenges to be addressed in demonstrators. This list includes items such as:

- Low-cost manufacturing methods for both concentrators and dense aperture arrays;
- Sensitive, cheap, highly integrated, uncooled receivers (Aperture Array, Luneburg lens, Cylindrical Reflector);
- Efficient, broadband, feeds with optical arrangements yielding minimum spillover (concentrators);
- Low cost, reliable, cryogenics (concentrators other than Luneburg lens and Cylindrical Reflector);
- Large, cheap, focal plane arrays with accurate, low-loss, beamforming networks operating to beyond 10 GHz (large concentrators);
- Economical, high bandwidth (Tb/s), fibre optic signal transmission links (compatibility with commercial standards and maximum bandwidth efficiency are dominant issues in distant and central SKA distance regimes, respectively); and
- Scaleable signal processors – including correlators – demonstrating processing power, flexibility and connectivity issues.

While much remains to be done, there have been some notable engineering achievements already. A sampling of the 2002–2003 engineering highlights noted in the report cited includes items such as:

- Broadband log-pyramidal feeds covering the range 0.5–11 GHz;
- New methods of accurately hydroforming aluminium paraboloids in the 6–12 m diameter range;
- Development and test of new pulse-tube cryogenic coolers operating at 70 K;
- Decade-band, low-noise, RF amplifier development, both cooled and uncooled, using technologies ranging from indium-phosphide to CMOS;
- Refinement of cost-effective, light-weight, dense focal plane arrays using Vivaldi end-fire elements;
- Active panel controls and new manufacturing methods for large reflectors (>100 m) diameter;
- Accurate, low-cost, feed positioning systems for large reflectors; and
- New low-loss artificial dielectric material ($\tan \delta < 10^{-4}$) for Luneburg lenses and other e.m. applications, together with an associated feed translation system for lenses.

Figure 4 shows some examples of recent SKA antenna-related prototyping.

Recent studies of SKA computing and software engineering requirements (e.g. Cornwell, 2005a,b) have highlighted the substantial challenges facing the project in these areas. Simply scaling up existing imaging techniques and using general-purpose computers (with projected computing powers adjusted for Moore's Law gains) appears unlikely to solve the SKA problem. In the worst-case estimate the cost of post-processing scales as the inverse eighth power of the concentrator diameter and the cube of the maximum baseline which, for geographically extended large-N concepts, implies 2015 costs well in excess of the entire SKA budget of USD 1B. At the same time, first simulations of imaging with large concentrators (small-N SKA) raise doubts about whether the required imaging quality can be achieved. Clearly, extension of the simulation program is essential if key SKA design questions are to be answered. In response to this imperative an international simulations working group has recently been established. The development of new imaging algorithms and methodologies, perhaps invoking a combination of dedicated hardware and general-purpose computers, will also be linked closely to the SKA simulation tools and facilities now being established.

5. Industry collaboration

As a matter of principle the international SKA project, and its associated national programs, welcome interest from potential industry partners. In general terms, any joint research and development is viewed as a shared risk endeavour, with SKA consortia and industry each contributing to defined activities. In some countries industry is able to offset or recoup its contribution via government funding programs or the tax system. The SKA has an agreed policy on intellectual property (IP) developed under its aegis. Broadly, industry partners exploit their own IP contributions

Figure 3. SKA concept visualizations. The left panel shows two large adaptive reflector ideas, one based on a 200 m diameter Arecibo-like spherical reflector (upper), the other using a much flatter reflector and an aerostat-mounted focus cabin. The middle panel (upper) shows offset paraboloids, the proposed 12 m versions being similar to the 6 m ATA antennas shown; the lower illustration shows an SKA station using aperture arrays. The right panel (upper) shows a station based on 64×7 m diameter Luneburg lenses; the lower visualization is of a station employing a single 110 m × 15 m cylindrical reflector. The concepts originate in China, Canada, the USA, Europe and Australia (both right hand panels). Not shown is a 12 m symmetrical paraboloid concept proposed by the Indian SKA group.

Figure 4. SKA technology prototypes. At left are components for large reflectors. The upper panel shows a prototype deformable surface for the FAST demonstrator while the lower panel illustrates a test aerostat for the suspended-feed LAR concept. The middle panel shows three 6 m dishes of the ATA (upper), together with a patch of an aperture array demonstrator based on Vivaldi-notch end-fire elements (lower). At upper right is a 0.9 m Luneburg lens with a two-arm feed translator. The lower right panel shows a prototype 12 m dish constructed using a mesh surface and pre-loaded radial members.

in arenas outside the SKA project but innovations are available to the SKA project free of any licensing charge. Advantages for industry participants include:

- The opportunity to hone the creative energies of their best professionals in a highly imaginative project;
- The ability to perfect leading-edge techniques and products in a demanding application, with technologically sophisticated users;
- The ability to generate and share information with R& D partners in a benign and commercially non-threatening environment;
- High visibility flowing from association with an innovative international mega-science project; and
- Potential for early involvement and favourable positioning in a USD 1B project spanning a range of engineering and computing disciplines.

An outline timescale for industry involvement is shown in Table II, and a summary of SKA consortia interests and expertise is also available (Hall, 2004).

TABLE II
SKA key dates and industry opportunities

Year	SKA project milestone	Industry opportunities
2003	Initial siting proposals received from four countries	Scope for continuing industry involvement in national site characterization
2004	Plans for national SKA demonstrators submitted	Possible links to national SKA technology development programs
2005	Final SKA site submissions	Possible involvement in compiling national proposals
2006	SKA site decision; critical review of technology demonstrator programs	Possible links in development of objective international methodology for site and technology selection, and for risk management
2009	Choice of SKA technology	
2010	Start construction of on-site SKA demonstrator (5–10% SKA area)	Likely participation in infrastructure provision, and instrument design and construction
2014	Start construction of SKA	Maximum involvement at levels of final design, project management, and construction contracts and sub-contracts
2017	Stage 1 SKA complete and operational	Opportunities in commissioning, operations and maintenance
2020	SKA complete	Continuing operations and maintenance role

Internationally, the first industry input may be via the ISPO and its interests in complex decision making formalisms and risk management strategies.

6. Conclusions

With many innovative concepts proposed for the SKA the selection process relies heavily on demonstration of key systems and components. While not uniform in scale, demonstrators currently being built will give insight into the feasibility of various approaches. Additional engineering development programs valued at more than USD 60M may soon be available, paving the way to a large-scale, on-site, demonstrator using selected technologies. This will most likely be built in the period 2010–2013, minimizing the "leap of faith" required to build the full SKA. At 5–10% of the SKA area, this telescope will be a formidable instrument in its own right, allowing a huge expansion of, for example, the database of galaxy red-shift measurements. This data could conceivably reveal the equation of state (pressure/density relationship) of the Universe, giving insight into the birth, and eventual death, of the Cosmos.

Acknowledgements

I thank many colleagues from a variety of international SKA groups for permission to use graphics and other material included in this paper. In particular, I thank R. T. Schilizzi and my associates on the SKA Engineering Working Group for their help in compiling this summary.

References

Cornwell, T. J. and Glendenning, B. E.: 2005a, 'Software Development for the SKA', *Exp. Astron.* **17**, 311–317.
Cornwell, T. J.: 2005b, 'SKA and EVLA Computing Costs for Wide Field Imaging', *Exp. Astron.* **17**, 331–345.
Carilli, C. and Rawlings, S. (eds.): 2004, 'Science with the SKA', *New Astronomy Reviews*, Elsevier, Amsterdam.
Hall, P. J. (ed.): 2003a, 'The SKA: Initial Australian Site Analysis', available at http://www.skatelescope.org/pages/p-docsandpres.htm
Hall, P. J. (ed.): 2003b, 'Report to the ISSC by the IEMT', SKA Memo 41, available at http://www.skatelescope.org/pages/p-docsandpres.htm
Hall, P. J.: 2004, 'The International SKA Project – Industry Interactions. Paper 1 – Background and Collaborative R&D', SKA Memo 52, available at http://www.skatelescope.org/pages/p-docsandpres.htm

Horiuchi, S., Chippendale, A. and Hall, P.: 2004, 'SKA system definition and costing – a first approach', SKA Memo 57, available at http://www.skatelescope.org/pages/p-docsandpres.htm

ISPO Concept Whitepapers: 2003, available at http://www.skatelescope.org/pages/p-docsandpres.htm

ISPO Site Whitepapers: 2003, available at http://www.skatelescope.org/pages/p-docsandpres.htm

Jones, D. L.: 2004, 'SKA Science Requirements – Version 2', SKA Memo 45, available at http://www.skatelescope.org/pages/p-docsandpres.htm

Schilizzi, R. T.: 2004, 'The Square Kilometre Array', *Proc. SPIE conf. (Ground based Telescopes)*, SPIE, Glasgow.

SKA DEMONSTRATORS AND RESULTS

THE ALLEN TELESCOPE ARRAY

DAVID DEBOER[1,*], ROB ACKERMANN[1], LEO BLITZ[2], DOUGLAS BOCK[2], GEOFFREY BOWER[2], MICHAEL DAVIS[1], JOHN DREHER[1], GREG ENGARGIOLA[2], MATT FLEMING[2], GIRMAY-KELETA[1], GERRY HARP[1], JOHN LUGTEN[2], JILL TARTER[1], DOUG THORNTON[2], NIKLAS WADEFALK[3], SANDER WEINREB[3] and WILLIAM J. WELCH[2]

[1]*SETI Institute, 2035 Landings Drive, Mountain View, CA 94043, U.S.A.;* [2]*Radio Astronomy Laboratory, University of California, Berkeley, CA 94720, U.S.A.;* [3]*JPL/Caltech, 4800 Oak Grove Road, Pasadena, CA 91109, U.S.A.*
(*author for correspondence, e-mail: ddeboer@seti.org)*

(Received 17 August 2004; accepted 4 October 2004)

Abstract. The Allen Telescope Array, a joint project between the SETI Institute and the Radio Astronomy Laboratory at the University of California Berkeley, is currently under development and construction at the Hat Creek Radio Observatory in northern California. It will consist of 350 6.1-m offset Gregorian antennas in a fairly densely packed configuration, with minimum baselines of less than 10 m and a maximum baseline of about 900 m. The dual-polarization frequency range spans from about 500 MHz to 11 GHz, both polarizations of which are transported back from each antenna. The first generation processor will provide 32 synthesized beams of 104 MHz bandwidth, eight at each of four tunings, as well as outputs for a full-polarization correlator at two of the tunings at the same bandwidth. This paper provides a general description of the Allen Telescope Array.

Keywords: array technology, radio astronomy, SETI

1. Introduction

Two primary efforts in astronomy today are to achieve a greater number of degrees of freedom available to a telescope and to increase the net surface area collecting signal from the cosmos. The first broad class of issues (the available degrees of freedom) is meant here to mean many receiving elements (e.g., feeds for a radio telescope or pixels for an optical telescope) as well as flexibility in observing modes (e.g. spectral diversity, spatial diversity, processing power,...). The second effort is just raw surface area which equates with improved sensitivity and the ability to detect and map ever fainter sources. As a "large-N" array (that is, comprising many antennas), the Allen Telescope Array addresses both of these efforts (Welch and Dreher, 2000; DeBoer and Bock, 2004).

The Allen Telescope Array is leveraging developments in low-cost, ultra-widebandwidth microwave electronics, high-speed digital electronics and "mid-quantity" production techniques. The goal is to lower the per element cost sufficiently to be able to afford to build a significant array on a relatively modest budget and, concurrently, make it better by incorporating additional flexibility (degrees of freedom) with these new technologies. These improvements come from utilizing many receivers (in radio astronomy, many receivers has meant on the order of 20–30 – we intend to emplace 350 or more antennas each with dual-polarization receivers), bringing back the entire ∼11 GHz of analog bandwidth, and incorporating many, concurrently-used radio-astronomical "back-ends" (the equipment that accepts the raw data stream and produces the scientific data-product). The net result is a telescope that will have unprecedented radio-imaging capability, significant sensitivity and very productive multitasking capability (several independent observing programs concurrently utilizing the entire array).

The Allen Telescope Array is a joint project of the SETI Institute (Search for ExtraTerrestrial Intelligence, Mountain View, CA) and the Radio Astronomy Lab at the University of California, Berkeley. Conceived in a series of meetings convened by the SETI Institute in 1997–1999, the instrument was originally called the One Hectare Telescope (1hT), after its 10^4 square meters of collecting area. Thanks to the generosity of the Paul G. Allen Foundations (Seattle, WA), the telescope has been funded through the research and development and initial construction phases. Named for its first and primary benefactor, the Allen Telescope Array is under construction, with the first three antennas currently in place (Figure 1) and a schedule that calls for 33 antennas in 2004, 206 antennas by the summer of 2006,

Figure 1. The three antennas currently under test at the Hat Creek Radio Observatory.

Figure 2. Rendering of the completed ATA-350 at the Hat Creek Radio Observatory.

and 350 antennas sometime later in the decade. Figure 2 shows a rendering of how the 350 dishes will look at the Hat Creek Radio Observatory.

The SETI Institute is a private, nonprofit organization dedicated to scientific research, education and public outreach. Founded in 1984, the Institute today employs more than 100 scientists, educators and support staff. The mission of the SETI Institute is to explore, understand and explain the origin, nature and prevalence of life in the universe. In pursuing this mission, the Institute is involved in numerous programs across a wide breadth of disciplines.

The Radio Astronomy Laboratory (RAL) at the University of California, Berkeley, has pioneered radio astronomy for more than 40 years, supported by the National Science Foundation, the State of California, and in its early days by the Office of Naval Research. During that time, the RAL has made major innovations in radio astronomy instrumentation and fundamental discoveries in radio astronomy. The first large filter bank spectrometer with a radio astronomical application was built for use on the 85-foot centimeter-wave telescope at Hat Creek. The RAL also built the first millimeter-wave array for astronomical observations. In the last 20 years, the RAL has concentrated its efforts on providing outstanding millimeter-wave radio imaging.

2. Antenna

The antenna is an offset Gregorian design, meaning that the primary reflector is a non-symmetric portion of a paraboloid and the large secondary reflector has an ellipsoidal figure that is positioned near the edge of the primary to reduce unwanted blockage. The hydroforming technology used to make these surfaces is the same technique used to generate low-cost satellite-TV antennas by Andersen Manufacturing (Idaho Fall, ID). Figure 3a shows the cross-sectional profile of the optics while Figure 3b shows the view looking down the optical axis. Note the small amount of blockage on the lower edge (where the illumination is more than 10 dB down), which was allowed in order to better optimize the feed and antenna mounts.

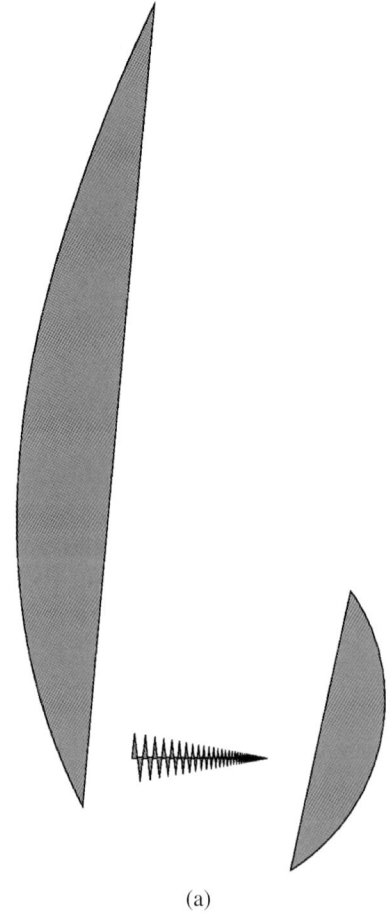

(a)

Figure 3. (a) Profile of antenna surfaces and log-periodic feed pointing at an elevation of approximately 30°. (b) View of antenna surfaces looking down the optical axis. The shaded portion is the total blockage at the low-illumination edge that was allowed to better optimize the feed and antenna mounts.

(*Continued on next page*)

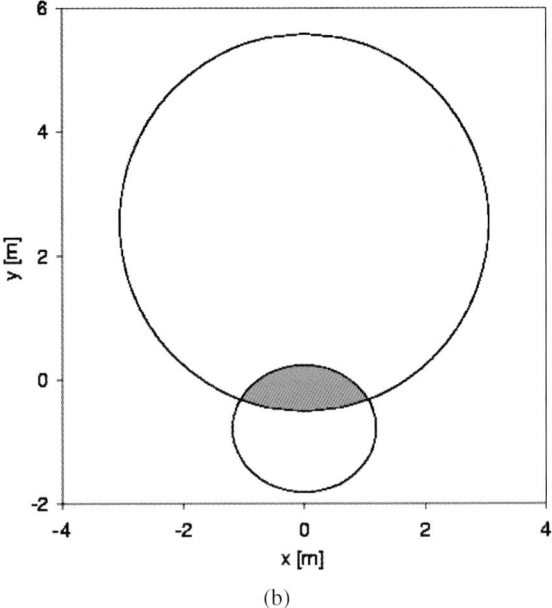

(b)

Figure 3. (Continued)

This offset design offers several benefits over the symmetric designs more traditionally used in radio astronomy. Firstly, it lets us incorporate a very large secondary to observe at low frequencies (at 500 MHz, the 2.4-m secondary is still about four wavelengths across). Secondly, it provides a field-of-view that is free from obstruction by the secondary so that the side-lobes (i.e., the responses from directions other than those desired) are minimized. This helps not only by maximizing the sensitivity in the desired direction, but also by reducing susceptibility to strong transmitters in other directions.

Since the primary and secondary antennas surfaces are single pieces of concave metal they have a good deal of rigidity (the "wok" effect) and the backup structure may be minimized to reduce cost. The ATA mount incorporates struts that support the reinforced rim and a center plate that allows axial displacement but provides torsional stiffness. Figure 4 shows the back of an ATA antenna, along with trenching for soon-to-be-emplaced antennas.

3. Receiver

The ATA feed is a pyramidal log-periodic feed (Figure 5) (Engargiola, 2002). The dual-linear polarization feed achieves about 12 dBi forward gain and is designed for an equivalent focal-length to antenna diameter ratio (f/D) of 0.65. This patented feed incorporates a central metallic pyramidal "spike" that allows a low-noise amplifier housed in a small cryogenic dewar to be placed directly behind the antenna terminals

Figure 4. ATA antenna support structure and trenching to soon-to-be-emplaced antenna (the three piers of the foundation may be seen at the end of the trench).

(at the "pointy" end) to yield a low receiver temperature. In transmission, the feed excites a waveguide mode at the terminals, which travels in the interior space until it meets a resonant condition (at about $\lambda/2$) and is then radiated in free space back towards the small end. In our case of receive-only, the reciprocal condition is relevant. The linear dimensions of the feed yield an operating range from 500 MHz to about 18 GHz.

This somewhat awkward geometry requires a pyramidal-shaped dewar with long, narrow twin-leads to connect the feed terminals to the end of the baluns. To fit in this space, a tapered-line balun was developed to match the balanced 240 Ω antenna terminals to the single-ended low-noise amplifier (Engargiola, 2004). Figure 6a shows the balun implemented on a crystalline-quartz substrate. Figure 6b is a close-up of the packaged amplifier interfaced to the balun. The balun and amplifier reside in the dewar, which is cooled by a pulse-tube refrigerator (PTR) developed for this feed by NIST and driven by a novel compressor designed and developed at the Radio Astronomy Lab (Lugten, 2003).

The low-noise amplifier (LNA) is a 0.1 μm Indium Phosphide monolithic microwave integrated circuit (MMIC) designed and packaged by Caltech (Weinreb et al., 2000; Weinreb and Wadefalk, 2003, personal communication). The current chips utilize the Northrop-Grumman foundry process (Grundbacher et al., 2002).

Figure 5. ATA log-periodic feed. The "zig-zags" are the arms that surround the interior pyramidal structure, which improves the electrical performance and also houses the dewar and cryogenics.

The measured and calculated performance of this LNA is shown in Figure 7. The amplifier feeds a thermally controlled, variable gain post-amplifier module utilizing wide-bandwidth RF Micro Devices gain blocks and Hittite variable attenuators. The entire bandwidth (two polarizations of 0.5–11.2 GHz) is brought back to a centrally located processing facility via analog fiber-optic links developed for this project by Photonic Systems Inc (Burlington, MA) (Ackerman et al., 2002).

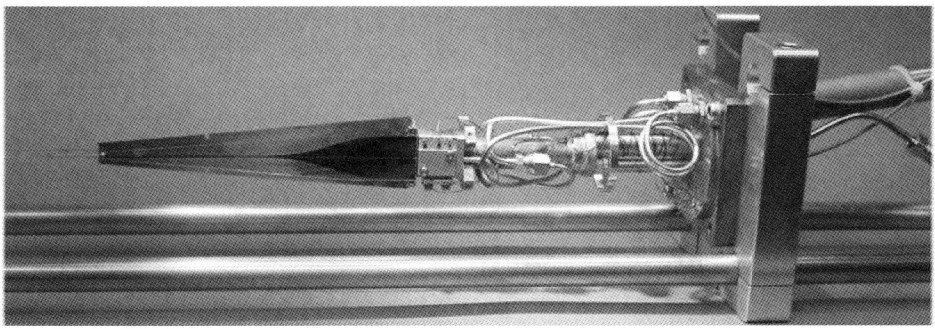

(a)

Figure 6. (a) Dewar interior. From left to right are the baluns, low-noise amplifiers, output coax to the dewar baseplate, from which the pulse-tube refrigerator protrudes. (b) Close-up of packaged amplifier interfaced to balun on the cold-head mounting block.   (*Continued on next page*)

26 D. DEBOER ET AL.

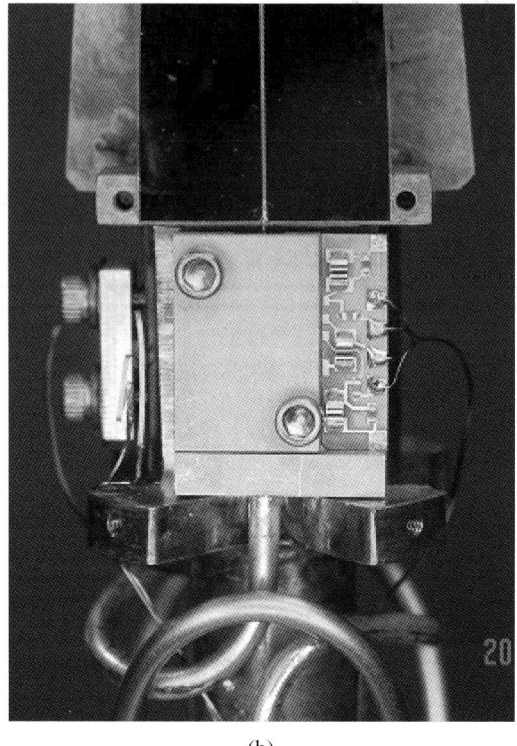

(b)

Figure 6. (*Continued*)

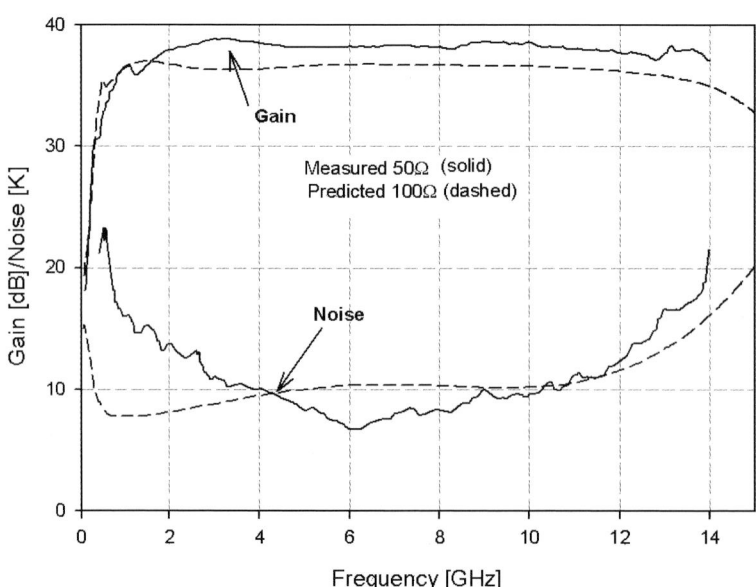

Figure 7. LNA performance. The top set of curves shows the measured gain at 50 Ω (solid) and predicted at 100 Ω (dashed). The bottom pair of lines depicts the noise temperature.

4. Signal path and observing

The accessibility of the entire bandwidth from every antenna back at the processor building allows a large degree of flexibility. Figure 8 shows a block diagram of the system, with its myriad options and abilities. The RF converter board accepts the two linear polarization fibers from an antenna and transmits four independent dual-polarization IF channels (400 MHz each at 1.56 GHz) to the digitizer, which produces 8-bit I&Q digital datastreams at 155 Msamples/s. These streams are transported via fiber to the IF Processor (IFP). The IFP then combines the streams to produce four independent dual-polarization beams per tuning (for a total of 32 beams) at a data rate of about 104 Msamples/s which are then routed to back-end processors. In addition, all data for two tunings are fed to the correlator at 104 Msamples/s. The digital hardware is based primarily on custom boards utilizing field-programmable gate arrays (FPGAs). The bandwidth of the digital processor has been kept relatively low, leveraging existing commodity components, so the cost may be minimized. We intend to upgrade the digital portion in fairly short timescales, taking advantage of the fast development pace of digital electronics. Control and communications of the entire array leverages the ubiquity and performance of the in-built TCP/IP and Java capabilities of microcontrollers.

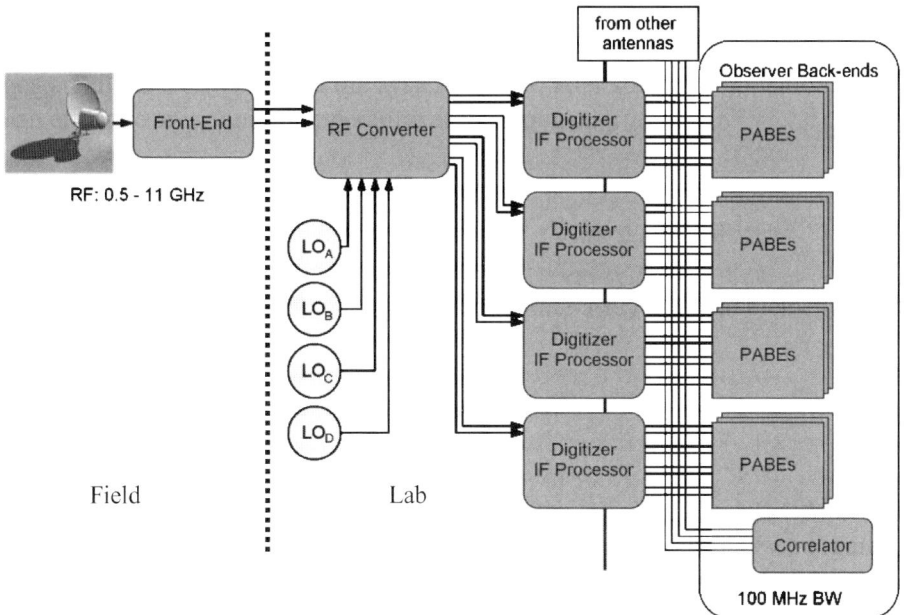

Figure 8. Signal path block diagram of the ATA. To the right of the vertical dashed line are items back in the lab. Analog optical fiber links transport about 11 GHz of both polarizations back to the lab, where the signal is split to allow for different frequencies and pointing.

The outputs from the IFP's are then passed to one of several types of backends. The phased array back ends (PABEs) and correlator may all be used concurrently. The typical observing mode will have the astronomers at the Radio Astronomy Lab stepping through large-scale surveys using the correlator, while those at the SETI Institute will use two tunings and several of the beams to simultaneously conduct multiple observations of stars. Other astronomers may concurrently use other beams to conduct other observing programs, for example, pulsar searches and monitoring.

As a joint project intended for large surveys, this commensal use of the array is paramount and is enabled by the large field-of-view, wide bandwidth brought back to the lab, and flexible digital hardware. This allows the SETI Institute to conduct the full-time, full-sensitivity targeted search for extraterrestrial technological signals concurrently with Radio Astronomy Lab-conducted wide surveys of HI using the correlator. In addition, other back-ends (e.g., a Beowulf-cluster spectrometer) can utilize other available beams. The "scarce" resources will be the pointing direction of the antennas (of which there is one, unless sub-arrays are used, which is possible down to arbitrary sub-arrays of one antenna) and the tunings (of which there are four). The data-streams from all the antennas for both polarizations for two tunings are also sent to the correlator, allowing images of the entire field-of-view at two independent frequencies at full polarization to be formed. This constrains four of the 32 beams, which are still available to other back-ends but are phased to point to boresight.

Figure 9 indicates the multiple-use capability. The large circle circumscribed over the Andromeda galaxy is the 3-dB field-of-view of the ATA at 1.420 GHz (contrasted with the Arecibo field-of-view) and the 32 smaller circles are the independent ATA beams (contrasted with the Arecibo resolution, which is the same as its field-of-view). These beams can point anywhere in the sky, however they will typically point within the main beam of the primary (although allowing beams to follow strong RFI sources out of the field-of-view may have benefits for RFI mitigation). In addition, as stated, the correlator can image the entire field-of-view at two independent frequencies. The moon is also shown for reference.

5. Array configuration

Given the large number of antennas, the imaging of the ATA will be unprecedented in radio astronomy. Figure 10 shows the intended configuration of the full ATA and some of the existing infrastructure of roads and buildings at Hat Creek. The filled circles are the positions of the first 33 antennas and the "X'ed" circles are the existing three antennas. The resulting beam is shown in Figure 11. The minimum baselines are about 9.8 m (notice the close groups of three antennas) and the maximum baseline is about 870 m. The beam is round at $+10°$ declination (about $78''$ with natural weighting) and then increases to about $100'' \times 78''$ at $90°$ and $200'' \times 78''$ at $-30°$.

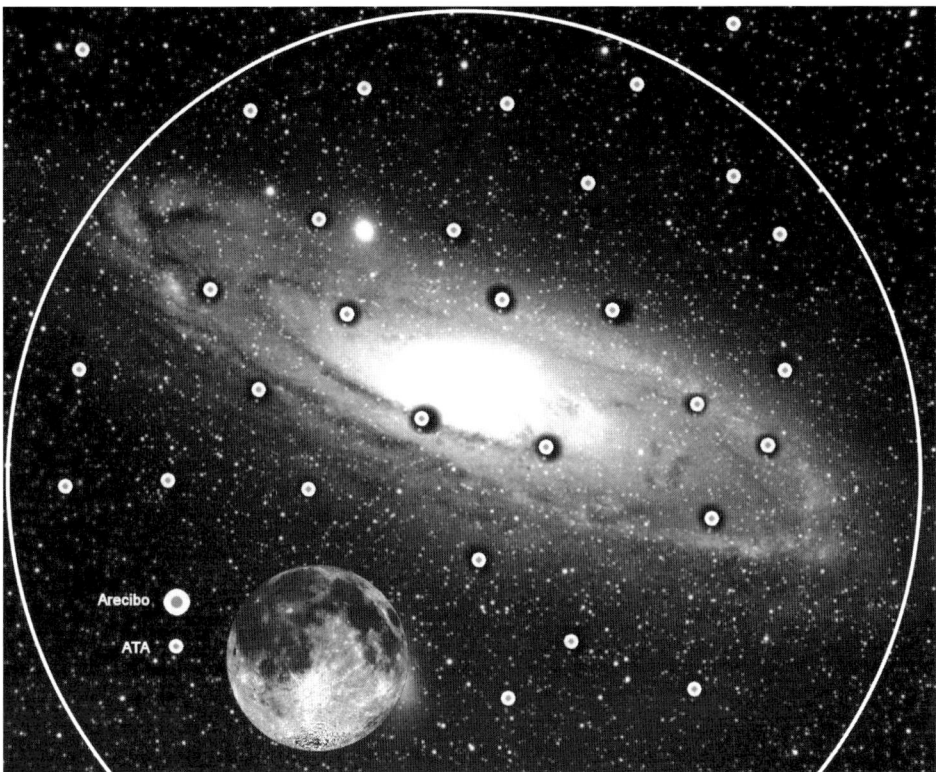

Figure 9. Flexibility of the ATA. The large field-of-view and the distributed array allow effective multiple concurrent use of the array. The small, filled circles within the inscribed 2.5° field-of-view at 1.42 GHz are the 32 individual beams – four dual-polarization pairs at each of four tunings. The correlator may also image the entire field-of-view with full polarization at two tunings. The Arecibo field-of-view/beam is shown for comparison. The background image is the Andromeda galaxy and the moon is also shown for reference.

The *uv* distribution of this configuration is shown in Figure 12, which shows the radial density of *uv* points relative to the target Gaussian and the azimuthal positional angle relative to the target uniform distribution. The primary constraints were accommodating the terrain, which is dominated by a lava flow, and to keep shadowing to an acceptable level (about 15% for a 2 h track at $-30°$). The other goal was to suppress the sidelobes. Figure 13 shows the peak and rms distribution of the sidelobes provided by this array distribution, with and without the antenna primary pattern included.

6. The science of radio astronomy

The science of radio astronomy seeks to explain the universe through the measurement of the electromagnetic spectrum from meter to sub-millimeter wavelengths.

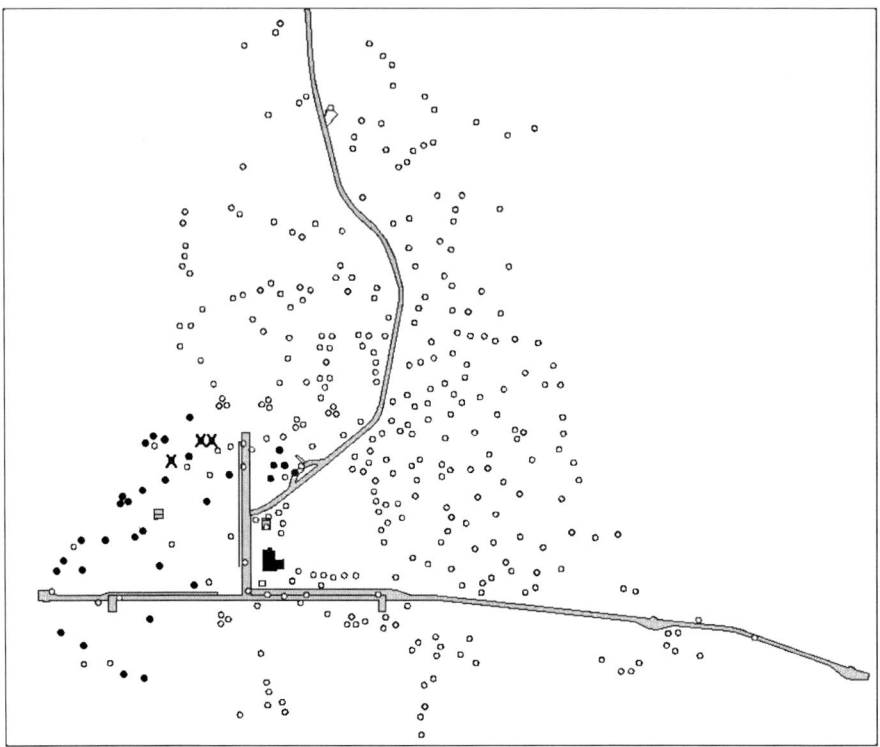

Figure 10. ATA configuration, roads and buildings at the Hat Creek Radio Observatory. The filled circles are the first 33 that are currently being emplaced and the three with the 'X's are the existing antennas. Existing accessways and buildings are also shown.

These may come from nearby, from Jupiter or the Sun or other stars, from galaxies, or from the most distant parts of the universe. The principal types of radio emission are *thermal, non-thermal* and *line* radiation. Thermal emission is the black body radiation described by the Planck function. At radio wavelengths this emission is commonly seen from hot gas in star-forming regions, and most notably as evidence of the Big Bang that marked the beginning of the universe. The latter is implied by a residual background radiation in every direction corresponding (in accordance with theory) to a physical temperature of 2.725 K. Non-thermal emission comes from the synchrotron radiation of very high-speed electrons spiraling in magnetic fields. This radiation is polarized, and allows us to "see" magnetic fields and electrons in, for example, the interstellar material shocked when stars explode at the end of their lives, nebulae powered by spinning neutron stars that often remain after the explosion, or in the spiral arms of galaxies. Emission (or absorption) of line radiation occurs at specific frequencies corresponding to the energies of certain quantum-mechanical state transitions. Most well-known of these is neutral hydrogen, which radiates at 1420.41 MHz. However, there are many other emission and

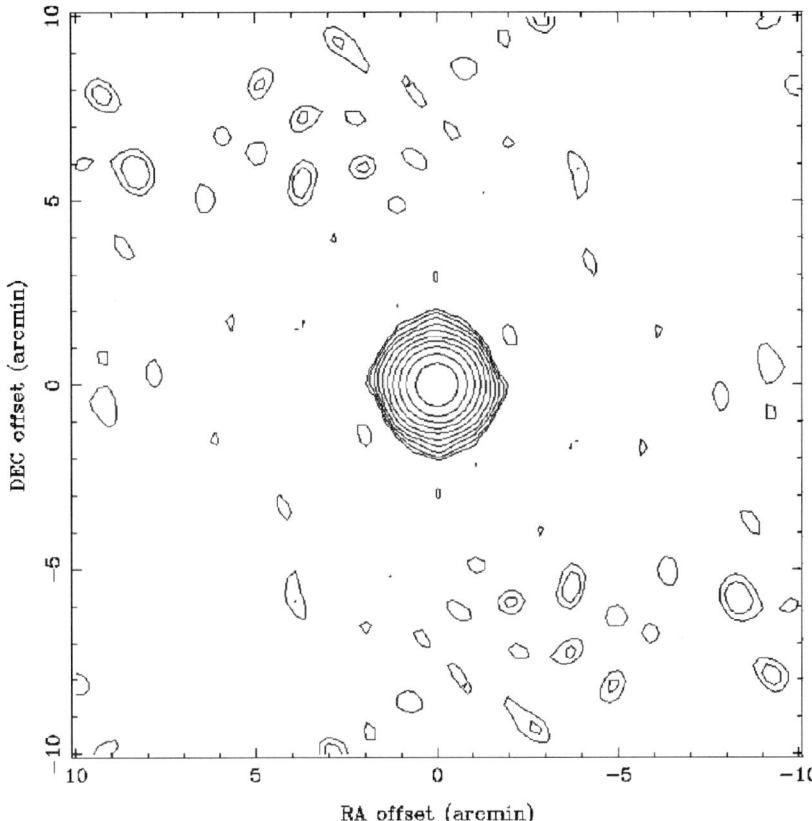

Figure 11. Naturally weighted snap-shot beam of the ATA at 1420 MHz (+10° declination). Contours are 0.3, 0.5, 0.9, 1.6, 2.9, 5.3, 9.5, 17, 31 and 56%.

absorption lines, particularly from molecules at millimeter wavelengths. Measurements of line emission have the advantage that they can tell us the velocity of the emitting material through the Doppler effect, which in turn can often provide the third dimension of an image.

The Allen Telescope Array will be nearly as sensitive as the most sensitive current radio telescopes, because it will have a large collecting area, about 10,000 sq. m. However, the ATA will be unusual in several respects: it will have a wide field-of-view (about 5 square degrees, at 1420 MHz – compare the Sun and the moon at about 0.2 square degree each), continuous frequency coverage from 0.5 to 11.2 GHz (with simultaneous observations possible throughout the band), and a very large number of antennas. These features lead to some scientific investigations for which the ATA will be uniquely well suited.

The wide field-of-view and the continuous frequency coverage are both capabilities that will make the ATA a powerful survey instrument. In three years, the ATA will do an all-sky survey for neutral hydrogen that will measure all galaxies

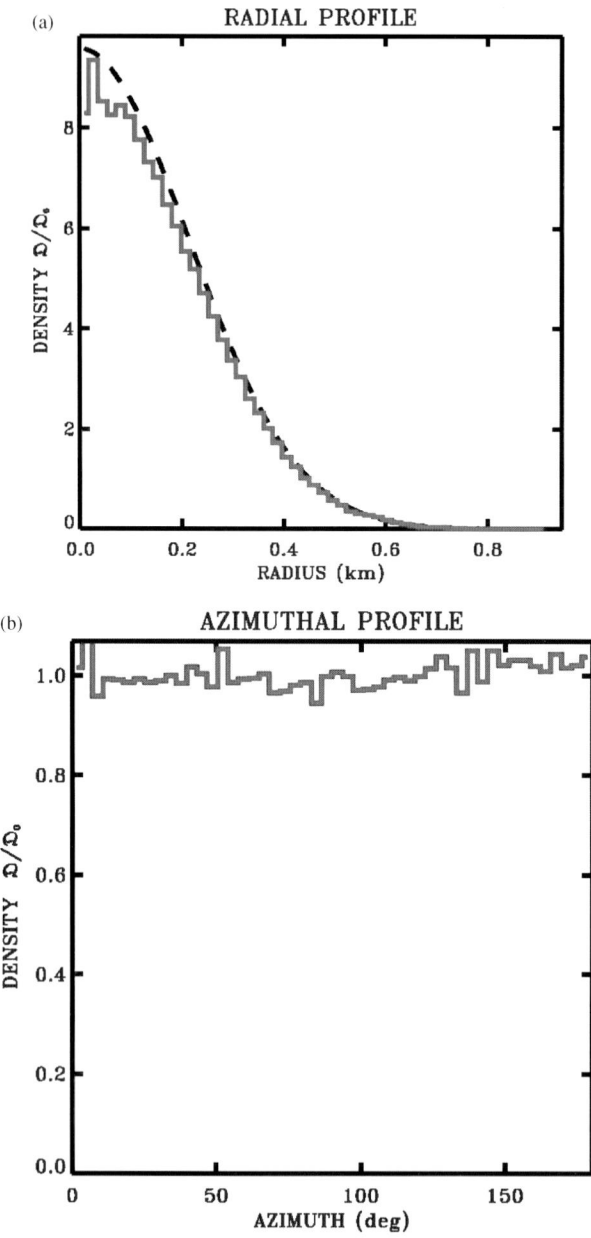

Figure 12. (a) Density of *uv* points relative to the radial Gaussian target distribution. (b) Density of *uv* points relative to the uniform azimuthal target distribution.

with masses at least as large as the Milky Way out to a billion light years. At the same time it will search for unknown transient and variable sources, from relatively nearby rapidly rotating "dead" stars (pulsars) to the most energetic explosions in the universe, gamma-ray bursts, to signals that might come from extra-terrestrial

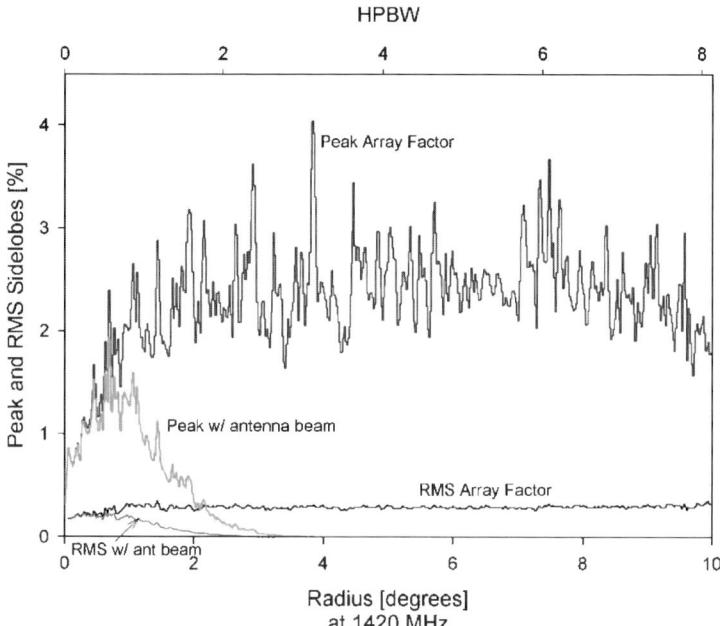

Figure 13. Sidelobe distribution of the 350-element ATA array with and without the antenna primary beam included. For the array factor alone, the rms is well below 1%, while the peak sidelobes are typically between 2–3%.

intelligent beings. The continuous frequency coverage will allow the first sensitive surveys for line radiation in parts of the radio band not previously explored.

The great number of antennas will enhance surveys of large areas. As described above, this is because the ATA will measure the sky with high fidelity in a short observation, whereas current radio telescopes must synthesize over several hours for high-quality observations. The large number of antennas will also provide another benefit: flexibility for rejecting interference. The ATA will have uniquely many degrees of freedom that can be manipulated to filter out interferers, either during the observation (by steering nulls in the array pattern) or afterwards in post-processing. The ATA will be used to test and validate these techniques, which help enable observations despite the increasing spectral "fill factor." Of course, the best protection is to steer clear of areas with many transmitters: the Hat Creek Valley where the facility is sited has few transmitters today. However, encroachment will continue, and satellites are always up there, transmitting directly towards these sensitive receivers.

7. Conclusion

The first three array elements (as seen in Figure 1) are installed and have been under test since early spring of 2003. Given that an array works even when not "completely finished," the construction of the Allen Telescope Array is staged to yield scientific output at the earliest possible point. The first phase is to produce another 30 antennas (with total equivalent collecting area of a 34-m antenna) by the end of 2004, with observations by both the SETI Institute, which will conduct a SETI survey of the galactic plane, and the Radio Astronomy Lab, which will use a correlator (recently funded by the National Science Foundation and currently under construction) to image the sky commencing shortly thereafter. Some of the first projects will be to map the neutral hydrogen in dwarf galaxies and thus measure the dark matter through the effect of its gravity on the dynamics of the system, to look for polyatomic molecules in interstellar molecular clouds, and to try to detect emission from primordial deuterium. The next stage is to build about 200 antennas by mid-2006 and before the end of the decade it is expected that the full complement of 350 antennas will be productively probing the skies. Future funding will then be sought to enhance the telescope still further, so that this flexible instrument will always remain on the cusp of discovery.

References

Ackerman, E., Cox, C., Dreher, J., Davis, M. and DeBoer, D.: 2002, 'Fiber-Optic Antenna Remoting for Radioastronomy Applications', URSI 27th General Assembly, Maastricht.

DeBoer, D. and Bock, D.: 2004, *IEEE Microwave Theor. Tech. Magazine* June, 46–53.

Engargiola, G.: 2002, 'Non-planar log-periodic antenna feed for integration with a cryogenic microwave amplifier', *Antennas and Propagation Symposium Digest*, June 16–21, 2002, San Antonio, TX.

Engargiola, G.: 2004, 'Tapered microstrip balun for integrating a low noise amplifier with a nonplanar log periodic antenna', *Rev. Sci. Instr.* **74**(12).

Grundbacher, R., Lai, R., Barsky, M., Tsai, R., Gaier, T., Weinreb, S., Dawson, D., Bautista, J. J., Davis, J. F., Erickson, N., Block, T. and Oki, A.: 2002, 'InP HEMT Devices and MMICs for Cryogenic Low Noise Amplifiers from X-band to W-band', *Indium Phosphide and Related Materials Conference, 2002. IPRM. 14th*, 12–16 May 2002, pp. 455–458.

Lugten, J. B.: 2003, *Adv. Cryog. Eng.* **49**, 1367–1372.

Weinreb, S., Gaier, T., Fernandez, J. E., Erickson, N. and Wielgus, J.: 2000, 'Cryogenic Low Noise Amplifiers', GaAs2000 Symposium, Paris, October 2.

Welch, W. J. and Dreher, J. W.: 2000, *Proc. SPIE* **4015**, 8–18.

SKY NOISE LIMITED SNAPSHOT IMAGING IN THE PRESENCE OF RFI WITH LOFAR'S INITIAL TEST STATION

STEFAN J. WIJNHOLDS*, JAAP D. BREGMAN and ALBERT-JAN BOONSTRA
ASTRON, Oude Hoogeveensedijk 4, 7991 PD Dwingeloo, The Netherlands
(*author for correspondence, e-mail: wijnholds@astron.nl)

(Received 12 July 2004; accepted 18 October 2004)

Abstract. The initial test station (ITS) is the first full scale prototype of a low frequency array (LOFAR) station. It operates in the 10–40 MHz range and consists of 60 sky noise limited dipoles arranged in a five-armed spiral structure offering an instantaneous synthesized aperture of almost 200 m diameter. We will present all sky snapshot images demonstrating sky-noise limited imaging capability in the presence of a strong RFI source that exceeds the all sky power by 27 dB. This result is obtained with a two stage self-calibration procedure. First, the RFI source near the horizon is used as calibrator and then subtracted, after which Cas A shows up at a level that is a factor 2000 lower and then dominates the picture with its side lobes. A second self calibration on Cas A then reveals the same extended galactic emission as found in a RFI free adjacent spectral channel. This demonstrates that a single 10 kHz channel of a 6.7 s snapshot of a single LOFAR station already provides a dynamic range of over 10^4.

Keywords: CLEAN, LOFAR, peeling, phased array, RFI, RFI mitigation, snapshot imaging

1. Introduction

The low frequency array (LOFAR) is a radio telescope currently being designed and planned. Its targeted observational frequency window lies in the range 10–250 MHz. LOFAR will consist of a number of stations spread over an area of 350 km diameter. Each station consists of order 100 receiving elements which operate together as a phased array. The initial test station (ITS) is the first full scale prototype of a LOFAR station and became operational in December 2003. In this paper we will demonstrate that this system may be used to do sky noise limited snapshot imaging in the presence of RFI. This is an important milestone towards sub-milliJansky imaging at these low frequencies. As pointed out by Boonstra and Van der Tol (accepted for publication) and Van der Veen et al. (2004) RFI signals need only be reduced to the noise level of an individual snapshot image. Further, integration to improve the SNR implies rotation of the sky relative to a hemispheric image. This has the consequence that after derotation the sky sources improve in SNR while point sources appearing at various fixed locations, like RFI, decrease in SNR and will become invisible.

In the next section we will provide a short description of ITS before proceeding with a mathematical description of the processing done to obtain the image and to perform the calibration. Once the processing is clearly described we will demonstrate the sky noise limited performance by presenting a plot of the differences between two neighboring channels after calibration on a RFI source in a nearby channel. Thereafter, we will show that Cas A appears when the RFI source is self calibrated and removed by standard CLEAN techniques. Finally, we will show that after removal of the RFI source, the calibration result can be improved further by another calibration on Cas A.

2. ITS system overview

ITS consists of 60 sky noise limited inverse V-shaped dipoles with East–West orientation. The dipoles are arranged in a five-armed spiral configuration as shown in Figure 1. The instantaneous or snapshot (u, v)-coverage and the beam pattern or PSF are also displayed in this figure.

Each antenna samples the same single polarization and is connected to a receiver unit where the signal is filtered by a 10–40 MHz bandpass filter and digitized by a 12 bit A/D converter before it is transmitted over a high speed optical link to a twin input module (TIM) board. The TIM board is a data acquisition card having two high speed optical link receivers, 2 GB of RAM buffer and a PCI interface for read-out and control, which was developed at ASTRON.

The on-site data acquisition and processing system consists of 16 standard PC's (2.4 GHz Intel Xeon, 1 GB RAM, Linux 2.4.20) for data acquisition and initial processing, 1 industrial PC for central processing and data storage (dual 2.0 GHz Intel Xeon, 512 MB RAM, Linux 2.4.20) and 1 router (2.4 GHz Intel Xeon, 1 GB RAM, Linux 2.4.21). The PC's are interconnected by a 100 Mbit control network and a 1 Gbit data network.

Each data acquisition PC is equipped with two TIM boards, giving the full system a capacity of 64 input channels. On these machines the raw data from the individual antennas $x_i[n]$, where the subscript i denotes the antenna number, can be

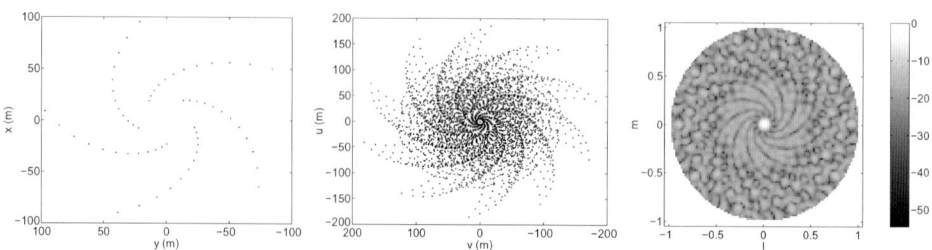

Figure 1. The left hand plot shows the configuration of the ITS antennas. Positive x is pointing north, positive y is pointing west. The middle plot shows the (u, v)-coverage for a single snapshot. The baselines are depicted in meters instead of wavelengths for convenience since ITS covers a factor 4 in frequency (10–40 MHz). Finally, the beam pattern of ITS at 30 MHz is shown in the right hand panel. The sensitivity is depicted in dB, 0 dB being the sensitivity of the main lobe pointed towards the zenith.

acquired from the TIM boards. These signals can be sent to the central processing and storage machine directly, but they also offer some initial processing options, such as statistics computation and Fourier transform. The central processing and storage machine can either store the incoming data or do some additional processing, such as beam forming or correlation.

3. Data reduction theory

For our observations we used the data acquisition PC's to do a fast Fourier transform on the acquired time series $x_i[n]$. For this operation the data was split up in non-overlapping blocks of N_{FFT} samples each forming a single snapshot. A Hanning taper $w_{han}[n] = \frac{1}{2}(1 - \cos(2\pi n/N_{FFT}))$ was applied to each snapshot (with index m) before doing the actual Fourier transform to obtain

$$y_{i,m}[k] = \sum_{n=0}^{N_{FFT}-1} w_{han}[n] x_i[n + mN_{FFT}] e^{-2\pi jnk/N_{FFT}}. \tag{1}$$

Since the rest of the discussion focuses on operations on a single frequency channel, we will drop the explicit statement of the dependence on k, i.e. $y_{i,m} = y_{i,m}[k]$. On the central processing and storage machine, first a full correlation matrix for every individual snapshot

$$\mathbf{R}_m = \mathbf{y}_m \mathbf{y}_m^H \tag{2}$$

is calculated. In this equation $\mathbf{y}_m = [y_{1,m}, y_{2,m}, \ldots, y_{p,m}]^T$ where the number of used elements is p and the superscript H denotes the Hermitian transpose. Before storage this result is integrated over M snapshots to arrive at the integrated array correlation matrix (ACM)

$$\mathbf{R} = \sum_{m=0}^{M-1} \mathbf{R}_m. \tag{3}$$

Beam steering in phased array telescopes can be done by applying complex weights w_i to the individual elements such that signals from the selected direction are added coherently. The voltage output of the beam former can thus be described as

$$A_m = \mathbf{w}^T \mathbf{y}_m \tag{4}$$

where $\mathbf{w} = [w_1, w_2, \ldots, w_p]^T$. The power output of the beam former after integration over M snapshots will be

$$|A|^2 = \sum_{m=0}^{M-1} A_m \bar{A}_m, \tag{5}$$

where $\bar{A}_m$ denotes the complex conjugate of A_m. This result can also be obtained using the ACM, since

$$\bar{\mathbf{w}}^H \mathbf{R} \bar{\mathbf{w}} = \mathbf{w}^T \left(\sum_{m=0}^{M-1} \mathbf{y}_m \mathbf{y}_m^H \right) \bar{\mathbf{w}} \tag{6}$$

$$= \sum_{m=0}^{M-1} (\mathbf{w}^T \mathbf{y}_m)(\mathbf{y}_m^H \bar{\mathbf{w}}) \tag{7}$$

$$= \sum_{m=0}^{M-1} A_m \bar{A}_m \tag{8}$$

$$= |A|^2 \tag{9}$$

The theory presented up to this point assumes a perfectly calibrated array where all elements have intrinsic gain 1, so the output of the beam former is completely determined by the input signal and the beam forming weights. In practice element i will have intrinsic gain g_i and receiver noise power of σ_i^2. These properties can be summarized in vector $\mathbf{g} = [g_1, g_2, \ldots, g_p]^T$ and vector $\boldsymbol{\sigma}^2 = [\sigma_1^2, \sigma_2^2, \ldots, \sigma_p^2]^T$, respectively. If the ACM as would be measured at an ideal noiseless array is denoted by $\mathbf{R_0}$, then the measured ACM will be

$$\mathbf{R} = \mathbf{g}\mathbf{g}^H \odot \mathbf{R_0} + \mathrm{diag}(\boldsymbol{\sigma}^2) \tag{10}$$

where $\odot$ denotes element-wise multiplication of two matrices and $\mathrm{diag}(\boldsymbol{\sigma}^2)$ is the diagonal noise matrix. The elements of $\mathbf{R}$ represent the sky brightness spatially integrated by the antenna elements.

If the system is sky noise limited, the receiver will contribute less to the autocorrelation power than the sky. Therefore the second term in Equation (10) can be neglected. Thus

$$\mathbf{g}\mathbf{g}^H = \mathbf{R} \oslash \mathbf{R_0} \tag{11}$$

where $\oslash$ denotes element-wise division of two matrices. The gain vector $\mathbf{g}$ is now the Eigen vector corresponding to the largest Eigen value of $\mathbf{g}\mathbf{g}^H$.

4. Sky noise limited snapshot imaging

In this section, we will demonstrate the near sky noise limited behavior of the ITS after calibration. The starting point will be a 6.7 s observation done on February 26, 2004 at 3:50 AM. A 8192 point FFT was used and integrated ACMs were computed based on the first 4096 Fourier coefficients providing ACMs for 9.77 kHz wide frequency channels covering the full range from 0 to 40 MHz.

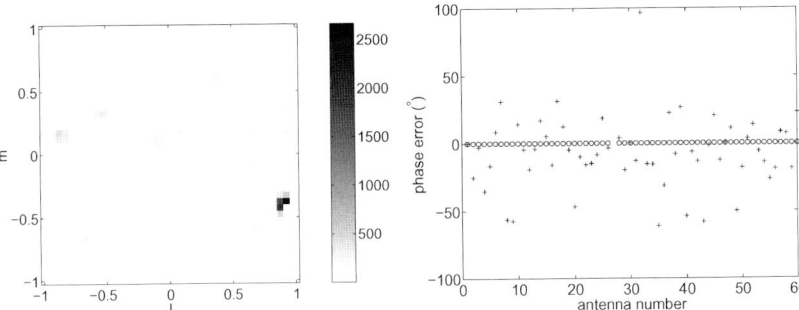

Figure 2. The full sky image shown in the left panel is based on a full correlation matrix for a 9.77 kHz wide frequency channel around 18.916 MHz from a 6.7 s observation on February 26, 2004 at 3:50 AM. The image is completely dominated by the RFI source on the south–eastern horizon. The image values are the direct result from Equation (6) after normalizing **R** such that all its diagonal elements are equal to 1. The right hand plot shows the phase errors of the individual signal paths before (+) and after (◦) calibration on the RFI source. The standard deviation before calibration is 26.5°, after calibration it is only 0.0011°.

As shown in the left panel of Figure 2 the 9.77 kHz band around 18.916 MHz is completely dominated by an RFI source on the south-eastern horizon. This source was used as calibration source following the procedure outlined in the previous section. The right hand panel of Figure 2 shows that this greatly reduces the phase errors in the individual signal paths (standard deviations before and after calibration are 26.5° and 0.0011°, respectively). The errors after calibration are the combined result from neglecting the finite sample effect and ignoring the sky and system noise contributions.

The phase errors before calibration are partly frequency dependent due to path length differences in the analog circuitry and partly frequency independent due to slightly different responses of the analog filters. Since there is only a gradual change in the phase over frequency, we can safely apply the calibration data at 18.916 MHz to two neighboring RFI free frequency channels centered around 18.760 and 18.779 MHz from the same observation. The resulting images are shown in the left and middle panel of Figure 3.

The right panel of the same figure shows the differences between these two sky images. The differences are expressed as fractions of the sky background level in these sky maps. The standard deviation of the image values is 0.0057. If the system is sky noise limited, one would expect the noise in the individual sky maps to be $\frac{T_{sky}}{\sqrt{B\tau}}$ where T_{sky}, B and τ denote the sky noise temperature, the bandwidth and the integration time. Based on the observational parameter, one would thus expect the noise to be a fraction 0.0039 of the sky noise. This should be multiplied by $\sqrt{2}$ for the difference image giving an expected value of 0.0055. From the agreement between the measured (0.0057) and the expected (0.0055) values we can conclude that the system has sky noise limited performance.

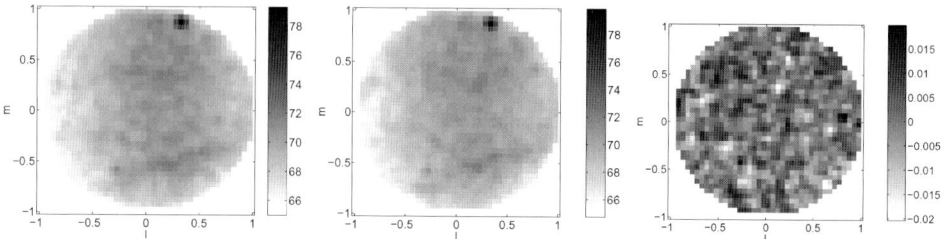

Figure 3. The left and middle panel show full sky images based on full correlation matrices for two 9.77 kHz frequency channels centered around 18.760 (left) and 18.779 MHz (middle) from the same 6.7 s observation as mentioned in the caption of Figure 2. The image values are the direct result from Equation (6) after normalizing $\mathbf{R}$ such that all its diagonal elements are equal to 1. Both images show the sky background with the north galactic spur on the southern hemisphere and Cas A near the northern horizon. The right panel shows the differences between the two maps. The image values are expressed as fractions of the sky background levels in the left and middle plots. The standard deviation of the image values is 0.0057.

5. Spatial RFI removal using CLEAN

The CLEAN algorithm removes a point source with power σ_{src}^2 in the visibility domain. In terms of spatial filtering this is called subtraction filtering (Boonstra and Van der Tol, accepted for publication) and can be described as

$$\mathbf{R}_- = \mathbf{R} - \sigma_{\mathrm{src}}^2 \mathbf{a}\mathbf{a}^{\mathrm{H}} \qquad (12)$$

where $\mathbf{a} = [a_1, a_2, \ldots, a_p]^{\mathrm{T}}$ describes the ideal array response to a point source of unit power at position $(l_{\mathrm{src}}, m_{\mathrm{src}})$.

When the source in the left panel of Figure 2 is subtracted from the image in the image domain before calibration, the image in the left panel of Figure 4 is obtained.

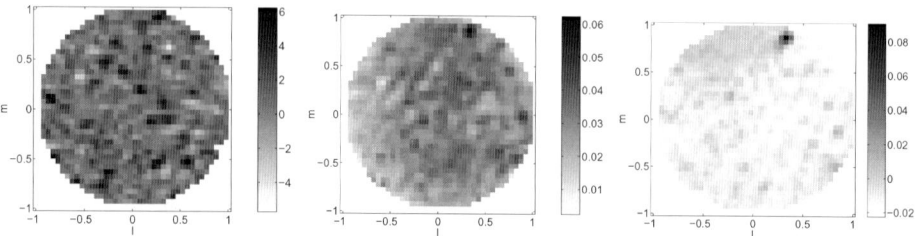

Figure 4. The left panel shows a full sky image of the RFI infested channel of Figure 2 after subtracting the RFI source in the image domain without calibration. The image values are represented as percentage of the peak value in the original image. This shows differences between the expected and the actual beam pattern as large as 5% of the maximum. The middle panel shows the result after subtraction of the RFI source in the visibility domain after calibration on the same scale. In this image Cas A becomes visible. The right panel shows a full sky image after another calibration on Cas A, an approach called peeling.

The values are expressed as percentage of the peak value in the original image. This shows that the differences between the expected beam pattern and the actual beam pattern are as large as 5% of the maximum.

When subtraction is done in the visibility domain using the procedure outlined above but deriving the necessary parameters (σ_{src}^2 and (l_{src}, m_{src})) from the image after calibration, the sky map in the center panel of Figure 4 is obtained. Although the power estimate was not perfect as can be seen from the fact that there is an "absorption" feature at the location of the RFI source, the image clearly shows Cas A on the northern horizon. This plot also shows that the ITS approximates the behavior of an ideal array within 0.05% otherwise we would not be able to see significant features at this level.

Finally, an attempt was made to refine the calibration parameters by recalibrating the array correlation matrix obtained after subtraction of the RFI source using Cas A. This approach, which is called peeling (Noordam, 2004), resulted in the sky map shown in the right panel of Figure 4. This image clearly shows extended emission along the northern horizon which can be identified with the galactic plane. This shows that even with a single LOFAR station significant features of 0.01% of the strongest source in the field can be detected indicating a usable dynamic range of over 10^4. It also shows blobs of similar power in other parts of the sky which can not be identified with astronomical sources. We therefore conclude that these are artefacts due to imperfect subtraction of the RFI source.

This image also shows a shortcoming of the point source calibration approach in that it assumes an isotropic element beam pattern, which does not hold for the inverse V-shaped dipoles of the ITS. This becomes apparent when comparing sky noise intensity distribution in the image in the middle panel of Figure 4 with that in the right panel of the same figure. In the middle panel it clearly decreases toward the western horizon due to the reduced sensitivity of the individual elements at low elevation. The calibration routine assumes that a single point source is the only structure on the whole sky and will thus flatten out any other structure that may be present. This causes the sky noise to look completely homogeneous in the right panel.

6. Conclusions

Based on a 6.7 s observation in a 9.77 kHz band it was shown that a single LOFAR station closely resembles the behavior of a sky noise limited system in a single snapshot. This has been achieved by a simple spectral Hanning taper for the whole 10–40 MHz band sampled by a 12 bit ADC.

We have also demonstrated that astronomical observations can be done in the presence of RFI using standard astronomical approaches. To this end an RFI infested image from a 6.7 s observation in a 9.77 kHz band around 18.916 MHz was taken. After selfcal on the RFI source and using CLEAN to remove it from the image Cas A and some emission from the galactic plane became visible indicating a dynamic range of over 10^4.

A suitable point source calibration approach for sky noise limited systems was presented in this paper. The results obtained in the paper implicitly show that this calibration technique attains the required level of accuracy; strong RFI signals can be used for initial complex gain calibration to a level of 3×10^{-6}, limited by the noise fluctuations in the system. For other positions on the sky we obtain a dynamic range of only 10^4, which is however enough for the strongest sky source.

References

Boonstra, A. J. and van der Tol, S. : accepted for publication, *Spatial filtering of interfering signals at the initial LOFAR phased array test station*: Radio Science, Special section on Mitigation in Radio Frequency Interference (RFI) in Radio Astronomy.

Noordam, J. E.: 2004, *LOFAR Calibration Challenges*, SPIE conference on astronomical telescopes and instrumentation, Glasgow (UK), June 21–25, 2004.

van der Veen, A. J., Leshem, A., and Boonstra, A. J.: 2004, *Signal Processing for Radio Astronomical Arrays*, SAM workshop, Barcelona (Spain), July 18–21, 2004.

THE LOFAR CENTRAL PROCESSING FACILITY ARCHITECTURE

KJELD VAN DER SCHAAF*, CHRIS BROEKEMA, GER VAN DIEPEN
and ELLEN VAN MEIJEREN
ASTRON, P.O. Box 2, 7990 AA Dwingeloo, The Netherlands
(*author for correspondence, e-mail: schaaf@astron.nl*)

(Received 23 July 2004; accepted 28 February 2005)

Abstract. Reconfiguration is a key feature characteristic of the LOFAR telescope. Software platforms are utilised to program out the required data transformations in the generation of scientific end-products. Reconfigurable resources nowadays often replace the hard-wired processing systems from the past. This paper describes how this paradigm is implemented in a purely general-purpose telescope back-end. Experiences from high performance computing, stream processing and software engineering have been combined, leading to a state-of-the-art processing platform. The processing platform offers a total processing power of 35 TFlops, which is used to process a sustained input data-stream of 320 Gbps. The architecture of this platform is optimised for streaming data processing and offers appropriate processing resources for each step in the data processing chains. Typical data processing chains include Fourier transformations and correlation tasks along with controlling tasks such as fringe rotation correction. These tasks are defined in a high level programming language and mapped onto the available resources at run time. A scheduling system is used to control a collection of concurrently executing observations, providing each associated application with the appropriate resources to meet its timing constraint and give the integrated system the correct on-line and off-line look and feel.

Keywords: cluster, correlator, high performance computing, HPC, parallel processing, pipeline, streaming data, supercomputer

1. The LOFAR central processor facility

LOFAR is an acronym for the *LOw Frequency ARray*, a phased-array radio telescope operating over the very low frequency range from 20 to 240 MHz. The LOFAR telescope is built as a distributed sensor network, with 100 stations distributed along five spiral arms. The stations hosts two antenna systems for the frequency bands from 20–80 MHz and 110–240 MHz, respectively. Each station contains 100 antennas of each type. At the stations, signal filtering and beamforming are done on a frequency window of 32 MHz for each antenna channel. Multibeaming is provided through up to eight independent field of views per station. Along with antennas for radio astronomy, other sensor types are located at the station sites to provide data for other scientific fields such as geophysics. For a more extensive description of LOFAR, see Butcher (2004) and de Vos et al. (2001).

The station-oriented processing is performed almost completely on digitised signals. Most of the processing tasks are executed on Field Programmable Gate

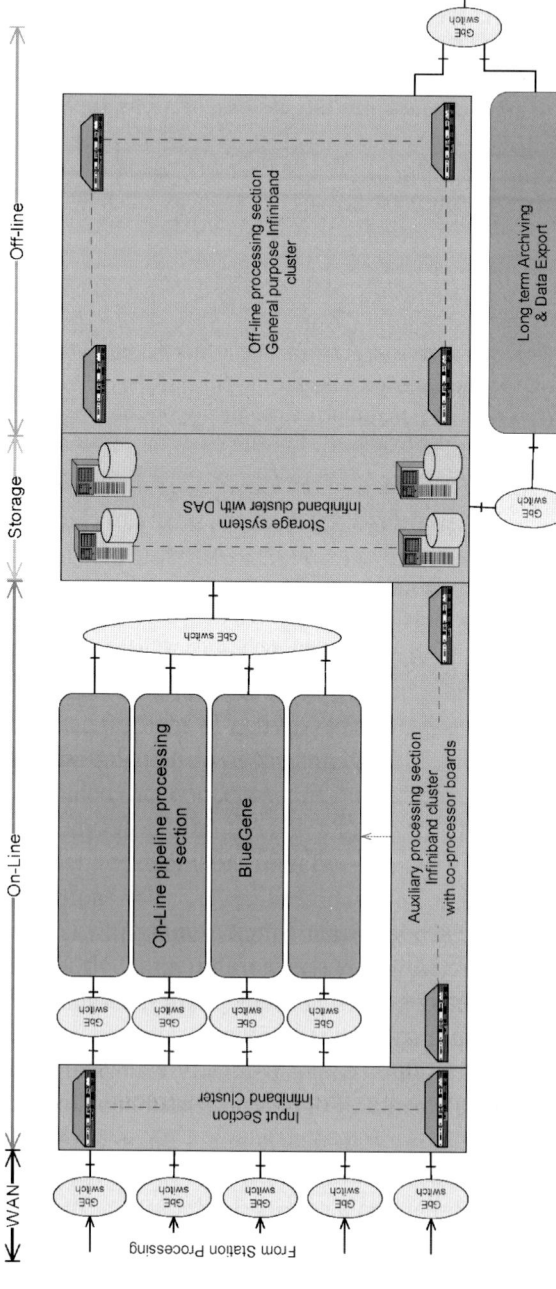

Figure 1. Central Processor Facility hardware architecture overview. An IBM BlueGene supercomputer is embedded in a Linux cluster computer. The cluster is connected internally with Infiniband; the connections between the cluster and the BlueGene IO nodes is through Ethernet switches connected to a selection of the cluster nodes. The sizes of the various parts of the cluster are scalable.

Arrays (FPGA) using re-configurable firmware. The main station processing data transformations are filtering, frequency separation and beamforming. Signals for each voltage beam are transported to the central site for further processing. This paper describes this *CE*ntral *P*rocessing facility (CEP) in more detail. We will use the abbreviation CEP in the remainder of this paper.

The aim of the CEP architecture is to provide enough processing power, data transport bandwidth and storage capacity to allow for continuous operation of the observation facility. Along with that, the operation of the facility must allow for fast reconfiguration and multiple concurrent observations. LOFAR can be reconfigured to observe in a wide range of observational modes, of which the key modes are:

- For sky imaging applications, the central processor is used for correlation, beamforming, integration and, after short-term storage, calibration and image creation.
- For pulsar observations, an "extended core" of up to 50% of the stations produces data for a central beamformer. The major processing tasks to be executed on the CEP are beamforming and de-dispersion measurement.
- For transient detection and analysis on time scales of a second or more, a series of snapshot images can be analysed using Fourier transforms. For shorter time scales, detection algorithms operate directly on the station beams data stream. In this mode the main tasks to be executed on the CEP are correlation and on-line calibration.

Since the LOFAR stations can produce multiple digital beams simultaneously, it is possible to observe in multiple observing modes concurrently. We will see later how this feature is supported in the CEP software model.

The end products of the CEP are typically standard final products such as images, spectra and source lists. For other users calibrated data cubes will be delivered. The data product can be specified for each individual user or observation. For surveys all data reduction will be performed automatically. At the other extreme, complete datasets can be provided if needed. In general, the idea is to perform the largest bulk of processing since the CEP provides very high volumes of processing power. Data reduction is done at least to a point were the data volume is decreased enough to allow for transport and further processing on more moderate processing facilities.

The hardware architecture of the central processor is based on cluster computer techniques with a high bandwidth switch fabric interconnect system. Multiple clusters with different hardware configurations are combined, each cluster being optimised for the specific tasks in the complete data processing flow. An IBM BlueGene (Adiga et al., 2002) supercomputer is integrated in the cluster environment to provide the bulk processing power needed particularly for correlation and beamforming tasks. The BlueGene supercomputer provides massive processing power of 30 Tflops peak in a 6144 node cluster with a three-dimensional interconnect topology. This architecture is schematically shown in Figure 1.

2. Operational model

The available processing power available in the CEP system can easily be distributed over multiple applications running in parallel by the use of advanced schedulers, as explained further on. The input section receives data from the LOFAR stations and routes these data streams to the appropriate processing applications that implement the required processing tasks. Multiple tasks can be connected to the input section concurrently, using a switch fabric to distribute the data to the appropriate processing resources.

In Figure 2 we see a typical example of processing tasks during operation of the CEP. Four concurrent on-line observations are running, each storing their data through storage access tasks running in the storage section. The input section and auxiliary processing section are typically running supporting observations only. Multiple off-line processing pipelines are running concurrently, accessing multiple datasets on the storage cluster. The multi-beaming facilities of LOFAR result in typically two to eight concurrent observations being processed concurrently. This results typically in 15–20 concurrently running applications on the CEP resources.

The storage cluster is also operated in a service mode. Storage applications run on this section that receive data from the processing tasks and implement the distributed storage. In the same way, storage servers run that provide storage services to the off-line processing applications.

3. Subsystems

The CEP system is schematically shown is Figure 1, which shows the main sections and their connections. Data enters the CEP system at the left side of the figure and flows through the system, resulting in exported data products leaving at the right side. The major tasks and data transformations applied to the data are input data handling, back-end processing chains, data merge and storage, off-line processing pipelines and final data product generation and export. Those tasks are mapped onto the different sections of the system, as described below. For each section the main requirements are stated and we will explain how these are fulfilled in the architecture.

Four of the sections in the CEP systems are implemented as cluster computers. Each of these clusters contain hundreds of cluster nodes. These clusters use Infiniband[1] for high bandwidth communication both between the nodes within the clusters as well as in between the subclusters. The BlueGene system is connected to two of those subclusters using Ethernet connections.

The subsections of the CEP are described next. All information is based on the current status of the design and implementation efforts, which are close to the final designs (see also Section 6).

---

[1] The Infiniband Trade Association http://www.infinibandta.org/.

THE LOFAR CENTRAL PROCESSING FACILITY ARCHITECTURE 47

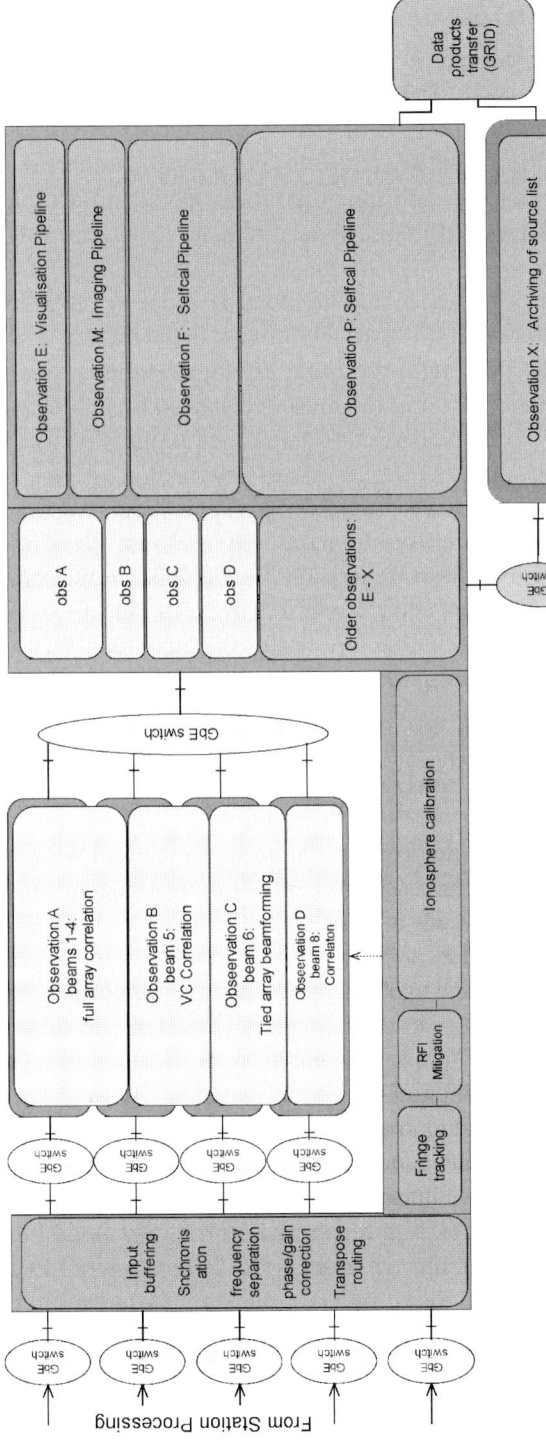

Figure 2. Typical mapping of applications and data files on the hardware resources in the operational system. Multiple observations are processed concurrently. The green applications show basic processing services that typically have a lifetime longer than the running observations. The observation specific applications are shown in yellow. These applications will have different lifetimes reflecting the observation schedule. The light blue applications operate on data observed in the past.

3.1. INPUT SECTION

This part of the central processing facility contains the connections to the stations (through the wide-area network). The total data stream originating from all LOFAR stations and entering the input section of the CEP is 320 Gbps. This is the total data content from the 100 stations, each observing a frequency band of 32 MHz with 1 kHz resolution with some additional overhead for identification and error detection. The data is transported in 16 bits complex samples for each frequency channel. Multiple voltage beams may be produced at the stations in which case the frequency bandwidth per beam would be lower in order to keep the total data production per station the same.

The input section must be able to receive these input streams. Furthermore, a re-sort of the complete data stream is required and station-wise filtering needs to be applied. This filtering is specified as one complex multiplication to each sample in the stream, which equals roughly 0.5 TFlops of processing power.

The data sent by the stations is sent in logical packages, each containing a time-frequency window of the signals from a single voltage beam. Depending on the number of beams produced at the station, such a voltage beam will contain up to 32 000 frequency channels. Each input node will receive data from the stations and run a data handling application that will buffer the input data and synchronise its output stream with the other input nodes based on the timestamps contained in the data. The connections to the WAN are implemented in 1 Gigabit Ethernet technology with two connections per input node.

The correlator architecture of LOFAR requires a transpose operation on all input signals at the central site. All 100 stations provide data for a spectral band of 32 MHz at 1 kHz resolution, yielding 32 000 channels per station. For the transpose operation this is a gigantic routing task. In the central processor design this transpose function is executed in two steps. In the first step, packages of multiple frequency channels are routed between cluster nodes receiving data from the stations and nodes connected to BlueGene. The receiving nodes memory can efficiently execute the remaining fine-grained transpose operation in cache. The transpose operation requires a bandwidth of order 2 Gbps between all nodes in the input section. For this subsystem, the internal Infiniband connections are heavily used to execute the transpose tasks, which re-sorts the station based data streams in the appropriate order for correlation. This resort interchanges the single station, all frequencies, data stream at the input side into all stations, and multiple smaller windows, at the output side. This transpose function is performed on the Infiniband switch fabric using an all-to-all connection scheme. Great care is taken to optimise this connection scheme for the particular switch topology and to prevent pathological breakdown of the switch performance through collisions or hotspots.

CPU power is available for various tasks. To begin with, the headers of received data packages are checked for time ordering and flags. Furthermore a combined correction is applied on the station data stream which corrects for fringe rotation, delay tracking and ionosphere calibration; these correction factors are produced by on-line processes running on the auxiliary processing subsystem and merged into a single phase/gain correction factor per data sample.

Of the order of 4GB of RAM per node is available to implement cyclic buffers of effectively 10–20 s of data. These buffers are used to correct for transport time differences between the stations. Moreover, the buffer length is extended to separate the hard real-time constraints of station data reception from the soft real-time operation mode of the downstream CEP applications. We have chosen for this soft real-time model in order to profit maximally from available software packages, in particular software available for high performance computing (HPC) and Linux clusters. This allows the application developer to use a greater range of libraries and tools as would be available in a hard real-time environment. For example, the message passing interface (MPI Snir, 1998) and CORBA[2] libraries are used for most data transport within the CEP.

The soft real-time regime allows for resource sharing between concurrently running applications and allows for the separation of back-end processing and so-called tuning applications that generate input parameter streams to the back-end tasks. These tuning applications run separately from the back-end processing tasks and are shared between multiple observations where possible. They execute typically on the auxiliary processing subsystem. Examples are the detection and mitigation control of interference signals and calibration of the phase rotation caused by the ionosphere.

3.2. BACK-END PROCESSING CHAINS

This section is the heart of the central processor facility. This is where the correlator is implemented in the case of imaging observations. The LOFAR operational requirements lead to the concept of multiple concurrently running observations in this part of the system. The observations can be of different types, implying for instance that a correlator is needed for some of them and a tied array beamformer for another observation. This motivated the choice for a re-programmable processing platform. We have analysed the various combinations of observational modes that must be available simultaneously and evaluated the processing needs for these combinations. This showed that deep-imaging observations with all stations and 32 MHz frequency bandwidth demand the highest processing power in this section. This is caused by the quadratic character of the correlator; the number of outputs is proportional to the square of the number of inputs. For this observation type, the back-end processing section must provide 10^{12} complex multiplications per second (1 trillion complex multiplications per second) on 8 bit complex samples.

This processing power is in the range of state-of-the-art supercomputers. The equivalent processing power on a supercomputer is in the order of 10–15 TFlops,

---

[2] See the OMG website: http://www.corba.org/.

depending on how efficiently this task can be mapped onto the hardware. Currently, this would place the CEP within the top 10 of the top 500 list of largest supercomputers in the world.[3] We have chosen to use supercomputer hardware to reduce the development effort for this section and, indeed, we have designed and tested the production prototype for this subsystem in about 5 man-years of work including the software development framework (see Section 5). This prototype contains a complete back-end for imaging observations with data handling, correlator and storage.

Large amounts of processing power and internal interconnection bandwidth are provided through the BlueGene supercomputer. This supercomputer contains 4096 compute nodes organised in a three-dimensional torus interconnect structure. The compute nodes each contain a dual-core processor with double floating point pipelines. These FPUs can perform two multiplications per clock cycle; so four multiplications can be performed on each compute node per clock cycle. This corresponds to the required multiplications of a complex multiplication, which is the basic operation of the correlation task. See Adiga et al. (2002) for an elaborate description of the BlueGene architecture.

The BlueGene resources are used to execute the heart of the back-end processing tasks. These are typically correlation and/or beamforming tasks for the main observational modes. Processing chains receive data streams from the input nodes. The output data streams are sent to the storage section for temporary storage. This communication is implemented using a communication library (e.g. Sockets or MPI) in order to isolate the application from the file system. Both these input and output applications communicate with the processing tasks via FIFO buffers. Therefore, the back-end processing applications can be simple data driven applications. This way, the programming model for the BlueGene applications is very simple and allows for the use of parallel programming models such as functional decomposition and master/slave operation (also called task farming). These standard paradigms offer well-established solutions to the application developer (Silva and Buyya, 1999). The basic parallelisation paradigm used for the back-end processing tasks is data decomposition, also known as data parallelism or single-program multiple-data (SPMD). The inputs data streams are split and sent to multiple processing task that each perform the same data transformation on their data stream. The basic dimensions that are used to split the data are beam direction (field-of-view) and frequency. So an individual correlation process executing on a particular BlueGene compute node will calculate all cross- and self-correlation products from the signals of a particular voltage beam for all stations for a limited frequency window. The neighbouring node will do the same, but on a different frequency windows and/or beam directions.

Since we will primarily use data parallel operation in this section, without communication between these processes, we program individual "compute cells" within the BlueGene supercomputer. Such a compute cell consists of one IO node and eight

---

[3] top500 supercomputer sites November 2004, http://www.top500.org/lists/2004/11/.

compute nodes. The eight compute nodes each run one process in the distributed MPI application, which typically consists of a few thousand processes. The MPI library is used for communication between these processes within the BlueGene.

The IO node provides the data communication between the BlueGene application and the Linux cluster environment. In hardware terms, this data transport into and out of this section is through bi-directional Gigabit Ethernet connections to a switch tree. Unix sockets are used for this communication between applications on the BlueGene and applications on the surrounding clusters.

3.3. AUXILIARY PROCESSING

This section provides processing power for additional on-line processing tasks such as ionosphere calibration and detection tasks – e.g. interference and transient phenomena on the sky. Such tasks are intended to tune the back-end processing applications (indicated by the dotted arrow in Figure 2). Low-bandwidth data transport connections are available to send these tuning parameters to the back-end processing tasks.

This part of the system will also be used for additional back-end processing tasks with specific needs that may be hard to implement for the BlueGene architecture or are not available at all for that platform. For example, the BlueGene applications must be single-threaded and some legacy libraries won't be available. Whether such tasks will execute on this cluster or on the BlueGene depends on the implementation details for the algorithm. At the current state of the project it is not yet clear if execution on BlueGene will be feasible. This auxiliary processing cluster provides the possibility to increase the processing capabilities for these applications if needed.

For these applications high-volume processing resources are provided by adding co-processor boards to the PC cluster nodes. Currently two types of processing boards are identified and tested: FPGA processor or graphics cards on the PC extension bus. This subsystem is especially scalable for an optimal adjustment of the implementation to the user needs.

3.4. INTERMEDIATE STORAGE

The storage section is responsible for the storage of complete measurements until the observation data are fully processed and the final data products are exported or archived. The collecting nature of this storage provides the transformation of streaming data into a coherent dataset. These coherent datasets are needed for calibration and analysis. Large parts of the observations, such as surveys, can be processed automatically, which reduces the use of storage volume. The required storage capacity is intended to hold the equivalent of two days of data from imaging type observations. This equals to at least 500 TByte of storage, which can be implemented in the storage cluster nodes. Each storage cluster node is equipped with

RAID disk arrays with a total volume of 5 TByte per node. The storage volume can be increased by adding new nodes to the storage cluster. Additional long-term storage and archiving are provided in LOFAR, but are outside of the CEP scope.

3.5. OFF-LINE PROCESSING CLUSTER

This section offers general-purpose processing power and high-bandwidth interconnections to the off-line processing applications. This section is used for all off-line processing applications. Important examples are self-calibration and image creation. The self-calibration application implements a minimisation algorithm with a model containing of order 1 00 000 parameter values which have to be fitted to a 25 TByte large dataset using iteration. This is a very complex minimisation tasks with many constraints. Image creation requires two-dimensional FFTs on slices of the complete data cube. Both these applications will require substantial processing power. However, these algorithms have not been used for routine data reduction at this scale before. Algorithm development is ongoing, but based on initial results we expect that the required processing power is of order of 2–5 Tflops, which corresponds to 200–500 cluster nodes. The exact processing power is not accurately known yet since the efficiency of the various applications is still increasing and the processing demands depend on the actual data quality. The architecture of this subcluster is very commonly used for high-performance computing calculations and can be scaled to thousands of nodes if required.

The largest part of this cluster is a standard Linux cluster computer optimised on cost per flop. Such a cluster uses a shared-nothing architecture which means that the cluster nodes have separated memory and file systems and are connected through a network. On such a cluster we apply message passing based programming. The cluster nodes will probably be dual CPU symmetric multiprocessing (SMP) nodes, allowing for small-scale shared memory programming. Dual CPU machines generally perform much better in a high-bandwidth network. Portability of applications to hardware resources available at the users' institutes will be guaranteed by installing the common general-purpose hardware available at the time of purchase.

A few multi-processor nodes with shared memory can be added for special processing tasks such as the solve task used in the Self-Calibration algorithm (Noordam, 2004), the size and number of these nodes will depend on the algorithms. Other node types can be added if needed for specific applications.

3.6. INTERCONNECT SYSTEM

Multiple high-bandwidth interconnect systems for clusters are available nowadays. These interconnect systems are supported by advanced middleware libraries such as MPI, parallel virtual machine (PVM) and shared memory libraries. With such software packages available one can develop programs with high performance communication patterns with relatively little development effort. The basic functional implementation is straightforward and one can directly focus on the performance optimisation. We have adopted Infiniband (see footnote 1) for our design, mainly because of the relatively low price. The design uses the $4\times$ specification which provides 10 Gbps per connection in four wire pairs and non-blocking crossbar switches.

A cluster wide switch fabric will be constructed from smaller switches, typically 32, 64 or 96 channels each. The topology of the Infiniband switch fabric is adjusted to the data transport needs in the various stages of processing.

In the input section, the transpose operation requires an all-to-all communication scheme, which is implemented in a fat-tree switch network. Such a fat-tree is also known as Clos or Constant Bisectional Bandwidth network (Clos, 1953; Leiserson, 1985). It is a switch topology in which smaller non-blocking crossbar switches are used to build a non-blocking switch topology supporting a much larger number of endpoints. Since we do not need the full point-to-point Infiniband bandwidth, we alter the fat-tree topology somewhat to spare switch capacity.

The combination of multiple special purpose clusters interconnected by an Infiniband network results in a very flexible and scalable system. The scalability is guaranteed by the division in smaller subsystems, for which the internal communication through the Infiniband switch fabric is scalable while staying well below any potential upper bound. Making use of the main dataflow direction, and therefore constructing a non-symmetrical switch fabric between the subsystems, can combine these subsystems together without scalability limits. This scalability has been demonstrated by detailed measurements on prototype processing applications; see also Schaaf et al. (2003a).

4. Software model

The overview of applications in the operational model, as shown in Figure 3, closely matches the hardware structure of the central processor facility. This is a result of the combined HW/SW design effort. All applications utilise multiple processing resources through parallel execution. Since we use a shared-nothing architecture, parallel operation is performed using message passing between the processes. Typically between 2 and 8 observations will run concurrently; these correspond to one processing application for each measurement. The measurement specific processing chains typically are mapped onto partitions in the BlueGene system.

All operation in the LOFAR system is first scheduled and then initiated by the Monitoring & Control system. The scheduling starts with a long-term volume scheduler, which is followed by a short-term scheduler that takes all scheduling constraints into account. Amongst these constraints is of course the visibility of sources. Additional constraints like predictions about interference and weather can be taken into account. The short-term scheduler takes the actual instrument

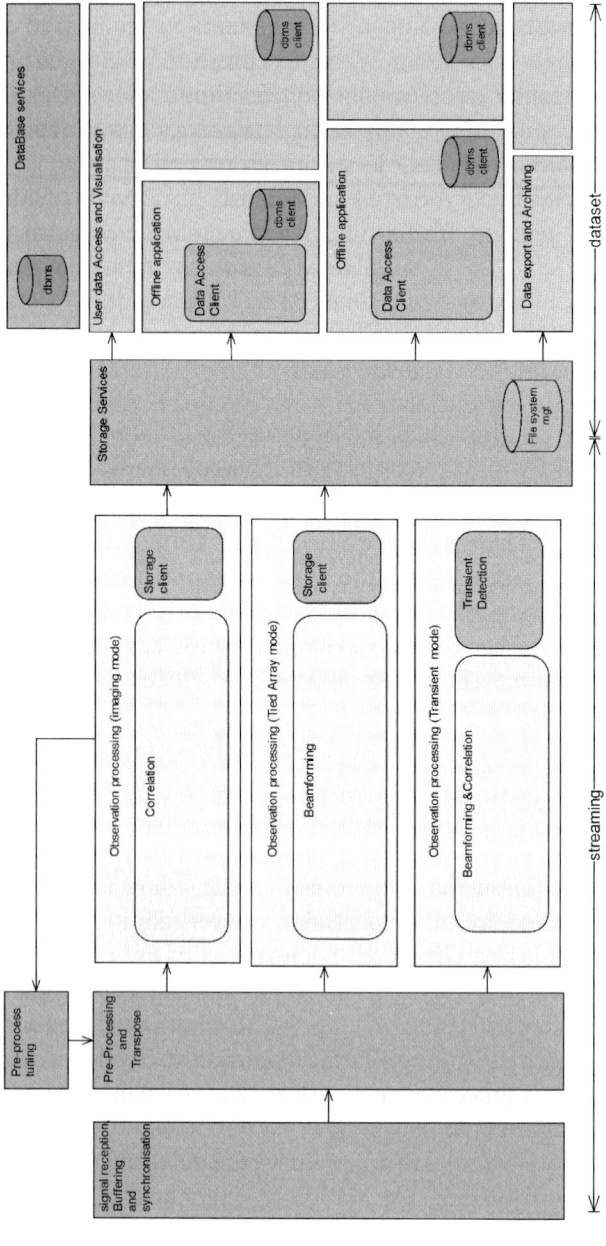

Figure 3. Overview of applications in the CEP system during typical operation. The input and storage applications run continuously, offering basic services to the remaining measurement or user specific applications. Multiple measurement applications run continuously executing different processing tasks and having different lifetimes.

availability provided by the monitoring system into account. A short-term schedule is generated for the complete LOFAR instrument. Upon execution of this schedule, the control system will use additional task schedulers to control the deployment of applications on the CEP cluster resources. These cluster schedulers implement load balancing on the cluster level.

All applications will operate as clients of the LOFAR Monitoring & Control (MAC) system, thus being started and controlled by the MAC system and providing monitoring data. MAC will start application controller tasks on the front-end nodes that will serve as interface to the distributed applications running on BlueGene and the Linux cluster.

5. Application development framework

All applications running on the central processing facility will be developed using the CEPFrame libraries. These libraries are used both on the Linux clusters and on the BlueGene, thus facilitating portability and data communication between these platforms. Three libraries exist with increasing functionality. The `Transport` library offers basic data transport functionality between *DataHolder* objects. The user will specialise that DataHolder class to reflect the required message structure. The transport library provides abstracted access to multiple often used communication mechanisms. The most important ones are MPI, Sockets and (shared) memory. The Transport library allows application programmers to specify the actual transport mechanism to be used for each separate connection. These mechanisms can be specified at run time. In this way applications, or components thereof, can be developed independently of middleware libraries. This allows for flexible operation since the optimal communication mechanisms for a particular combination of components with a particular deployment can be adjusted to the actual situation at run time.

The `tinyCEPFrame` library adds the concept of WorkHolders. A WorkHolder implements the data transformation from input data into output data. This gives the application developer the basic building stones to implement processing networks; see Figure 4 for an example processing chain. The tinyCEPFrame package is ported to the BlueGene platform.

The full `CEPFrame` package gives the application developer more tools to define the structure of the processing network at hand. Also the dynamic behaviour of the data transport can be defined better. For example, non-streaming data and for the communication of parameters between processes. The full CEPFrame is not available on the BlueGene platform, but applications developed with CEPFrame can communicate with those developed with tinyCEPFrame.

There is a huge difference in the nature of data between on-line and off-line applications. Most on-line application operates on streaming data and all off-line applications operate on datasets. This difference is reflected by using different

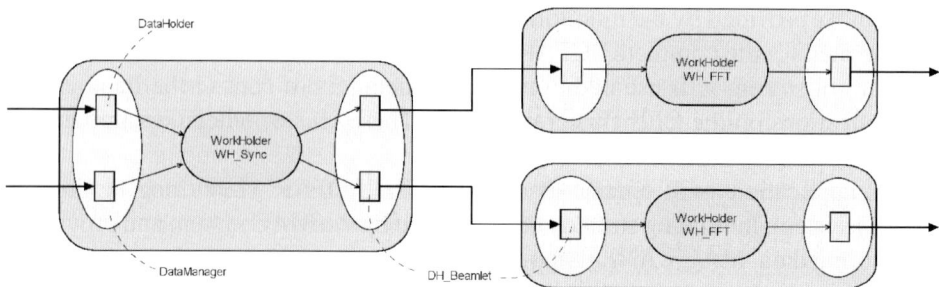

Figure 4. Example usage of the CEPFrame basis classes. This example shows a processing pipeline consisting of three processing blocks called "Steps." Each Step implements the data transformation from input data objects (called "DataHolders") into output data objects through user defined WorkHolders.

branches of the CEPFrame software for the application development. The current CEPFrame version includes a few features for non-streaming applications and fully supports the development of streaming data applications. More features for non-streaming applications are being added.

6. Status and further work

The architectural design of the LOFAR central processor facility is based on experiments on a small Linux cluster with a SCI switch fabric (Hellwagner and Reinefeld, 1999).[4] The switch fabric has been tested for SCI-specific performance, but was also used to test the performance of the CEPFrame application development framework in high bandwidth interconnects in general. Some of these results are presented in Schaaf et al. (2003b). On these resources we have developed the CEPFrame application development framework and tested the operational model. The tests included the co-operation of multiple applications, even with different lifetimes.

A new test cluster has just been installed which consists of 32 dual Xeon nodes with Infiniband interconnects. Additionally, two storage nodes with direct attached storage devices (DAS) are available to test the storage cluster concept and especially the interaction of storage servers with processing applications. The first results on this new hardware show good data communication performance. Point-to-point bandwidths of 700 MByte/s have been measured. A basic measurement for streaming data processing is shown in Figure 5. In this measurement a chain of three computers is used. The first one generates a random data stream that is sent to the second node. There we perform some floating point operations on each data sample and sent the output to the third node. From this figure we can conclude that

---

[4] SCI information on the web: http://www.dolphinics.com/ and http://www.scizzl.com/.

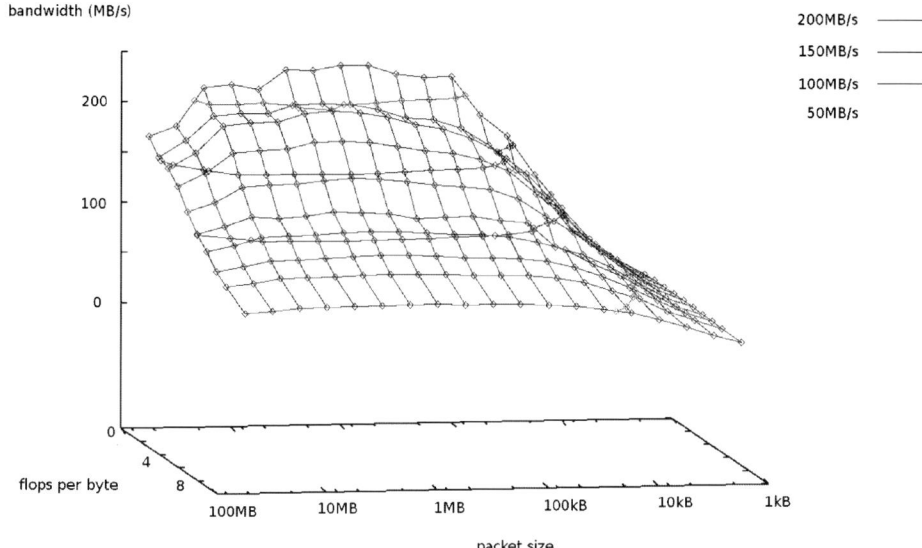

Figure 5. Initial results of streaming data processing throughput measurements. The throughput bandwidth through a cluster node is measured as function of package size and amount of processing that is performed to the data stream. The cluster node is a Dual Xeon running Linux and is connected to the input and output nodes using Infiniband 4×. A throughput of 200 Mbyte/s is observed when 2 floating point operations are applied to each input byte (equivalent to slightly more than 2 complex multiplications per complex<float> input sample).

basic data processing chains can be constructed with a throughput of 200 MByte/s (1.6 Gbps) per node. We expect some additional performance improvements for this particular hardware and MPI library.

A first implementation of a parallel correlator application for the BlueGene architecture has been developed. The current implementation provides over 50% of the theoretical available processing power per BlueGene compute node for the correlation function and possible improvements have been identified. This correlator demonstrator proofs the correlation on high performance computing concept and gives a first performance impression for the complete on-line processing system. The `tinyCEPFrame` application development framework has been ported successfully to the BlueGene platform and was used for the correlator demonstrator.

7. Conclusion

The LOFAR central processing facility architecture and operational model describe a pure software and HPC based telescope back-end. The presented architecture provides great flexibility and scalability of all processing resources. This flexibility will be especially used in the operational model of the instrument where we can run multiple observations of different kinds concurrently, even with different life times and in reaction to events that require quick follow-up observations.

The architecture is being tested on prototype implementation systems which show good performance so far. Scalability proofs towards the final scale are ongoing.

References

Adiga, N.R., et al.: 2002, *An Overview of the BlueGene/L Supercomputer*, Proceedings of SC2002, http://sc-2002.org/paperpdfs/pap.pap207.pdf.

Butcher, H.R.: 2004, *Proc. SPIE* **5489**, 537.

Clos, C.: 1953, *Bell Syst. Tech. J.* **32**, 406.

de Vos, C.M., Schaaf, K.v.d. and Bregman, J.D.: 2001, Cluster Computers and Grid Processing in the First Radio-Telescope of a New Generation, *Proceedings of IEEE CCGrid*.

Hellwagner, H. and Reinefeld, A.: 1999, Sci: Scalable Coherent Interface: Architecture and Software for High-Performance Compute Clusters, Springer-Verlag, ISBN 3540666966.

Leiserson, C.E.: 1985, October, Fat-Trees: *IEEE Trans. Comput.* **34**, 892.

Noordam, J.E.: 2004, *Proc. SPIE* **5489**, 817.

Silva, L.M.E. and Buyya, R.: 1999, *Parallel Programing Models and Paradigms*, chapter 1 of *High Performance Cluster Computing, Vol. 2*, R. Buyya, Prentice Hall PTR 1999, ISBN 0130137855.

Snir, M.: 1998, MPI: *The Complete Reference*, MIT press, ISBN 0262692163.

Schaaf, K.v.d., Bregman, J.D. and de Vos, C.M.: 2003a, *Hybrid Cluster Computing Hardware and Software in the LOFAR Radio Telescope*, Proceedings of PDPTA, CSREA Press.

Schaaf, K.v.d., Bregman, J.D. and de Vos, C.M.: 2003b, *Hybrid Cluster Computing Hardware and Software in the LOFARRadio Telescope*, Proceedings of IEEE PDPTA.

HEMISPHERIC IMAGING OF GALACTIC NEUTRAL HYDROGEN WITH A PHASED ARRAY ANTENNA SYSTEM

STEFAN J. WIJNHOLDS*, A. GER DE BRUYN, JAAP D. BREGMAN
and JAN-GERALT BIJ DE VAATE
ASTRON, Oude Hoogeveensedijk 4, 7991 PD Dwingeloo, The Netherlands
(*author for correspondence, e-mail: wijnholds@astron.nl)

(Received 28 August 2003; accepted 4 October 2004)

Abstract. The thousand element array (THEA) system is a phased array system consisting of 1 m^2 tiles having 64 Vivaldi elements each, arranged on a regular 8-by-8 grid, which has been developed as a demonstrator of technology and applicability for SKA. In this paper we present imaging results of Galactic neutral hydrogen with THEA. Measurements have been taken using a dense 2-by-2 array of four tiles as a four tile adder. The results are compared with results from the Leiden-Dwingeloo Survey, showing qualitative agreement, but also indicating that further studies are needed on the instrumental characteristics.

Keywords: Galactic neutral hydrogen, imaging, LOFAR, phased array, phased array antenna, SKA, Square Kilometre Array

1. Introduction

The international radio astronomy community is currently making detailed plans for the development of a new radio telescope: the Square Kilometre Array (SKA). This instrument will be a hundred times more sensitive than telescopes currently in use. ASTRON is developing one of the options for this new synthesis telescope, using antenna stations with phased array technology consisting of over one million receiving elements with a mixed RF/digital adaptive beam former. The thousand element array (THEA) (Kant et al., 2000) is one of the demonstrator systems that has been built during the SKA development program and is officially continued as the THEA experimental platform (THEP), but is usually still named THEA.

THEP is an out-door phased array system which was originally designed for 16 tiles, but currently consists of only four 1 m^2 tiles having 64 broadband Vivaldi elements arranged on a regular 8-by-8 grid each. Within each tile, beam forming is done at RF level before the normal and quadrature components of the signal are each sampled as 12-bit signal. This system is implemented twice in each tile allowing THEA to make two beams over the sky simultaneously.

The complex signals from the beams from all tiles are fed into a powerful digital backend where beam forming at array level is done. This backend provides an FX correlator having 1024 frequency channels and 20 MHz bandwidth allowing us to produce the power spectra of two signals or determine the cross correlated power of two signals. Both results can be integrated over time and stored on hard disk.

By scanning the beam of a single tile, a sky image can be formed at low resolution (Vaate et al., 2002; Vaate and Kant, 2002). In this paper we will show that such a system in phased array technology can be used successfully for astronomical observations by presenting results from observations on Galactic neutral hydrogen made with a four tile adder of 2-by-2 tiles. The results will be compared to results from the Leiden-Dwingeloo survey (LDS) (Hartmann, 1994; Hartmann and Burton, 1997).

2. Preparations on data from the Leiden-Dwingeloo survey

The Leiden-Dwingeloo Survey has provided an all sky map of Galactic neutral hydrogen visible from Dwingeloo. The Dwingeloo telescope is a 25-m single dish telescope having a half power beam width (HPBW) at 1420 MHz of 0.6°. The data is collected on a 0.5° grid or at 60% of the spatial Nyquist criterion to reduce the number of measurements. The RMS noise of the resulting maps is 0.07 K per frequency channel.

The LDS data cube is expressed in Galactic longitude l and latitude b and the velocity with respect to the local standard of rest (LSR), while the THEA data cubes are expressed in the directional cosines l and m and the velocity along the line-of-sight (LOS). Furthermore the spatial as well as spectral resolution of the two instruments differ. The following steps were taken to convert the data to the same coordinate system for comparison.

1. The spectral resolution of the LDS data is 1.03 km/s, while THEA operates with a spectral resolution of 4.22 km/s at 1420 MHz. Therefore the LDS data was divided in consecutive series of four frequency planes and averaged over the frequency planes within each series. This step also reduced the number of data points by a factor 4.
2. It was concluded that the first and last frequency planes only contained some small high velocity clouds, which would not be detectable by THEA. The relevant frequency planes were all found between the 70th and the 150th plane. Therefore the amount of data was reduced further by selecting only those planes for further processing.
3. The coordinate transformation between Galactic coordinates and the (l, m)-plane poses the problem that a regular grid in one coordinate system is mapped on a distorted grid in the other. Since the phased array beam pattern has a constant shape in the (l, m)-plane, convolution with the phased array beam pattern in the (l, m)-plane is algorithmically less complex than in Galactic coordinates. Therefore the convolution was done in the (l, m)-plane. Since a regular grid

greatly simplifies the convolution operation, a regular (l, m) grid with a spacing of 0.01 was defined and mapped on the (l, b)-plane. A bilinear interpolation was used to obtain the intensity values at the desired positions. This produces a $0.57°$ spacing in the (l, b)-plane in the zenith direction. This step results in a data cube of $201 \times 201 \times 81$ points, which is padded with zeroes on positions where $\sqrt{l^2 + m^2} > 1$.
4. The points in the resulting (l, m, v_{LSR}) data cube were weighted with $\sqrt{\cos(\frac{1}{2}\pi \sqrt{l^2 + m^2})}$ to account for the sensitivity pattern of the individual receiving elements.
5. The array beam pattern was calculated on a grid with $-2 \leq l, m \leq 2$ and a spacing of 0.01 assuming an ideal THEA tile with omnidirectional receiving elements aimed at the zenith. This assumption can be made since the sensitivity pattern of the receiving elements is already taken into account in the previous step. The seemingly very large grid is needed to account for the full side lobe and grating lobe structure for every point on the map.
6. The frequency planes of the (l, m, v_{LSR}) data cube were convolved with the array beam power pattern calculated in the previous step using the standard Matlab routine for two-dimensional convolution of two matrices. The resulting product of the element beam pattern and the array beam pattern serves a first-order approximation to the actual beam pattern of the instrument.
7. The final result was obtained by down sampling the convolved cube by keeping only the spectra at (l, m) coordinates at which the actual measurement were done. An example is shown in the left panel of Figure 1.

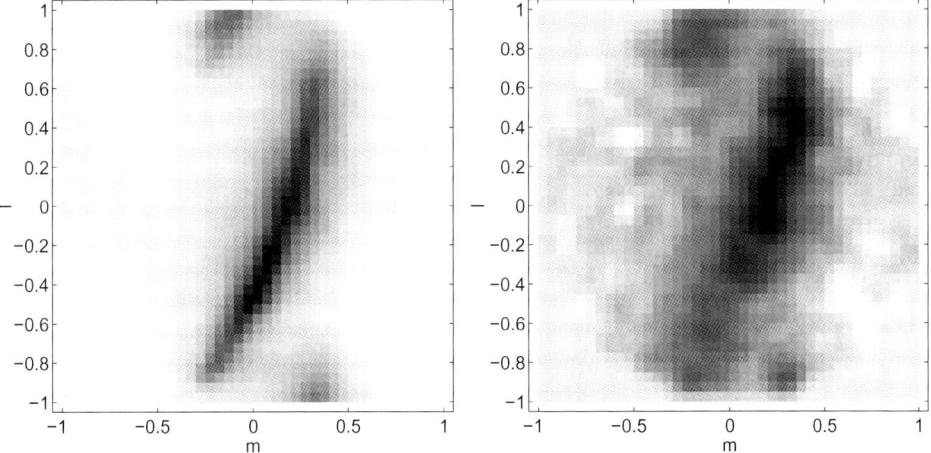

Figure 1. The left panel shows a total HI map based on LDS measurements after projection on the (l, m)-plane for April 8, 2003, 9:46.40 (GMT). The image is convolved with the array beam predicted for a 2-by-2 array of THEA tiles (see text for details). The right panel shows the total HI map actually measured with a 2-by-2 array of THEA tiles at the specified time. The image consists of 1246 points and was produced with 1 s integration per point.

3. Measurements with a 4-tile adder

For the observation of Galactic neutral hydrogen four THEA tiles were placed in a dense 2-by-2 configuration oriented parallel to the quarters of the compass. Beam steering was done by adjusting the phases of all 256 elements such that the array of THEA tiles acted as a single larger tile. Phase deviations of the tiles not due to the geometrical delay, for example phase differences caused by slightly different lengths of the LO cables, were corrected by calibration on Afristar at 1480 MHz. This calibration was also used to find the gain differences between the tiles. These phase and amplitude differences were corrected by multiplication of the signals from the individual tiles by appropriate factors in the digital backend before the adder stage. Using this setup a full sky scan was made on April 8, 2003, 9:46.40 (GMT) on a regular (l, m) grid with spacings of 0.05. Only the points above the horizon were actually measured, the spectra of the other points were set to zero. The scan consisted of 1246 points which were measured with 1 s integration. The complete scan was made in 23 min. Since the beam of the array of THEA tiles has a HPBW of 6°, the result may be slightly distorted due to sky rotation. However this will hardly be visible since neighboring points are measured within 1 min after each other.

RFI spikes were removed from the resulting data cube by comparing the value in each frequency channel to the median of the surrounding frequency channels and replacing its value by this median value if it deviated more than 20%. This procedure removes spikes very efficiently without modifying the more continuous features such as the shape of the pass band and HI profiles.

Baseline subtraction was done between 1420.1 and 1420.7 MHz by a linear interpolation between the lower reference band between 1419.9 and 1420.2 MHz and the higher reference band between 1420.6 and 1420.9 MHz. The choice for a simple linear interpolation was made based on a first visual inspection of the raw data. The resulting total HI map shown in Figure 1 was obtained by simply adding the intensity values between 1420.1 and 1420.7 MHz after baseline subtraction. The gray scale of the LDS image ranges from 0 K to the maximum in the map, while the gray scale of the THEA image ranges from the minimum of the fluctuations in the map to its maximum.

Comparison of the THEA result with the corresponding result from the LDS shown in the left panel of Figure 1 shows qualitative agreement, although there are some significant differences which require further explanation.

These fluctuations are about four times larger than expected based on the system temperature, bandwidth and integration time. This can be explained by imperfect baseline subtraction due to ripples in the baseline of the instrument on frequency scales smaller than the scale on which the linear interpolation is applied. This causes a slight over- or underestimation of the intensity along specific lines of sight. These errors are relatively large for low intensity values. This effect may also vary over the sky, since the baseline is also affected by low power intermodulation products from radio stations. Unfortunately we found that there are many of these intermodulation products around 1420 MHz and have not found an effective mitigation strategy yet.

An important aspect of a phased array system is mutual coupling between individual elements of the array. The mutual coupling impedances express the coupled voltages to other elements as a function of the actual current in an element. The latter depends on the actual loading condition of the array. If the sum of all coupled voltages differs 180° from the Thevenin voltage of an element, nothing is detected by that element. Since a THEA tile forms a regular array, this situation may occur simultaneously for a large number of elements, causing a sharp decrease in sensitivity for a few specific directions which are called blind angles. This could explain some of the white spots in the THEA image, but this needs further quantitative study. The last noticeable difference between the LDS and the THEA image is the shift in maximum intensity along the Galactic plane itself. This may be due to an asymmetric element beam caused by the asymmetry in the electromagnetic surroundings of the array and the fact that the tiles were not identical. Some simulation studies on the element beam have been done at ASTRON, but experimental verification has not been successful yet.

As a result of these difficulties and especially those concerning baseline subtraction, velocity profiles can differ strongly from the corresponding LDS results as shown in Figure 2. It has already been concluded that the noise is about four times as large as expected based on bandwidth and integration time, thus leading to a noise level of about 6 K per channel. There is however agreement in the sense that along lines of sight where the LDS velocity profile shows only a single peak, the THEA velocity profile also shows a single peak and when the LDS profile is more complicated, this holds for the THEA profile as well.

Therefore a sample of profiles can be compared to find the scaling factor relating the number of counts from the A/D converter to brightness temperature. This scaling

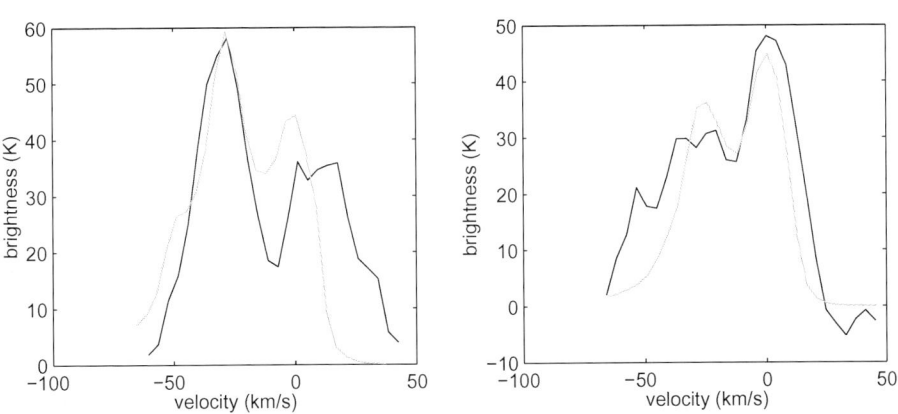

Figure 2. Comparison of velocity profiles from LDS (light gray) and THEA (black). The THEA profiles are scaled and shifted to match the LDS profiles.

factor was used to derive the noise temperature from the number of counts in the noise floor before the baseline subtraction. With this approach a system temperature of 186 ± 10 K was derived, consistent with independent tile system temperature measurements.

4. Conclusions

With the successful detection of Galactic neutral hydrogen we have demonstrated that astronomical observations are possible with telescopes in phased array technology. However, our measurements do indicate that further research is needed to deal with instrument properties typical for phased array antennas, such as blind angles and the fact that RFI from all directions is detected by the instrument. Since these issues have an effect on instrument calibration as well the calibration routines used should also be reconsidered.

On a qualitative level we have been able to explain the differences between measurements from the Leiden-Dwingeloo Survey and THEA. This demonstrates that we already have a considerable level of understanding of the operation of phased array telescopes and particularly the THEA system. On a quantitative level we have been able to use a sample of velocity profiles to derive the system temperature of the THEA system which was found to be 186 ± 10 K. It has become clear however that the aforementioned further research is needed if reliable astronomical measurements are to be done without using statistics, i.e. in a single measurement.

References

Hartmann, D.: 1994, *The Leiden/Dwingeloo Survey of Galactic Neutral Hydrogen*, PhD thesis, Leiden University, October.

Hartmann, D. and Burton, W. B.: 1997, *Atlas of Galactic Neutral Hydrogen*, Cambridge University Press.

Kant, G. W., Gunst, A. W., Kokkeler, A. B. J. and Smolders, A. B.: 2000, *Proceedings of the SPIE Conference on Receiver Architecture of the Thousand Element Array (THEA)*, Vol. 4015.

Vaate, J. G. bij de, and Kant, G. W.: 2002, The Phased Array Approach to SKA, Results of a Demonstrator Project', in *Proceedings of the European Microwave Conference*, Milan, September.

Vaate, J. G. bij de, Kant, G. W., Cappellen, W. A. van and Tol, S. van de: 2002, *First Celestial Measurement Results of the Thousand Element Array*, URSI-GA, Maastricht, August.

Experimental Astronomy (2004) 17: 65–77

ELECTRONIC MULTI-BEAM RADIO ASTRONOMY CONCEPT: EMBRACE A DEMONSTRATOR FOR THE EUROPEAN SKA PROGRAM

A. VAN ARDENNE[1], P. N. WILKINSON[2], P. D. PATEL[1,*] and J. G. BIJ DE VAATE[1]
[1]ASTRON, P.O. Box 2, 7990 AA Dwingeloo, The Netherlands; [2]Jodrell Bank Observatory, University of Manchester, Cheshire, SK11 9DL, U.K.
(*author for correspondence, e-mail: patel@astron.nl)

(Received 18 August 2004; accepted 28 February 2005)

Abstract. ASTRON has demonstrated the capabilities of a 4 m^2, dense phased array antenna (Bij de Vaate et al., 2002) for radio astronomy, as part of the Thousand Element Array project (ThEA). Although it proved the principle, a definitive answer related to the viability of the dense phased array approach for the SKA could not be given, due to the limited collecting area of the array considered. A larger demonstrator has therefore been defined, known as "Electronic Multi-Beam Radio Astronomy Concept", EMBRACE, which will have an area of 625 m^2, operate in the band 0.4–1.550 GHz and have at least two independent and steerable beams. With this collecting area EMBRACE can function as a radio astronomy instrument whose sensitivity is comparable to that of a 25-m diameter dish. The collecting area also represents a significant percentage area ($\sim$10%) of an individual SKA "station." This paper presents the plans for the realisation of the EMBRACE demonstrator.

Keywords: dense arrays, EMBRACE, European demonstrator for SKA, multi-beam phased arrays, multiple fields of View (FOV), novel radio telescope concept

1. Introduction

The international radio astronomy community is currently making detailed plans for the development of a new radio telescope, namely the Square Kilometre Array (SKA) (Hall, 2005). This instrument will be two orders of magnitude more sensitive than telescopes currently in use and is required to operate from 100 MHz to 25 GHz. Various reflector antenna design activities are being pursued around the globe. However, a unique SKA concept based on 2-D dense phased arrays (Van Ardenne, 2004), which comprise the entire physical aperture and which provide multiple, independently steerable, fields-of-view (FOVs) is being considered as the European contribution. This dense "aperture array" concept has hitherto been unavailable for radio astronomy. We are in the process of establishing the viability of aperture array technology in Radio Astronomy, which is particularly attractive for the frequency range 100 MHz to $\sim$2 GHz, where up to four of the five Key Science Projects recently identified for the SKA may be carried out.

In this paper a brief background of the precursor project (ThEA) is provided followed by a detailed description of the EMBRACE demonstrator and how it is likely to be designed, implemented and tested. We also give an outline of its usefulness as a unique radio astronomy observational instrument, especially operating in the lower part of the SKA frequency band.

2. Historical background – Thousand Element Array (ThEA)

ASTRON has built and evaluated a small-scale phased array prototype known as the Thousand Element Array (ThEA) (Van Ardenne et al., 2000; Bij de Vaate et al., 2003; Woestenburg and Dijkstra, 2003; Kant et al., 2000). This prototype demonstrated the capabilities of the dense phased array principle. ThEA consisted of four active "tiles," each of 1 m^2 in size and containing 64 Vivaldi antenna elements, as depicted in Figure 1. After amplification and beamforming, the signals from each tile were down converted, digitised and transmitted to the back-end by means of an optical fibre link. A dedicated back-end performed digital beamforming on the four digital data streams. The snapshots of data, together with the weighting factors of the beamformer derived from an RFI mitigation algorithm were processed in the adaptive digital beam-former. The output data of the digital beamformer was subsequently time integrated to obtain the required sensitivity levels. In addition to the fast dedicated hardware at the back-end, a flexible PC-based data processing capability was used to analyse the data.

With this platform a large number of experiments have been performed although the total collecting area of 4 m^2 limited experiments to strong radio sources only. Interferometer fringes of the sun gave information on the beam pattern. Geo-stationary satellites could also have been used for this calibration but, instead, they were used

Figure 1. Photograph of the THEA tiles at the ASTRON test site.

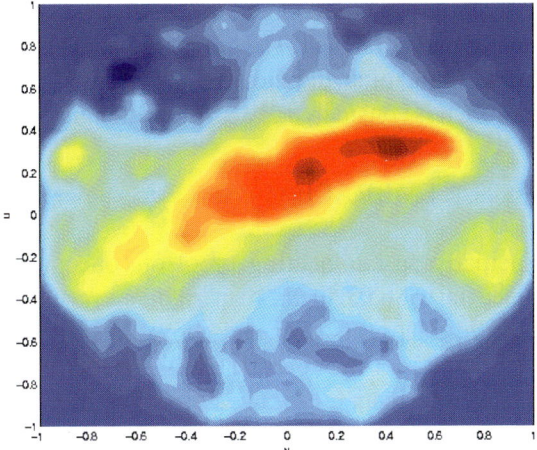

Figure 2. Full sky image of the intensity of HI radiation in our Galaxy measured with four THEA tiles plotted on the (u, v) plane with a resolution of 5.7 degrees near the zenith.

for beam-tracking as well for nulling experiments (Bij de Vaate et al., 2003). Of particular astronomical interest is that the HI spectral line from atomic hydrogen at 1420 MHz can be observed within the band. The "all-sky" HI survey map with the signals added from the four tiles is presented in Figure 2. A total of 1246 points have been measured to map the sky with an integration time of 1 s at each point. Very good agreement with observations performed with traditional reflector type telescopes has been found.

3. EMBRACE

3.1. Motivation for EMBRACE

Many institutes around the globe are considering the design of SKA antennas using parabolic and cylindrical reflectors as the collecting area, the properties of which are well known within the radio community. The novel antenna concept proposed by the Europeans for the SKA is the dense aperture array (Van Ardenne, 2004), in which elementary antennas are densely-packed and connected together with beamforming circuitry. This is the most innovative concept under investigation for the SKA, and it offers unique features and potential opportunities for future growth in radio astronomy. The encouraging result obtained from the ThEA project was the main reason leading the European radio astronomy community to adopt this concept.

EMBRACE will be much larger than ThEA, the goal being an array of area 625 m^2. It will operate in the band 0.4–1.550 GHz, similar to that used for ThEA, and will have at least two independent and steerable beams. This size of array can function as a radio astronomy instrument whose sensitivity is comparable to that of a standard 25-m diameter parabolic dish. The collecting area also represents a significant percentage area (~10%) of an individual SKA "station" (envisaged as 5000 to 10 000 m^2) and therefore it is considered as a most challenging step toward the SKA.

3.2. SCIENCE CASE FOR EMBRACE

EMBRACE is principally an engineering proof-of-concept and is not intended to be a major science instrument in its own right; nevertheless, it will provide some immediate scientific pay-offs over and above these short-term rewards, the long-term scientific potential of the aperture array concept is being held firmly in mind by the EMBRACE designers. First we note some immediate rewards of the programme:

- As a stand-alone instrument, with a collecting area comparable to that of a "standard" 25-m diameter radio telescope EMBRACE will, most crucially, demonstrate the ability to form at least two well-calibrated "fields-of-view" offering the full sensitivity of the array in very different directions on the sky simultaneously. This alone is sufficient rationale for the project and will create a paradigm-shift in respect of the astronomical community's perception of the observational potential of phased array radio telescopes.
- The sensitivity offered by a 25-m class instrument with two large fields-of-view will provide a new opportunity to examine the radio sky for bright transient sources, perhaps of unknown types. With the limitations imposed by the narrow fields-of-view of current radio telescope designs, little exploratory work has been done in this area for many decades.
- EMBRACE will also operate as an element in the Dutch WSRT array and, using aperture synthesis techniques, the combined system will be used to demonstrate the ability to form higher resolution images within separate fields-of-view. EMBRACE will enhance the imaging power of the WSRT by providing baselines in the N–S direction and allow it to observe more efficiently close to the celestial equator.

Turning to the long-term science potential of aperture arrays, the SKA's international Science Working Group (SWG) has recently agreed upon five Key Science Projects (KSPs), against whose requirements any technical design must be tested, (Jones, 2004). These are the following:

- Strong field tests of gravity using pulsars and black holes
- Probing the "Dark Ages" of the Universe
- The origin and evolution of cosmic magnetism
- The cradle of life
- The evolution of galaxies and large scale structure

Since the SKA will also be a tool for the next generation and indeed for users who are not yet born, the International SKA Steering Committee (ISSC) has therefore recommended that an additional Key Science Project be defined, namely:

• The exploration of the unknown

This places an onus on SKA planners to allow for the maximum flexibility in its design right from the start.

How are these science goals influencing the development of the aperture array concept? During the development of these KSPs it has become clear that much of the scientific potential of the SKA arises from its enormous surveying potential. In addition, new discoveries often arise from large-scale surveys during which rare discrete objects of a new type, or new large-scale emergent properties of the ensemble of objects, stand out. The surveying speed of a radio telescope is proportional to the square of the instantaneous sensitivity multiplied by the number of independent beams on the sky. Since all the SKA concepts now under development offer comparable sensitivity (roughly one hundred times greater than existing interferometer arrays), the sky coverage, *and the flexibility with which it can be deployed*, will be major factors in any assessment of their scientific potential. In our view the low-frequency aperture array concept, being pioneered in EMBRACE, has the greatest overall potential for meeting these scientific challenges. It offers:

- The largest possible fields-of-view ($>100\,\mathrm{deg}^2$). If this advantage can be realised then the aperture array SKA will have a surveying speed up to a million times faster than current instruments and up to two orders of magnitude faster than offered by other SKA concepts. These speeds may well prove to be vital for carrying out the enormous "all-sky" survey programmes involved in targeting the KSPs.
- The greatest possible number of design options for maximizing the flexibility with which the sky can be observed at any given time. In particular, we envisage many groups being able to observe with the full aperture simultaneously, thereby opening many "windows on the universe". More specifically the flexible multifielding capability, to be demonstrated by EMBRACE, generically incorporates:

  i. a science survey advantage – required for that range of key science programmes requiring large amounts of telescope time and which would be impossible with conventional systems;
  ii. a "community" advantage – many groups, including students and schools, can access the whole aperture simultaneously, allowing the operation of the SKA to resemble that of particle accelerators or synchrotron light sources;
  iii. a multiplex advantage – simply by increasing the volume of data which can be collected;
  iv. an adaptive beam forming advantage – "reception nulls" are steered to cancel out sources of radio frequency interference.

The different fields-of-view could, for example, be used for:

i. imaging a deep field – integrating for long periods for the ultimate in sensitivity;
ii. studies of time variable phenomena – seeking transient radio sources and responding instantly to transients discovered in other wavebands;
iii. pulsar timing – finding, and then picking out the unusual ones from 20 000+ pulsars;
iv. experimentation – not scheduled by standard peer-review.

Overall we believe that the advance offered by the aperture array concept promises a complete revolution in our ability to survey the radio sky – so far in advance of anything which has been done before that it truly can claim to take the SKA into an arena of "discovery science." But before we can move into this exciting new era we must first construct and operate a well-engineered proof-of-concept. This is the scientific motivation of the EMBRACE programme.

It must be noted, however, that the aperture array concept is currently only feasible at relatively low frequencies (<2 GHz) because of the fact that the number of antenna elements, and hence the overall complexity and cost, rises as (frequency)^2. In this respect the aperture array concept must look to other technologies based on shaped collector elements to meet the high frequency requirements (i.e. $\sim$2 to $\sim$20 GHz) of the SKA. We return to this issue later on.

3.3. Overview of the European collaborative proposal and its relationship with EMBRACE

To further investigate and provide solid information on the viability of the aperture array concept, a Consortium has been formed and a proposal submitted to the European Commission under the Sixth Framework research program (FP6). The Consortium (Square Kilometre Array Design Studies (SKADS) proposal, 2004), consists of 32 institutes, mostly from European countries, but Australia, South Africa and Canada are also partners.

The basic aim of SKADS is to establish cost-effective technologies appropriate for the Key Science Projects in the low frequency end of the SKA spectrum ($\sim$0.1–$\sim$2 GHz).

The SKADS program has been structured as a series of strategic *Design Studies* (DS) including both feasibility studies and technical preparatory work. The feasibility studies are "paper" exercises relevant to the design and costing of the SKA network and infrastructure, the overall assessment and a project plan. The technical preparatory work mainly involves hardware R&D associated with establishing the cost-effectiveness of our specific solution for an SKA "station." Each of these

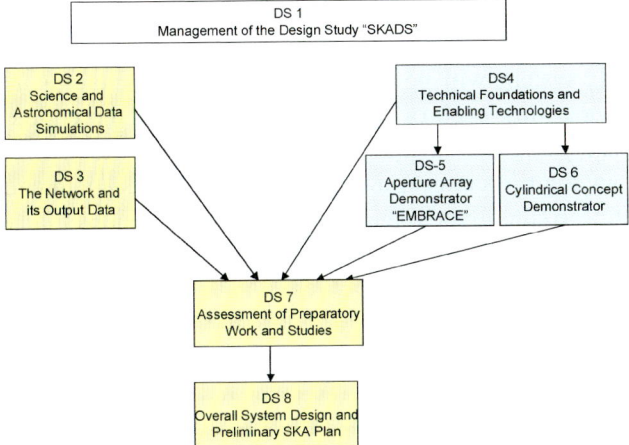

Figure 3. A schematic overview of the SKADS program.

Design Studies has been sub-divided into coherent *Design Study Tasks* with clear aims, milestones and deliverables. A schematic overview of the SKADS program is depicted in Figure 3.

The European demonstrator EMBRACE is embedded within the SKADS proposal and is described in detail in DS5.

Briefly, the Design Studies will address the following aspects of the SKA:

- The management of 'SKADS' (DS1);
- The development of technical specifications based on a quantification of key science drivers and a study of the technical requirements (DS2);
- An end-to-end study of the network data handling from collection to delivery to the astronomical users (DS3);
- The design of sub-systems for a cost-effective dual-polarised, all-digital, antenna system capable of forming multi-beams simultaneously and independently (requires a range of R&D to develop the technical foundations and enabling technologies appropriate for a final SKA design) (DS4);
- The construction of an engineering demonstrator low-frequency array resembling our vision of an SKA station and exploration of the practical issues involved in large-scale multi-beam acquisition and signal-processing concepts. This is the EMBRACE project (DS5);
- The application of phased array technology in the focal plane of mechanical flux concentrators (DS6);
- A continuous process of critical technical assessment and evaluation of *all* aspects of the SKADS programme (DS7);
- An overall system design and a preliminary, multi-component plan of how to realize the SKA (DS8).

3.4. EMBRACE SYSTEM CONCEPT

The EMBRACE system will be built on the same principle as ThEA. A large number of antenna tiles, each of area $\sim 1\,\text{m}^2$, will form the collecting area. The signals from the 50 to 250 elementary antennas from each tile will be amplified and initial RF (i.e. analogue) beamforming will be applied. It is not intended to perform any digital beamforming in the tiles but, due to the limitations imposed by the use of phase shift control with large instantaneous bandwidth requirements, a tile might be split in to quadrants. The outputs of these quadrants can then be combined with time-delay lines directly, or be sent to the backend for further digital beamforming.

A system level block diagram is shown in Figure 4, indicating the formation of two independent fields-of-view at the tile level and at the aperture array "station" level. Note that all the processing is done at the back-end using an analogue link from the aperture array. For the analogue link we are pursuing an RF-on-fibre approach where the RF signal modulates a laser directly. The analogue link from the tiles to the back-end processing simplifies the tile design and totally decouples the analogue antenna from the receiver thus improving, for example, cross-talk, since the local oscillator and clock signals now have to be distributed in the back-end cabinet only. The down-conversion and digitisation is also carried out in this back-end cabinet.

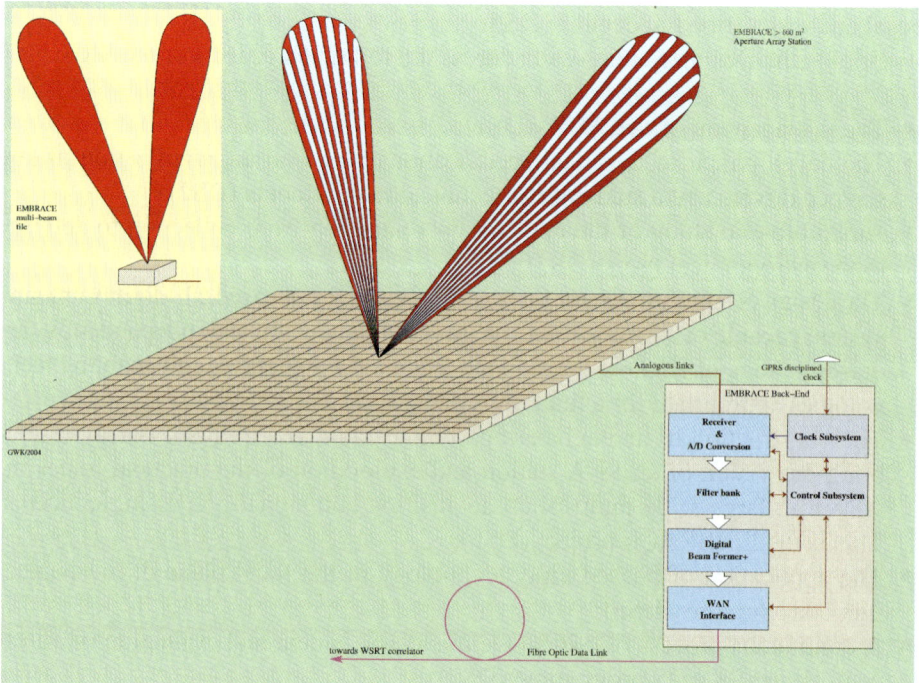

Figure 4. A system level block diagram of EMBRACE showing the formation of independent multiple fields-of-view at both the tile and station level, with processing at the back-end.

3.5. EMBRACE DESIGN APPROACH

In the design of the dense aperture array (Van Ardenne, 2004), the radiating antenna element spacing should not be significantly above $\lambda_0/2$, where λ_0 is the wavelength at the design frequency, otherwise grating lobes appear as the array is scanned over a broad bandwidth, reducing the gain, efficiency and effective aperture. We are therefore forced to design the array with an elemental separation derived from the high end of the frequency range. As previously mentioned in Section 3.2, observing frequencies above 2 GHz are not practical because of the large increase in the number of radiating elements required, necessarily driving up the cost.

Reiterating another point made in Section 3.2, the electronic steering and beam-forming creates the possibility of generating multiple independent fields-of-view, which makes the dense aperture array antenna a *multi-user* instrument, a very important aspect for a radio astronomy instrument of the size and cost of the SKA. The effective observing time therefore can now be expended without the loss of sensitivity simply by providing more electronics and computing power to create more beams.

3.6. EMBRACE SPECIFICATION

In Table I the specifications for the key parameters of the EMBRACE system are given. From the collecting area and the $\lambda_0/2$ spacing at 1 GHz, it is estimated that a total of 50 000 antenna elements will be required. The system will be built with tiles approximately of 1×1 m size although larger tiles, up to 2.5×2.5 m, will also be considered.

TABLE I
EMBRACE specifications

Frequency range of receiver chain	400–1550 MHz.
Element separation	$\lambda_0/2$ at 1 GHz
Polarization	Single polarisation
Physical collecting area	$\approx 25 \times 25$ m^2
Aperture efficiency	>0.80
Electronic scan range	Full hemispherical
T_{sys}	<100 K @ 1 GHz (aim for 50 K)
Antenna element phase control accuracy	3 or 4 bit
Instantaneous bandwidth	40 MHz
Dynamic range A/D converter	60 dB
Number of independently tuned FOVs (RF beams)	2
Number of digital beams	8 of 20 MHz per FOV

3.7. LOCATION AND TEST PLAN FOR EMBRACE

The EMBRACE instrument of 625 m² will be built close to the WSRT as this strategy offers several advantages:

- It will allow the use of existing infrastructure to speed up the validation process of EMBRACE.
- The WSRT correlator has four available spare signal channels allowing several configurations to be tested, such as:

  - 2 beams formed with 2 WSRT subarrays of seven telescopes correlated with the two independent FOVs of EMBRACE.
  - 1 beam of the complete WSRT correlated with 1 beam of EMBRACE, also in single polarisation.

- It will also allow access to a 1 GB/s fibre link for real time data transport to the Joint Institute for VLBI in Europe (JIVE) data processing centre, allowing Very Long Baseline Interferometer experiments.
- The WSRT is located inside a radio quiet zone.

A possible location for the siting of EMBRACE at Westerbork is shown in Figure 5. The positions of the existing telescopes are clearly visible; the EMBRACE proposed site is located approximately 1 km to the south of dish no. 10.

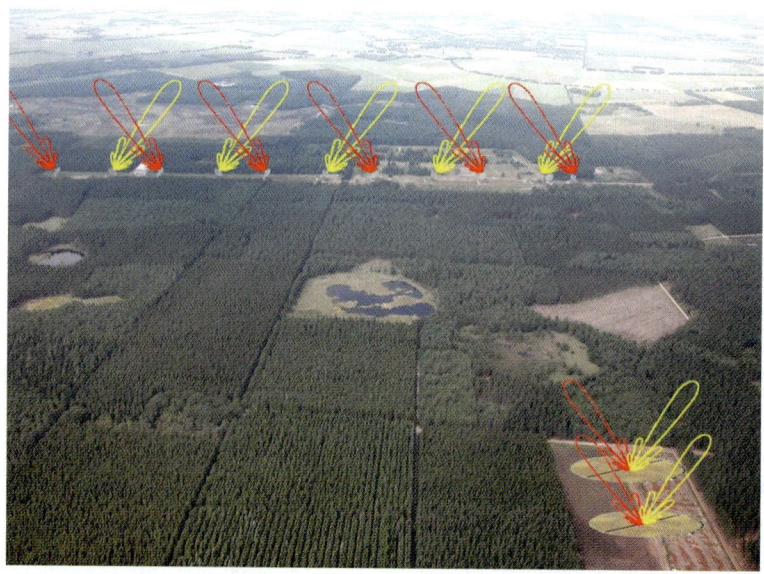

Figure 5. Possible location for EMBRACE siting at the Westerbork site.

The EMBRACE test field can also accommodate a LOFAR test station as is suggested with the two array depictions in Figure 5.

In conjunction with the Westerbork array possible experiments include low-declination imaging with EMBRACE as 15th element providing North-South coverage (and some instantaneous 2-D information) for snapshot imaging projects, including transients. Typical standalone projects with EMBRACE are HI surveys of the whole sky using all 16 (2 × 8) digital beams within the large field-of-view dictated by a tile, and very wide field searches for variable objects or transients.

3.8. From EMBRACE toward SKA

The purpose of the EMBRACE demonstrator is to show that a significant percentage of an SKA station can be realized, providing the necessary information regarding the maturity of technology and related cost issues. While the EMBRACE demonstrator will have only single polarization capability, a dual-polarisation tile design will be addressed and delivered within the scope of SKADS, in Design Study 4 (DS4). The EMBRACE design, plus the DS4 dual-polarized tile design, will allow the practical design for a full phased array SKA station in 2010.

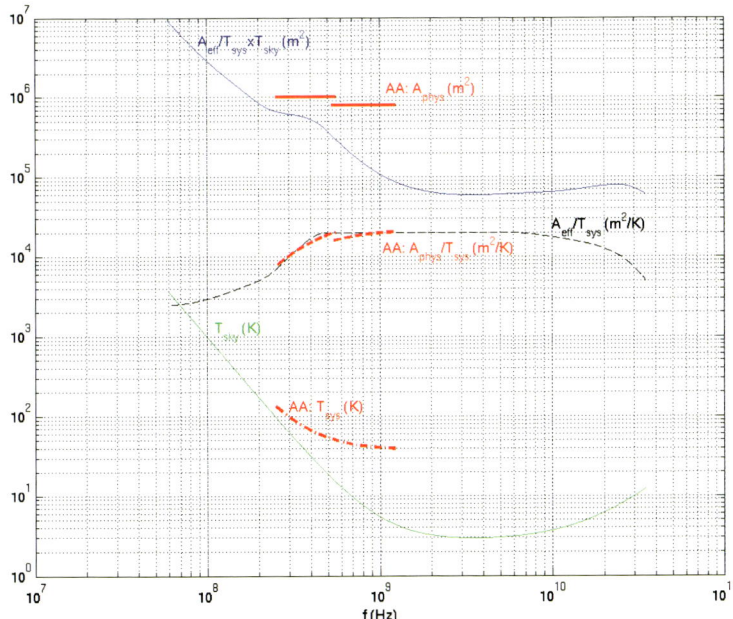

Figure 6. Possible two-band phased-array solution, in green the sky noise, in black the A_{eff}/T_{sys} as specified in SKA Memo 45 (Jones 2004), in blue the absolute minimum required effective area for an ideal system ($T_{sys} = T_{sky}$), in red from top to bottom for the aperture arrays: the physical Area (A_{phys}), the A_{eff}/T_{sys} and the T_{sys}.

On the basis of the DS4 and the EMBRACE demonstrator plus our previous submissions on the dense aperture array (van Ardenne and Butcher, 2003), we now outline the likely performance of the astronomical instrument built on dense aperture array at the SKA station level.

In Figure 6 the system characteristics for a two-band aperture array concept are presented. The frequency bands as plotted can be extended to cover 200 to 1500 MHz with further optimization and with limited decrease in sensitivity. The two suggested arrays, one of 1 km^2 and a slightly smaller one of 0.8 km^2, meet the SKA specification $A_{\mathrm{eff}}/T_{\mathrm{sys}} = 20\,000$. This two-band approach or some derivative of this concept, such as a three-band approach where the bandwidth requirement per array is somewhat less severe, maybe adopted by the European SKA Consortium. As can be seen in Figure 6, to meet the goals of the SKA Key Science Project for the full frequency band, a hybrid solution composed of a phased array system for the lower frequencies and another concept for the higher frequencies can form a very effective SKA. These two collector concepts, e.g. phased-arrays and small dishes, can share physical infrastructure, data transport and central processing infrastructure. As part of the European SKA engineering assessment studies, the cost effectiveness of such hybrids will be studied and evaluated.

4. Conclusion

Although ThEA had clearly demonstrated the principle of Aperture Array with its two independent and steerable multi-beams, it is still necessary to design, build and test a much larger array, to provide some concrete experience of large phased arrays. This type of knowledge is necessary to further extrapolate our experience up to a SKA station level.

This paper has presented the motivation, the science case, the system concept and the design approach for EMBRACE: the first large radio astronomy telescope based on dense phased array antenna technology. We have also suggested how EMBRACE might be extended to the size of an SKA station, which may be more than 100 m in diameter.

References

Bij de Vaate, J. G., Kant, G. W., van Cappellen, W. A. and van der Tol, S.: 2002, 'First Celestial Measurement Results of the Thousand Element Array', *Proceedings of the URSI GA*, Maastricht.

Bij de Vaate, J. G., Wijnholds, S. J. and Bregman, J. D.: 2003, 'Two Dimensional 256 Element Phased Array System for Radio Astronomy', *Proceedings of the IEEE, Phased Array Systems and Technology Conference*, Boston.

Hall, P. J.: 2005, *Exp. Astron.* this issue.

Jones, D. L.: 2004, 'SKA Science requirements: Version 2', SKA Memo 45, available at: http://www.skatelescope.org/pages/p_docsandpres.htm

Kant, G. W., Gunst, A. W., Kokkeler, A. B. J. and Smolders, A. B.: 2000, 'Receiver Architecture of the Thousand Element Array (THEA)', *Proceedings of the IEEE SPIE Conference*, 4015, Munich.

'Square Kilometre Array Design Studies (SKADS)', A proposal submitted to the European Commission under the Sixth Framework, European SKA Consortium, March 2004.

Van Ardenne, A.: 2004, 'The European Aperture Array SKA Demonstrator – Electronic Multibeam Radio Astronomy Concept: EMBRACE', presented at SKA 2004, Penticton, available at: http://www.skatelescope.org/pages/p_docsandpres.htm

Van Ardenne, A. and Butcher, H. R.: 2003, 'The Aperture Array approach for the Square Kilometre Array – Concepts Extensions and Response to Questions', SKA Memo no. 14, available at: http://www.skatelescope.org/pages/p_docsandpres.htm

Van Ardenne, A., Smolders, A. B. and Hampson, G. A.: 2000, 'Active Adaptive Antennas for Radio Astronomy; Results for the R&D Program Towards the Square Kilometre Array', *Proceedings of the SPIE conference*, 4015, Munich.

Woestenburg, B. W. and Dijkstra, K. F.: 2003, 'Noise Characterization of the Phased Tile', *Proceedings of the IEEE European Microwave Conference*, Munich.

THE AUSTRALIAN SKA NEW TECHNOLOGY DEMONSTRATOR PROGRAM

RAY NORRIS

CSIRO ATNF, P.O. Box 76, Epping, NSW 1710, Australia
(e-mail: ray.norris@csiro.au)

(Received 11 August 2004; accepted 4 October 2004)

Abstract. I present a brief overview of the SKA projects conducted under the Australian Major National Research Facilities program, and describe the largest of these projects – the SKA New Technology Demonstrator. The goal of this project is to construct an SKA technology demonstrator which will explore and evaluate a number of SKA technologies in a remote radio-quiet environment, while also achieving a restricted set of key science goals. Infrastructure and access to back-end facilities will also be provided for other international SKA groups who wish to evaluate or demonstrate technologies, or conduct science experiments, in a remote, radio-quiet, environment.

Keywords: radio astronomy, radio telescopes

1. Introduction

The location of the Square Kilometre Array (SKA) will not be decided until 2006, and Australia is one of the five candidate host countries (the others being Argentina, China, South Africa, and the USA). Within Australia, several sites are being evaluated, but a strong candidate is Mileura Station, near Meekatharra in Western Australia. Because of its radio-quietness, and other factors, this site was initially ranked as the preferred location for the international LOFAR telescope, although that telescope was ultimately located in the Netherlands for other reasons.

The technologies for the SKA are still being developed, and it is still unclear, for example, what type of antenna will be optimum, although there are several excellent competing technologies described elsewhere in this volume. Although the antennas are the most visible part of the SKA, equally important are the signal processing hardware and software, the control and monitor systems, data transport, etc. All these areas of technology are undergoing rapid development by the participants in the SKA project.

In Australia, much of this SKA work is funded by the Australian Astronomy Major National Research Facility (MNRF), which is a A$52 m (US$38 m) collaborative venture involving nearly all major astronomical institutions in Australia, with the aim of securing significant Australian participation in major new international astronomical facilities at both optical and radio wavelengths, represented by Gemini and the SKA, respectively. Roughly half of this funding was provided by the

Australian Federal government under its "Backing Australia's Ability" initiative, and the remainder was provided by the participants, including the Commonwealth Scientific and Industrial Research Organisation (CSIRO – Australia's largest research agency), Universities, industry, and the Western Australian State Government. The MNRF funding commenced in 2002, and will continue until 2007.

MNRF technology development projects are tightly integrated into existing astronomical instruments, so that technology developments for the SKA can be tested and demonstrated on currently operational telescopes, resulting in improved performance from those telescopes. MNRF projects relevant to the SKA include the following.

- NTD: development of an SKA New Technology Demonstrator, as described below.
- CABB: development of correlator and signal processing technology, including the broadband upgrade of the Australia Telescope Compact Array.
- MMIC: development of monolithic microwave integrated circuits, including integrated receivers and high-speed samplers.
- SKAMP: development of a cylindrical antenna SKA demonstrator by upgrading the Molonglo telescope.
- SKASS: development of high-speed supercomputer applications and SKA simulation software.
- WA Siting: investigation of Western Australian sites as potential locations for the SKA.

2. The New Technology Demonstrator

2.1. OVERVIEW

The MNRF New Technology Demonstrator (NTD) project has the explicit aim of developing and evaluating novel technologies for the SKA. The specific goal of the NTD project is to develop a wideband, multi-beam technology demonstrator consisting of a "mini-station" of a next-generation radio telescope, incorporating the core technologies of:

- wide field-of-view antennas;
- optical signal transport;
- digital signal processing techniques such as achromatic beam forming, wide-field imaging, and interference mitigation.

In the first two years of the NTD project, a significant amount of research has been devoted to evaluating a particular type of wide-field antenna – the Luneburg lens.

2.2. LUNEBURG LENS

One of the science requirements of the SKA is that antennas should have a wide field-of-view, or multiple beams. To achieve this with parabolic dishes, which have the fundamental limitation that they can point in only one direction at a time, is expensive. An innovative alternative is the Luneburg lens (Hall, 2002), which consists of a sphere of dielectric, with the dielectric constant graded radially such that a plane wave arriving at the lens is refracted to a focus on (or near) the opposite surface of the lens. By arranging a number of feeds around the lens, the antenna may observe several independent fields of view at once.

For the Luneburg lens to be suitable for the SKA, it was necessary to develop an inexpensive process to manufacture and assemble the dielectric, using new composite low-density, low-loss, dielectric materials. The SKA requirements presented a major challenge, which was tackled by bringing together expertise in several different areas: theory and measurement of the electromagnetic properties of materials, polymer foam technology, and materials science and manufacturing technology. The project successfully produced a working prototype lens, and patented a manufacturing process for a new artificial dielectric material. However, the available manufacturing processes were unable to produce a suitable low-density high-dielectric-constant material, within the available budget (Kot, 2004). Luneburg lenses have therefore now been withdrawn as an SKA concept, enabling the NTD project to focus on other areas of SKA technology.

2.3. REFOCUSSING THE NEW TECHNOLOGY DEMONSTRATOR

Since the NTD project was first conceived, there have been a number of significant developments which prompt a re-orientation of the direction of this project. These include a growing recognition of:

- the scientific importance of a wide field-of-view;
- the scientific importance of low frequencies, especially for cosmological studies;
- the advantages of extremely radio-quiet sites, both in permitting new types of science to be done, and in enabling more cost-effective technology to be used (e.g. lower interference levels permit a reduced number of bits of sampling).

These changes, together with the withdrawal of the Luneburg lens, have prompted a re-examination of the goals of the project. After a process of community consultation, the aims have been redefined to the following.

- Demonstrate a radio telescope prototype, which provides a very wide field-of-view ($\sim$50 square degrees) at the lower SKA frequencies.

- Demonstrate the feasibility of remote operation of a radio astronomy facility in a very radio-quiet site.
- Demonstrate other areas of SKA enabling technology, including signal processing, signal transport, data processing, control/monitor systems, infrastructure, and energy provision in a remote environment.
- Provide a facility for other groups to test SKA technologies in a remote radio-quiet site.
- Pave the way for key science outcomes.

A significant aspect of the project will be to investigate the establishment of a Radio-Quiet Zone at the Mileura site, to provide a site for future radio astronomy facilities of different types.

Collaborators in this project are expected to include CSIRO, the Universities of Sydney and Melbourne, the Massachusetts Institute of Technology (MIT), and the State Government of Western Australia. In particular, it is expected that work will begin in collaboration with MIT on the implementation of an additional antenna array at Mileura optimized for 80–300 MHz, which will share the development, backend, and infrastructure. This is a continuation of MIT's previous work on low-frequency radio astronomy, and the construction of this array is contingent on successful funding proposals by MIT.

2.4. SCIENCE GOALS

The primary goals of the NTD are to develop technology and explore the operation of a large instrument in a remote radio-quiet zone. However, these goals cannot be usefully demonstrated unless the instrument is also able to deliver significant science outcomes.

The potential key science goals of the NTD are as follows.

- Operation of the instrument at 1.0–1.4 GHz as a survey instrument for studying neutral hydrogen in galaxies in the local Universe. Depending on funding availability, we hope to be able to construct a survey instrument with significantly greater sensitivity than the Parkes HIPASS survey (Barnes et al., 2001).
- Operation of the instrument at 2.3 GHz as a Very Long Baseline Interferometry (VLBI) antenna, to be used together with the existing VLBI arrays in Australia, South Africa, and Japan. We expect most of these antennas to be equipped with broadband (>1 Gb/s) links within the lifetime of this project, enabling real-time wide-band VLBI.
- Operation of the instrument at 80–300 MHz to detect the cosmic signature of the Epoch of Reionisation (EOR) (Morales and Hewitt, 2004). As the frequency, strength, and width of this signal is currently unknown, this is necessarily a high-risk experiment, but worthwhile because of the high scientific impact if successful. We note that this instrument may be the only array in the world

capable of observing at 80–110 MHz, because of strong interference from FM stations at other sites.
- Operation of the instrument at 80–300 MHz to detect transient signals, such as stellar or supernova bursts, gamma-ray bursters, exploding black holes, and possibly other phenomena that may lie undiscovered in this unexplored region of observational phase space (e.g. Katz et al., 2003).

In addition, a strong case has been made in Australia for a specialised array (HYFAR) to observe neutral hydrogen at 700–1400 MHz, and which will be capable of using the galaxy distribution to determine the Equation of State of the Universe (Bunton, 2004). While the array to be constructed in the NTD project will have neither the sensitivity nor the frequency range to achieve this goal, we regard it as a technological precursor and prototype of HYFAR.

2.5. SAMPLE SYSTEM CONFIGURATION

Significant technology decisions on the system design are currently uncertain. In particular, it is not yet clear whether the requirement for a wide field-of-view is best satisfied by a cylindrical reflector (with a line feed) or a parabolic or shaped reflector (with a phased focal plane array feed). Furthermore, it should be noted that specifications are not yet firmly established, and all figures quoted in this paper should be regarded as provisional. However, a sample configuration, which is one of several being explored, is shown in Figure 1.

The signal processing hardware will include beamformers, digital filterbanks, and correlators, but the optimisation of these components will depend strongly on the antenna decision. It is hoped that as much as possible of this digital signal

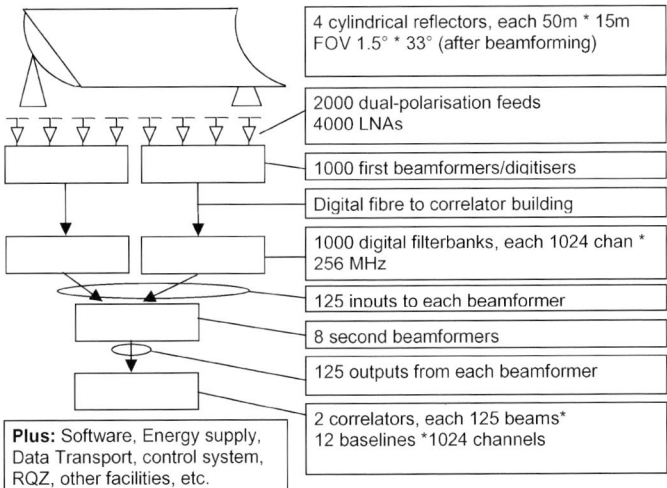

Figure 1. One possible system configuration for the New Technology Demonstrator.

processing hardware will be common to both the 1.0–2.4 GHz and the 80–300 MHz systems.

2.6. PROJECT MILESTONES AND TIMELINE

The project is divided into three phases. Completion of the array with the specifications described here is contingent on additional funding beyond that available from the MNRF, and the project will be de-scoped as necessary if this funding is not available. Furthermore, it is recognised that the timescales are necessarily ambitious, and so risk mitigation strategies are important. Construction of the 80–300 MHz array is contingent on a successful funding application by MIT.

Phase 1: July 2004–April 2005

- Early decision on concentrator configuration (cylinder/parabola, no. of elements in focal-plane array).
- Detailed project planning, risk assessment, risk mitigation strategy.
- Development of a concentrator + phased-array feed covering one octave around 1400 MHz.
- Development of signal processing, data transport, software, radio-quiet zone, etc.
- Overall system design and costing.
- Collaboration with MIT on phased array EOR tiles optimised for 80–300 MHz, sharing infrastructure and backend with 1 GHz array as far as possible.
- Installation of prototype MIT antennas and simple backend/infrastructure in December 2004.
- Ends with Preliminary Design Review (PDR) in April 2005.

Phase 2: May 2005–June 2006

- Architecture and configuration selected.
- Detailed design and costing.
- Prototype subsystems.
- Establish site.
- Start installing EOR tile antennas.
- Critical Design Reviews of component systems throughout Phase 2.
- Ends with Critical Design Review (CDR) of entire system.

Phase 3: July 2006–June 2007

- Construct small concentrator with array feed in WA with A <3000 m^2 (1000–2400 MHz?).
- Establish EOR/transient facility using MIT tiles (80–300 MHz).
- Establish facility with backend, infrastructure, for other groups wishing to use radio-quiet site.

- Paves way, subject to additional funding and SKA site decision, for either:

    o construction of 5% SKA Demonstrator/SKA Phase 1;
    o construction of an Australian niche instrument such as HYFAR.

3. Conclusion

The New Technology Demonstrator will provide a new focus for Australian SKA technology development while delivering on a restricted set of key science goals. It will also provide a facility to the international SKA project in which potential SKA technologies can be evaluated and demonstrated in a remote radio-quiet environment.

Acknowledgements

This brief paper summarises the work of many people, too numerous to be mentioned individually here, to whom I am indebted.

References

Barnes, D. G. et al.: 2001, *MNRAS* **322**, 486.
Bunton, J. D., Briggs, F. H. and Blake, C. A.: 2004, 'HYFAR: An Array of Cylindrical Reflectors for Precision Cosmology', *WARS'04 Conference*, Hobart, February 2004, see http://www.ips.gov.au/IPSHosted/NCRS/wars/wars2004/proceedings/final/j/bunton02i_f.pdf
Hall, P. J. (ed.): 2002, http://www.skatelescope.org/documents/SKA-AUS-CONCEPT-Luneburg-17072002.pdf
Katz, C. A., Hewitt, J. N., Corey, B. E. and Moore, C. B.: 2003, *PASP* **115**, 675.
Kot, J.: 2004, this volume.
Morales, M. F. and Hewitt, J.: in press, *ApJ* (astro-ph/0312437).

ANTENNAS

Experimental Astronomy (2004) 17: 89–99   © Springer 2005

LOW NOISE PERFORMANCE PERSPECTIVES OF WIDEBAND APERTURE PHASED ARRAYS

E. E. M. WOESTENBURG* and J. C. KUENEN

ASTRON, P.O. Box 2, 7990 AA Dwingeloo, The Netherlands
(*author for correspondence, e-mail: woestenburg@astron.nl*)

(Received 4 August 2004; accepted 17 November 2004)

Abstract. A general analysis of phased array noise properties and measurements, applied to one square meter tiles of the Thousand Element Array (THEA), has resulted in a procedure to define the noise budget for a THEA-tile (Woestenburg and Dijkstra, 2003). The THEA system temperature includes LNA and receiver noise, antenna connecting loss, noise coupling between antenna elements and other possible contributions. This paper discusses the various noise contributions to the THEA system temperature and identifies the areas where improvement can be realized. We will present better understanding of the individual noise contributions using measurements and analysis of single antenna/receiver elements. An improved design for a 1-m^2 Low Noise Tile (LNT) will be discussed and optimized low noise performance for the LNT is presented. We will also give future perspectives of the noise performance for such tiles, in relation to the requirements for SKA in the 1 GHz frequency range.

Keywords: low noise, phased array, radio astronomy

1. Introduction

The overall sensitivity of large aperture phased array antennas is a subject of growing interest. Maaskant et al. (2004) described a generalized method for modeling the sensitivity of array antennas, using S- and noise parameters of individual system blocks to predict overall performance at RF system level. In Woestenburg and Dijkstra (2003) an analysis of phased array noise properties and measurements was presented, resulting in a noise budget for a 1-m^2 tile of THEA (Bij de Vaate and Kant, 2002), which is used as the basis for this paper. Section 2 of this paper summarizes the results of the noise analysis and discusses the individual elements of the noise budget. One element of the noise budget is designated as an "unknown" uncorrelated noise source T_g, attributed to absorption losses in the antenna structure and noise through antenna side-lobes. On the basis of the measurement results from THEA tiles and individual antenna/receiver elements of such a tile, another, more likely, cause was identified and will be discussed in Section 3, giving a better explanation of the origin of this noise component.

As THEA was not designed as a low noise system, some of the contributions to the noise budget are higher than might be expected for a properly designed low noise system. Based on the knowledge from the THEA analysis in Section 3, an improved design was made for the antenna/receiver boards as elements of a low noise tile. Results of measurements for this design are presented and discussed in Section 4, leading to the design and modeled properties of a 1-m² Low Noise Tile (LNT). The noise budget of this tile shows the balance between individual noise contributions, as can be achieved with readily available receiver components using present day technology. The requirements for SKA dictate a reduction of all array noise contributions, of which the feasibility will be discussed in Section 5.

2. Phased array noise characterization

The analysis in Woestenburg and Dijkstra (2003) is based on the Y-factor measurement, where the Device Under Test (DUT) noise temperature is determined using reference loads at two different temperatures (T_h and T_c) at the DUT input and measuring the ratio of the available powers (the Y-factor) at the DUT output. The DUT noise temperature is then given by the general formula

$$T_{\text{DUT}} = \frac{T_h - Y T_c}{Y - 1}$$

In the proposed measurement procedure, a cold load is presented with the antenna main beam looking at the cold sky in zenith (antenna far field). Antenna losses and side-lobes may also contribute noise during the cold load measurement and are described here by noise sources with equivalent noise temperature T_{g_i}, with index i for the individual elements of the array. During the cold load measurement there also is a contribution T_{nc} from noise coupling between antenna elements. This contribution can be considered as a correlated one and may be incorporated in T_c. For the hot loads an array of V-shaped room temperature absorbers is used, completely covering the antenna array in the near field, presenting uncorrelated noise sources and effectively eliminating the noise coupling between antennas. The existence of T_{g_i} will not change the hot load temperatures from ambient temperature, if we assume that the antennas and loads are in thermal equilibrium. This measurement situation is schematically depicted in Figure 1 for an N-element array.

In the analysis we identify four contributions to the array system temperature:

- T_{array}, defined as the weighed average of the noise temperatures of the individual receiver channels T_{rfc}, including losses between antennas and receiver RF chains;
- T_{g_i}, due to absorption losses in the antenna and noise through antenna side-lobes from nearby warm objects
- T_{nc}, due to noise coupling between array elements, to be calculated from the active reflection coefficient and the amplitude of the noise wave emanating from the receiver input (T_r) towards the antenna (Weem and Popović, 2001)

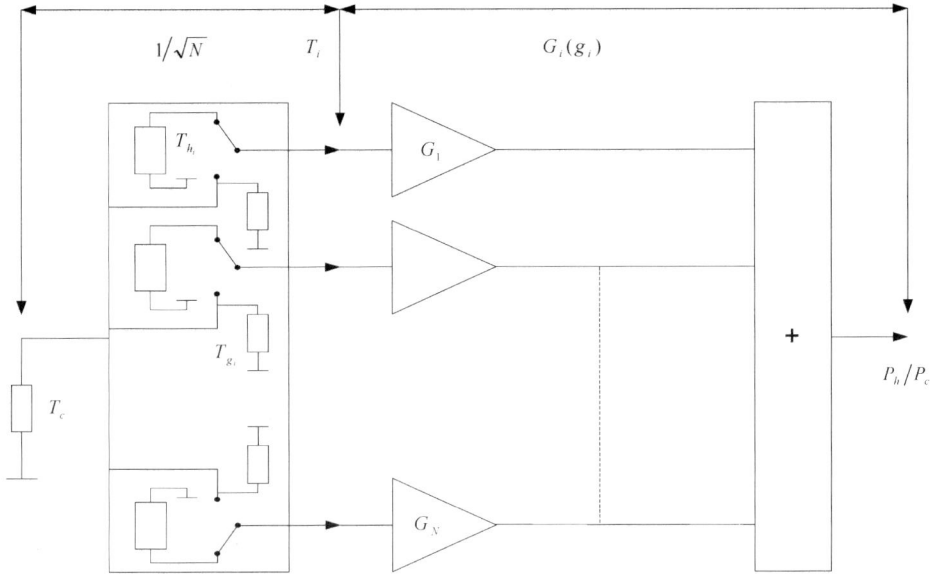

Figure 1. N-element array with noise sources at the N inputs.

- T_c, the cold sky temperature, presenting correlated noise at the inputs of the individual antenna array elements.

The Y-factor for this measurement set-up, assuming that the correlated contributions are all added in-phase at the system output, after some manipulation, results in an effective array noise temperature at the input:

$$T_{\text{array}} = \frac{\sum_n T_i G_i}{\sum_n G_i} = \frac{T_h - Y T_g - Y T_c Q}{Y - 1} \tag{1}$$

assuming that $T_{h_i} = T_h$ and $T_{g_i} = T_g$ (for an infinite array of equal elements) and

$$Q = \frac{\left(\sum_n \frac{g_i}{\sqrt{N}}\right)^2}{\sum_n G_i},$$

($0 < Q \leq 1$), a measure for the equality of the gains of the individual array elements. In the formula for Q, g_i and $\sqrt{N}$ are voltage coefficients (see Figure 1), that must be used for the summation of the correlated contributions. In formula (1) G_i is the power gain, corresponding to the voltage gain coefficient g_i, while T_i is the noise temperature of an individual receiver.

If we define the system temperature T_{sys} as the sum of T_{array}, T_g, T_{nc} and T_c, assuming Q is close to 1 and that neglecting it does not influence the accuracy of

the Y-factor measurement, then

$$T_{sys} = \frac{T_h - T_g - T_{nc} - T_c}{Y - 1} \tag{2}$$

It can be shown that (1) and (2) also represent the general situation where there are additional correlated sources or combinations of correlated and uncorrelated noise sources at the antenna inputs. The nature of these sources does not affect the result of the measurement if the quality factor of the array is close to 1. Otherwise a correction should be applied to the result, involving quality factors for the corresponding noise sources. In general the array noise properties reflect the average noise properties of the individual array elements. Further details about the array noise analysis and measurements can be found in Woestenburg and Dijkstra (2003).

3. Discussion of THEA noise contributions

Figure 2 shows a picture of one of the THEA boards with four Vivaldi antennas, each with LNA and second stage amplifiers, as well as the vector modulators

Figure 2. Picture of one of the THEA boards, showing four Vivaldi antennas and electronics.

TABLE I

Values of THEA noise contributions at the measurement frequencies

Frequency (MHz)	Y (dB)	T_{rfc} (K)	T_{array} (K)	$T_{nc} + T_c$ (K)	T_g (K)	T_{sys} (K)
724	4.6	72	126	8	15	149
1058	4.4	80	135	11	15	161
1354	3.8	110	169	12	17	198
1600	2.7	170	238	10	42	290

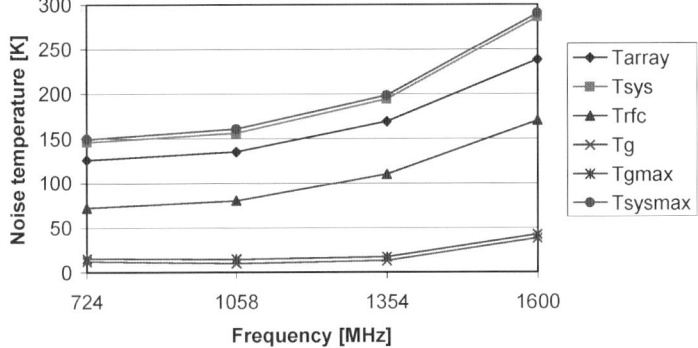

Figure 3. THEA noise results as a function of frequency.

for beam control. The LNA is a GaAs PsHEMT with 15 dB gain at 1 GHz, the second stage amplifiers are SiGe HBT gain blocks. The results of the noise analysis for a THEA tile by Woestenburg and Dijkstra (2003) are presented in Figure 3 and Table I, showing a system noise temperature ranging from 150–200 K. The largest contribution (T_{array}) comes from the receiver electronics (e.g. 135 K at a measurement frequency of 1058 MHz) and is composed of LNA and second stage noise (80 K), as well as the noise increase due to an 11-cm long, 50 Ω strip-line on the FR4 board with a calculated and measured loss of 0.6 dB, connecting the LNA to the feeding point of the antenna.

All three components of the receiver contribution can be reduced using a better LNA in combination with the second receiver stage and a low loss connection between antenna and LNA. A second contribution (T_{nc}) is due to noise, which emerges from the LNA inputs towards the antennas and couples to adjacent antenna elements. The exact (correlated) noise increase at the output of the array is determined by the amount of noise coming from the LNA, the coupling factors and the phasing of the array. In this analysis the array beam is phased in such a way that it looks at the zenith (broadside of the array), giving a calculated contribution of 11 K for $T_{nc} + T_c$, including 3 K for the cold sky. The third contribution, T_g with a value of 15 K, represents a noise budget element, attributed to noise through antenna side lobes and antenna losses. The idea behind this was that the antenna measurement setup with a limited size of 1 m² (thus not representing an infinite array) did not provide sufficient shielding from nearby warm objects at ground level. At the time of the first analysis this was considered to be an acceptable explanation for approximately 10% of the system temperature. However, further measurements with a configuration of five, closely spaced, THEA tiles providing better side-lobe rejection (visible by reduced interference levels), still showed a remaining excess noise. An estimate of absorption losses in the antenna could also not provide a proper explanation. Moreover, measurement of improved single THEA antenna/receiver boards in a well-shielded environment (using the surface of the Dwingeloo 25 m reflector) again showed the remaining unexplained noise contribution. This lead to the conclusion that "ground noise" and thus array beam properties, with the improved shielding and beam properties in the later measurements, could no longer be considered as a cause for the observed excess noise and that the remaining noise component had to originate from another source.

Comparing the results from the measurements on a set of tiles with those of a single board in a well-shielded environment, lead to the conclusion that the measurement results on a single board give a good indication for their noise performance in an array environment with one or more tiles. The same noise contributions as in a tile are present in a single board, with possibly a slightly different value for the noise coupling contribution. This hardly changes the overall result, because for the present system the noise from the receiver chain is dominating system performance.

The above conclusions have lead to a reassessment of possible causes for the higher than expected system temperature, based on a single antenna board as shown in Figure 2. The noise temperatures for the receiver components and overall system had been determined using a calculation model for the receiver electronics and measurements with standard noise measurement equipment for 50 Ω systems. This was done assuming that the antenna impedance was sufficiently close to 50 Ω and did not cause a significant increase in LNA noise, based on general information about the LNA noise parameters. In case the underlying condition is not satisfied, this may result in two errors. The first one could give an unexpected increase in LNA noise temperature, due to sensitivity of LNA noise to input impedance mismatch. The second could result in a measurement error due to differences in impedance seen at the LNA input during the hot and cold load measurements. The latter can be corrected if the impedance values during hot and cold load measurements are known, as well as the noise parameters of the LNA. After assessment of the noise parameters and loads by noise and impedance measurements, it was concluded that the THEA noise measurement results needed correction. The impedance at the antenna input indeed changed considerably, when moving the absorbers of the hot load closer to the antenna board. Kuenen (2004) showed that for a measurement frequency of 1058 MHz there is an increase in the LNA noise temperature of 38 K due to impedance mismatch in the cold load situation, with the antenna impedance significantly deviating from 50 Ω. This should be compared with an LNA noise temperature of 51 K for the matched situation (the difference of 29 K with the value

TABLE II
Corrected and original values for the THEA noise budget at 1058 MHz

Noise temperature (K)	Original value	Corrected value
T_{rfc}	80	118
T_{array}	135	179
$T_{nc} + T_c$	11	11
T_g	15	0
T_{sys}	161	190

Note. Frequency = 1058 MHz, $Y = 4.4$ dB.

for T_{rfc} of 80 K in Table I is due to the second stage contribution). With correction for the measurement error due to unequal hot and cold load impedances, this results in a THEA system noise temperature of 190 K (Kuenen, 2004). Table II shows the resulting noise budget with the corrected values at 1058 MHz, compared to the original ones. The increase in noise temperature due to mismatch at the LNA input (seen in T_{rfc}) is significant and indicates where major improvement can be found. For the specific case of the THEA design, T_{array} may be lowered further by reducing the loss between antenna and LNA, as well as decreasing the second stage receiver contribution.

It should be noted that the effect of LNA noise mismatch is not limited to array systems, but is in general relevant for the sensitivity of conventional low noise receiver systems for radio astronomy. It emphasizes the importance of control over the antenna impedance and proper noise matching between the LNA and antenna, to achieve low-noise system performance. For the case of a phased array the situation seems more complicated than e.g. for a parabolic dish with a single antenna/receiver system, but this is not really the case. The various noise contributions to the system temperature are limited in number and behave similarly for both types of system. The dominant noise contribution comes from the receiver electronics in both cases. As far as the extra noise due to input impedance mismatch is concerned, this is not influenced by the scan angle of a phased array, because the phasing of the array is done after the LNA stages, isolated from their inputs. The determining parameter here is the impedance seen from an individual LNA input towards its antenna input, with the other array antenna elements loaded with their LNAs (the passive array impedance).

A parabolic dish in general has a noise contribution from feed spill-over, which is not present in an aperture phased array. However, such an array has the noise coupling effect, which in principle is scan angle dependent through the active reflection coefficient. This will generally result in minor changes of the system temperature as a function of scan angle, but will have considerable effect in case of blind scan angles.

4. Design of a Low Noise Tile

For the design of a low noise system the various contributions to the system noise budget should be properly balanced. It does not make sense to put large efforts in trying to minimize one contribution to a negligible level, without reducing other contributions to comparable levels. If we take the measured system temperature of 190 K of THEA as a starting point, we could aim for a system temperature improvement with a factor of two ($T_{sys} < 100$ K) for the design of a Low Noise Tile (LNT), as an intermediate step towards the realization of the SKA noise specification, which would require another factor of two improvement. The first factor of two could be obtained using the same technology as for the THEA antenna/receiver boards with a proper design for low system noise performance. The second factor of two has to be obtained using better low noise technology and further design improvements.

For the LNT design we considered the use of the same LNA and antenna design as for THEA (Vivaldi antennas), with a low loss connection between antenna and LNA (0.1 dB instead of 0.6 dB in the THEA design, due to a somewhat shorter line, but mainly better loss properties of the printed circuit board material). For a proper low noise design the second stage receiver contribution should be limited to a maximum of 10% of the LNA noise, but this was not yet achieved for the present LNT design. Improvement of the second stage gave a significant reduction (down to 11 K) with respect to the original THEA design, although further reduction will be possible. The contribution from mismatch between antenna and LNA should ideally also be limited to a maximum of 10% of the LNA noise. However, with the use of a slightly modified version of the design of the THEA antennas, the contribution due to mismatch reduced considerably, but is still around 30%, increasing the LNA noise temperature to 65 K. These numbers were considered to be acceptable to obtain the goal for the LNT design. Figure 4 shows the envisaged results as a function of frequency, with the various expected noise contributions to the system temperature, based on measurements of a single antenna/receiver chain, which are indicative for the array behavior (see Section 3).

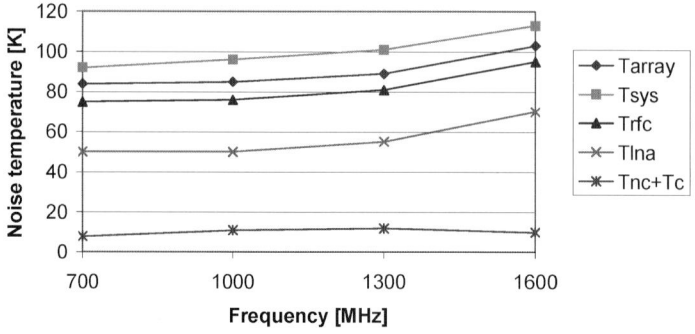

Figure 4. LNT noise contributions as a function of frequency.

TABLE III

Noise budget of the LNT with measurement result on a single antenna/receiver chain at 1058 MHz

Noise temperature (K)	Calculated value	Measured value
T_{rfc}	76	
T_{array}	85	
$T_{nc} + T_c$	11	
T_{sys}	96	97

Note. Frequency = 1058 MHz, $Y = 5.6$ dB.

The noise budget of the LNT is as presented in Table III, which also shows the measured system temperature of a single antenna/receiver chain on a board of similar design as in Figure 2, being the basic element for the LNT. The LNT will consist of 64 antenna elements in total, in the same configuration as THEA. There is close agreement between the calculated and measured noise temperatures, even though the measured result has not been corrected for the difference in hot and cold load impedances. Realization and measurement of a complete 1-m^2 LNT system are the next steps in the process of demonstration of the low noise capabilities of flat antenna aperture phased array systems.

The results show that a reduction factor of two for the noise temperature with respect to THEA may be obtained, even with the same LNA as used in THEA, by reducing the other noise contributions to the level of the LNA noise. With some improvements in the design a further reduction is possible, but would be limited by the LNA noise. For the present LNT design the system temperature is approximately two times the minimum LNA noise, consistent with our starting point of a properly balanced design in terms of system noise contributions. Reducing the LNA noise should be the next step in lowering system noise towards the requirements for SKA.

5. Perspectives for SKA

The results for the LNT design show that the LNA noise contribution accounts for roughly half of the system temperature. Reduction of LNA noise temperature, as well as the total of the other contributions to a comparable level, is required to reach the goal for SKA of 35 K system temperature in the 1 GHz range. Using a factor of two between LNA and system noise temperatures as a 'rule of thumb' for the noise budget of SKA, gives a specification for the LNA-noise that could be achieved, using 0.1 micron PsHEMT or MHEMT processes in III/V-technology. A noise temperature of 16 K for the LNA, extrapolated to the 2010 time frame, would be feasible. Noise increase through antenna mismatch, input losses and second stage contribution may then account for a maximum of 9 K, leading to an array

TABLE IV
Required noise budget for SKA in the 1 GHz range

Noise temperature (K)	Calculated value
T_{rfc}	19
T_{array}	25
$T_{\text{nc}} + T_{\text{c}}$	10
T_{sys}	35

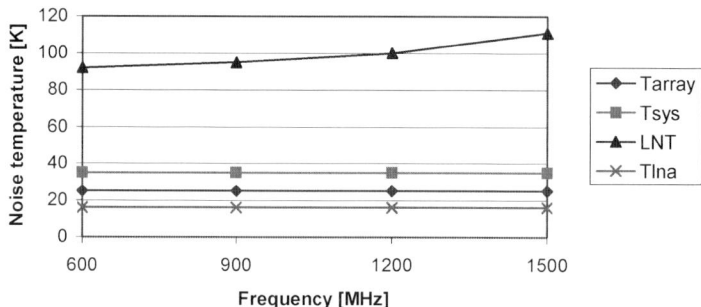

Figure 5. Desired SKA noise contributions for a 35 K system temperature in the 1 GHz frequency range, requiring a reduction of a factor of two to three compared to the LNT system temperature.

temperature of 25 K. With a cosmic background radiation of 3 K, this leaves 7 K for the noise contribution through noise coupling at the antenna inputs, giving a total for $T_{\text{nc}} + T_{\text{c}}$ of 10 K. This noise budget is presented in Table IV, where 10% second stage and 10% input noise mismatch contributions are taken into account in T_{rfc}. From T_{array} a maximum loss between antenna and LNA of 0.08 dB may be derived. These numbers set severe specifications for the SKA-receivers, but will be achievable with a future state-of-the-art low-noise system design, based on a noise temperature of 16 K for a room temperature LNA.

The desired result as a function of frequency is illustrated in Figure 5, which also gives the LNT result for comparison. Improvement with a factor of two to three with respect to the present LNT design is necessary to obtain the SKA design goal, emphasizing that for a low-noise system with the specifications for SKA, every "K" counts.

6. Conclusion

The noise budget for low noise phased array systems has been presented, on the basis of a general noise analysis. The main contributions to the system noise are due to the LNA, second stage noise, antenna connection losses and noise coupling. It was shown that for the Low Noise Tile (LNT) a system noise temperature below

100 K can be realized, using readily available components giving a factor of two improvement with respect to THEA. Reaching the SKA design specification requires a reduction of another factor of two to three with respect to the LNT design. Design goal values for the individual contributions to the SKA noise budget were defined. It was concluded that the realization of these values is feasible on the timescale of the SKA development trajectory.

References

Kuenen, J. C.: 2004, 'Impact of Antenna Matching on THEA Noise', ASTRON Report SKA-00939.

Maaskant, R., Woestenburg, E. E. M. and Arts, M. J.: 2004, 'A Generalized Method of Modeling the Sensitivity of Array Antennas at System Level', *Proceedings of the 34th European Microwave Conference,* Amsterdam, The Netherlands, pp. 1541–1544.

Vaate bij de, J. G. and Kant, G. W.: 2002, 'The phased array approach to SKA. Results of a demonstrator project', *Proceedings of the 32nd European Microwave Conference*, Milan, Italy, pp. 993–996.

Weem, J.P. and Popović, Z.: 2001, 'A method for determining noise coupling in a phased array', *IEEE MTT-S Digest*, pp. 271–274.

Woestenburg, E. E. M. and Dijkstra, K. F.: 2003, 'Noise Characterization of a Phased Array Tile', *Proceedings of the 33rd European Microwave Conference*, Muenich, Germany, pp. 363–366.

EFFECTIVE SENSITIVITY OF A NON-UNIFORM PHASED ARRAY OF SHORT DIPOLES

W.A. VAN CAPPELLEN*, J.D. BREGMAN and M.J. ARTS

ASTRON, P.O. Box 2, 7990 AA Dwingeloo, The Netherlands
(*author for correspondence, e-mail: cappellen@astron.nl*)

(Received 25 August 2004; accepted 17 January 2005)

Abstract. Short dipoles are a key element in new low frequency array antennas as proposed for LOFAR and other astronomical applications. Unfortunately standard texts on short dipole antennas are based on the effective area and do not lead to an astronomically useful sensitivity formulation in a straightforward manner. The concept of maximum effective area is applied to arrays of short dipoles and allows expressing the sensitivity as the ratio of this area over the effective sky brightness temperature as long as the output noise power is dominated by the antenna input radiation. For both quantities we only need to know the array directivity pattern that includes the mutual coupling effects when the actual loading conditions of the array elements are taken into account. Short dipole elements have a constant directivity pattern for frequencies below resonance, but they exhibit strong complex impedance variations that provide only narrow band performance when power matching is applied as required in transmit applications. However, in receive applications voltage or current sensing can be realized, for example with an active balun. Assisted by the steep increase of the sky brightness with wavelength for frequencies below 300 MHz, this can provide sky noise dominated performance over at least a three to one frequency range. Still the low frequency limit is determined by the amplifier noise contribution and the losses in the antenna and in the dielectric ground surrounding the elements. We show that for a sparse array with the elements non-uniformly distributed according to an exponential shell model, a constant sensitivity can be obtained over a frequency range of at least two octaves. In addition, such a configuration has a factor of six greater sensitivity than a rectangular array for a large part of the frequency band.

Keywords: active antenna, maximum effective area, mutual coupling, phased array, radiation efficiency, short dipole, spiral array

1. Introduction

The development of a new generation of large low frequency array antennas has become possible by the application of new technology (Bregman et al., 2000). Active short dipoles are such a key element for which conventional antenna paradigms are not practical to guide the design and to evaluate the system sensitivity (Konovalenko et al., unpublished; Tan and Rohner, 2000). Bregman (2000) showed that a fractal sparse array can have a relatively constant aperture efficiency over a three octave frequency band. In this paper we give a more detailed analysis taking into account antenna losses and the mutual coupling effects that reduce the effectiveness of closely spaced elements in an array.

The sensitivity of a radio astronomical observing instrument is defined as the ratio of the effective collecting area over the system temperature. For a short dipole antenna with an active balun, a complicated evaluation procedure is required to obtain these figures. Kraus (1988) introduced the maximum effective area, which can be derived from the antenna directivity in a straightforward manner. We need some estimate of the antenna efficiency, which can tell us whether the system output power is dominated by the incoming radiation such as the sky brightness temperature. Then the system sensitivity can be evaluated as the ratio of the maximum effective area over the sky brightness temperature averaged over a hemisphere by the directivity pattern.

We apply this concept to two different arrays of short dipoles and compare their maximum effective area (including mutual coupling) to a model without coupling.

2. System temperature

First, an expression for the antenna noise temperature is derived. Although the exact geometry varies, the radiating elements of all large low frequency systems are short horizontal dipole antennas (with respect to the wavelength λ) because of their low manufacturing costs. The radiation efficiency η of such a radiator is given by

$$\eta = \frac{R_{\text{rad}}}{R_{\text{rad}} + R_{\text{loss}}} \tag{1}$$

where R_{rad} is the radiation resistance and R_{loss} is the loss resistance, both measured at the terminals of the antenna. The loss resistance consists of both ohmic losses and dielectric losses of the ground below the antenna. For frequencies well below the resonance frequency of the dipole R_{loss} is constant (typically several ohm). At the low frequency end, the radiation resistance of dipoles above a ground surface is in the first order proportional to λ^{-4} (Balanis, 1997) leading to radiation efficiency also proportional to λ^{-4} (the exact exponent depends on the ground characteristics).

The antenna temperature T_a, as a measure for the noise power available at the terminals, is given by

$$T_a = \eta T_{\text{sky}} + (1 - \eta) T_0 \tag{2}$$

where T_{sky} is the galactic background noise and T_0 is the environmental temperature. At resonance the dipole radiation efficiency is close to unity, which means that its antenna noise temperature is sky noise dominated for all practical cases.

At frequencies below 600 MHz the sky background is proportional to $\lambda^{2.55}$ (see Figure 1 and Tan and Rohner (2000)) while the radiation efficiency is proportional to λ^{-4}. This implies that for frequencies far below resonance the second term in

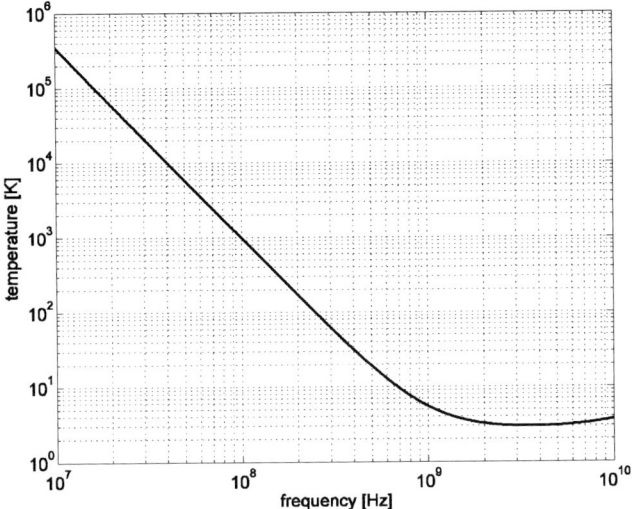

Figure 1. Model of the sky noise temperature.

Equation (2) eventually will dominate, so a short dipole antenna itself is then not sky noise dominated anymore.

The system temperature T_{sys}, defined at the physical terminals of the antenna, is equal to

$$T_{\text{sys}} = T_{\text{a}} + T_{\text{rec}} \tag{3}$$

where T_{a} is the antenna noise temperature and T_{rec} is the receiver noise temperature both referred to the antenna terminals. It has been shown by Konovalenko et al. (unpublished) that it is possible to design a receiver system such that $T_{\text{rec}} < T_{\text{a}}/10$. In that case, the system noise is dominated by the antenna noise and T_{rec} can be neglected. Also, the second term in (2) can be neglected in the proposed applications, resulting in a sky noise dominated system

$$T_{\text{sys}} \approx \eta T_{\text{sky}} \tag{4}$$

3. Effective area and maximum effective area

The IEEE (1993) defined the effective area of an antenna, in a given direction, as the ratio of the available power at the terminals of a receiving antenna to the power flux density of a plane wave incident on the antenna from that direction, the wave being polarization matched to the antenna. There is a direct relation between the antenna gain (in transmit situation) and the effective area A_e

$$A_e = \frac{\lambda^2}{4\pi} G \tag{5}$$

In this context, the maximum effective area A_{em}, described by Kraus (1988) and Balanis (1997) should be considered, which is defined as $A_{em} = A_e/\eta$ where η is the radiation efficiency of the antenna (note that Kraus (1988) uses the term maximum effective *aperture*). Since the gain G and directivity D are related as $G = \eta D$, the maximum effective area can be expressed as

$$A_{em} = \frac{\lambda^2}{4\pi} D \tag{6}$$

and is equal to the effective area A_e if the antenna losses are neglected. This is an attractive formula, since D can often be estimated from normalized antenna patterns or by using physical or geometrical arguments, without recourse to extensive electromagnetic analysis.

The system sensitivity A_e/T_{sys} can now be evaluated by substituting the expressions for T_{sys} and A_{em}, which leads for the sky noise dominated situation to

$$\frac{A_e}{T_{sys}} \approx \frac{A_e}{\eta T_{sky}} = \frac{A_{em}}{T_{sky}} \tag{7}$$

This situation is depicted in Figure 2, which gives an alternative approach for the evaluation of the sensitivity of a short dipole than the one presented by Tan and Rohner (2000). Electronic engineers concentrate on the available power at the terminals of the antenna to make sure that the receiver noise is indeed much smaller than the sky noise. For astronomical sensitivity calculations we need an effective sky temperature T_{skyeff} which results from averaging over a hemisphere by the complete directivity pattern of the antenna. Bregman et al. (2000) gives a simplified analysis for this situation using the main beam efficiency of the antenna.

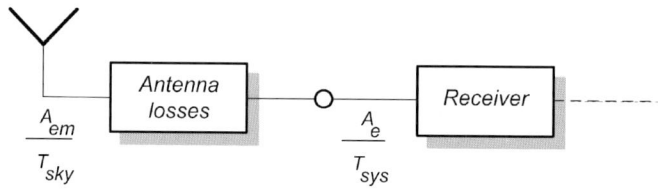

Figure 2. For sky noise dominated systems, A_{em}/T_{sky} approximates A_e/T_{sys}.

4. Maximum effective area of an array

We could extend the concept of effective area from a single port antenna to a multi-port array. According to de Hoop (1975) we then need to connect the array to a multi-port network with an impedance matrix that is the conjugate of the antenna mutual impedance matrix to extract the available power and then derive an effective area. Practical receive arrays, however, are not terminated by such a network with conjugate mutual coupling coefficients, but with fixed loads per element, which results into a different effective array antenna pattern. Therefore, it is more appropriate to determine a maximum effective area from this actual pattern. In practice, however, it is more convenient to evaluate the array directivity in transmit than in receive situation, but still with the same loading condition. In transmit we use a voltage source with a series load resistor, which means that for receive we need to use the signal current in that same load to get the appropriate reciprocal situation as defined by de Hoop (1975).

5. Arrays with short dipoles

We start with a simplified analysis to get an impression of the expected maximum effective area as function of frequency. These results will be compared with electromagnetic simulation in the next section.

First, a single short dipole element is considered, which has a constant directivity for frequencies well below resonance. This implies that its maximum effective area increases for lower frequencies proportional to λ^2 (see Equation 6). Although its physical area is constant (and very small) one could imagine the maximum effective area as an equivalent (virtual) surface around the element that is growing with wavelength and interacts with the incoming waves.

Now suppose N more elements are added and an array is formed. If the element spacing (in wavelengths) is large, the total effective area of all elements is equal to N times the effective area of a single element. However, if the element spacing is decreased, the equivalent surfaces around the elements start to overlap, indicating that less field power will be available for each element (see Figure 3). In

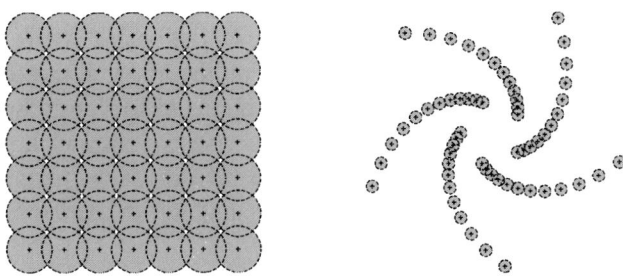

Figure 3. Rectangular and spiral array configurations with equivalent surfaces.

electronic terms, the mutual coupling reduces the induced current in the matched loads and the total effective area will decrease and will be less than N times a single element.

If the element spacing is smaller than $\lambda/2$, we are in the dense array regime. From the law of conservation of energy it can de derived that in an infinitely large dense array, consisting of units cells with area a, the effective area per unit cell is equal or less than the value of a projected on the incoming (broadside) wave plane. Therefore, the maximum effective area of a large finite dense array, of which the edge effects can be neglected, in broadside direction is approximately equal to its physical area.

6. Maximum effective area of arrays of short dipole configurations

Numerical simulations of several configurations are performed to evaluate their effective area as well as their maximum effective area, both at broadside (zenith) direction. The basic element is an inverted-V shaped, 2 m dipole, loaded with 50 ohm, resonant at a frequency f_0 of $\sim$75 MHz and 1.4 m above a conducting ground plane, which results in a lobe free antenna pattern for frequencies below 90 MHz. Two examples are considered in this paper, a rectangular array and a spiral array of dipole antennas. The rectangular array has 7×7 elements on a rectangular grid, with an element distance equal to half a wavelength at the resonance frequency f_0. The spiral configuration has its elements arranged along 5 arms. The smallest element spacing on an arm (near the center) is also equal half a wavelength at frequency f_0. The element spacing increases exponentially along the arms (see Figure 3).

First, the directivity and the mutual impedance matrix of the arrays were calculated with an electromagnetic simulation package for the transmit situation (all elements were excited equally with a 50 ohm Thevenin source). The mutual impedance matrix was used to calculate the available power and from there the effective area. The directivity was used to calculate the maximum effective area. There are only marginal differences between the two results, which can be attributed to the varying effective element excitation conditions over the frequency band. The effective area calculation assumes a conjugate mutual impedance matrix, which is a function of frequency, while the maximum effective area calculation assumes a constant load. The results presented in this article are based on the maximum effective area derived from the transmit situation. Figure 4 shows the results of the spiral array and the square array. For reference, the maximum effective area of a single element multiplied by 50 is shown as well. This equals the theoretical maximum effective area of 50 elements without mutual coupling and shows the expected quadratic behavior.

For f/f_0 below 1, the maximum effective area of the rectangular (dense) array is almost constant. It can be seen that for f/f_0 below 0.3 it increases and converges

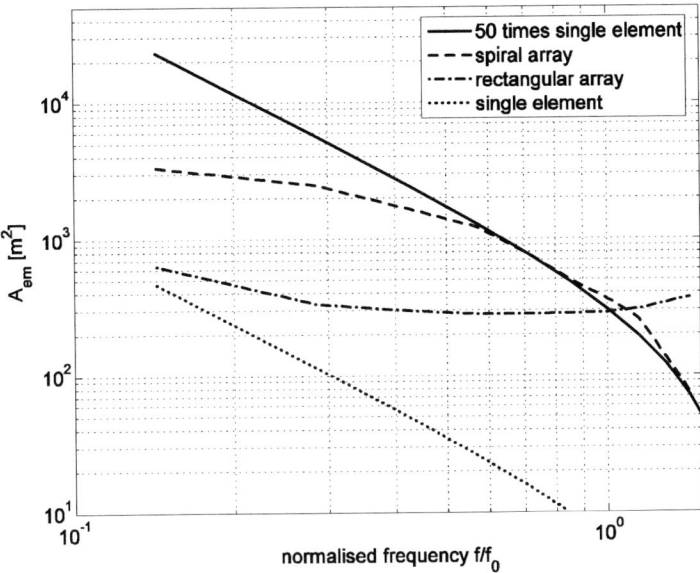

Figure 4. Simulated maximum effective area as a function of normalized frequency.

towards the single element result. A_{em} of the spiral array follows the 50 single elements curve for $f/f_0 > 0.6$. Within the band $0.14 < f/f_0 < 0.7$, the maximum effective area of the spiral configuration is approximately a factor 6 larger than that of the rectangular array, while both have almost the same number of elements.

It is interesting to note that for $f/f_0 > 1$ the maximum effective area of the arrays is larger than of the multiple single-element array, where mutual coupling is ignored. This can be explained by in-phase coupling between the central elements.

7. Sensitivity of short dipole arrays

The array sensitivity A_{em}/T_{sky} is calculated and presented in Figure 5. Since T_{sky} is proportional to $\lambda^{2.55}$ (for $\lambda > 1$ m) and A_{em} is proportional to λ^2 an almost flat sensitivity over a large frequency range is expected for isolated elements. This behavior is indeed shown by the multiple single element curve and for the spiral configuration over the range $0.5 < f/f_0 < 1.3$. The rectangular array configuration shows a large variation of A_{em}/T_{sky} over the band because of its constant A_{em}.

It should be noted that for normalized frequencies below 0.2, the $(1-\eta)T_0$ term in the expression for T_a, can no longer be neglected. In addition, for practical realizations the receiver noise determines the system temperature at these frequencies. Therefore, the presented figures suggest a too large bandwidth.

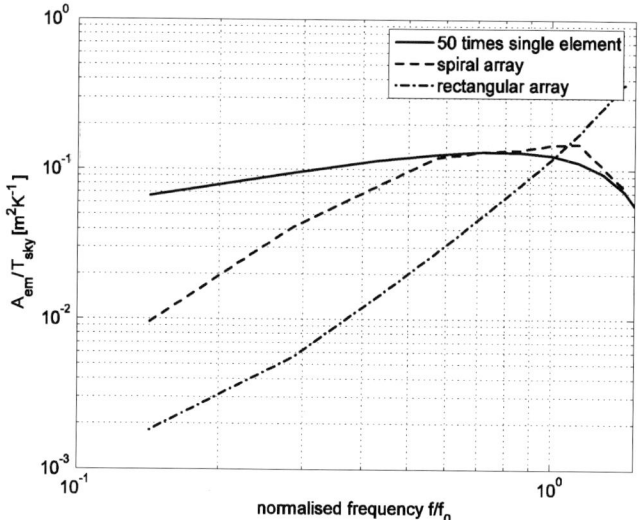

Figure 5. Simulated A_{em}/T_{sky}.

8. Conclusions

The results of our work can be summarized as follows:

- The sensitivity of sky noise dominated systems can be evaluated more conveniently from the maximum effective area A_{em} and the effective sky brightness temperature than from the effective area and the antenna temperature, eliminating the need for the calculation of the antenna efficiency.
- We have extended this concept to arrays with mutual coupling and have shown that even for arrays of short dipoles there are only marginal differences with sensitivities derived from the effective area, due to differences in the applied loading conditions.
- The effective sky brightness temperature to be used in the sensitivity calculation is the hemispheric temperature averaged by the directivity pattern of the array.
- Arrays of sky noise dominated receiving elements, operating at frequencies below 300 MHz, can be designed with a nearly flat sensitivity over a frequency band of up to three octaves, by using a sparse configuration with an exponential shell distribution with a constant number of elements per shell.

References

Balanis, C.A.: 1997, *Antenna Theory, Analysis and Design*, John Wiley & Sons, Inc., New York.
Bregman, J.D.: 2000, 'Concept Design for a Low Frequency Array', *SPIE Proc. Astron. Telesc. Instrum.*, **1415**, 19–32.

Bregman, J.D., Tan, G.H., Cazemier, W. and Craeye, C.: 2000, 'A Wideband Sparse Fractal Array Antenna for Low Frequency Radio Astronomy', *IEEE Antenn. Propag. Soc. Int. Symp.* **1**, 166–169.
de Hoop, A.T.: 1975, *Philips Res. Rep.* **30**, 302–315.
IEEE Standard Definitions of Terms for Antennas, IEEE Std 145–1993, The Institute of Electrical and Electronic Engineers Inc., New York.
Konovalenko, A.A. et al.: 2003, 'Thirty-element Active Antenna Array as a Prototype of a Huge Low-Frequency Radio Telescope', *Exp. Astron.* **16**, 149–164.
Kraus, J.D.: 1988, *Antennas*, McGraw-Hill, New York.
Tan, G.H. and Rohner, C.: 2000, *SPIE Proceedings Astronomical Telescopes and Instrumentation*.

LOW FREQUENCY END PERFORMANCE OF A SYMMETRICAL CONFIGURATION ANTENNA FOR THE SQUARE KILOMETRE ARRAY (SKA)

GERMÁN CORTÉS-MEDELLÍN
National Astronomy and Ionosphere Center, Cornell University, New York, U.S.A.
(e-mail: gcortes@astro.cornell.edu)

(Received 16 July 2004; accepted 9 December 2004)

Abstract. The frequency specifications of the Square Kilometre Array (SKA) call for an optimum operation of the antenna elements from 25 down to 100 MHz. The current 12 m diameter US-SKA design is specified from 500 up to 25 GHz, with an upper goal of 35 GHz. At the low frequency end of the band (i.e., 100 MHz), a 12 m reflector antenna is about four wavelengths in diameter. Then, the question is: how well can you do, at this low frequency end of the specified band of operation for the SKA, with a symmetric reflector configuration using an ultra-wide-band prime focus feed? This paper presents the analysis of the antenna performance, in terms of A_{eff}/T_A, of three symmetric configurations of the 12 m US-SKA antenna design between 100 and 200 MHz.

Keywords: low frequency, SKA, Square Kilometre Array, symmetric Gregorian antenna

1. Introduction

The Square Kilometre Array (SKA) is the next generation of radio telescope aiming to attain one square kilometre of collecting aperture. While there are several ongoing international design efforts and proposals for the SKA, the specifications required by the science missions for the SKA, and in particular the frequency coverage (Jones, 2004), from 0.1–25 GHz, make it very challenging to find a single design solution.

The US-SKA consortium proposes the use of approximately 4500 small dual reflector antennas, known as the *Large-N-Small D* concept. The current antenna element design is based on an offset Gregorian configuration, with a nominal frequency coverage from 0.5–35 GHz. In order to cover this very large band, the design proposes the use of three ultra wide band feeds.

Nevertheless, a symmetric antenna configuration is still a very attractive proposal for its inherent lower cost of fabrication, as well as reduced structural and system control requirements.

In this paper, we explore the use of a 12 m symmetric reflector antenna and analyze three different prime focus antenna configurations to address the issue of the performance at the lower end of the frequency band specification for the SKA, between 100–200 MHz. Two of these configurations include the use of a noise shield to improve the overall $A_{\text{eff}}/T_{\text{Sys}}$. Also, of particular interest is the evaluation of the matching efficiency when a prime focus feed is used with these antenna configurations.

2. Antenna geometry and simulation

Figure 1 shows three different reflector geometries considered in the simulation: first, a regular 12 m paraboloid dish used as reference, (*No-Mesh*); second, a 12 m parabolic reflector with an axial shield of 2 m in length at the edge of the main reflector (*Mesh-A*), and finally, a 12 m dish with a parabolic shield extension from 12 m–16 m diameter reflector, (*Mesh-B*).

We use a prime focus feed to illuminate the antenna with a radiation pattern equivalent to those of the TRW (Ingerson's) feed or Chalmers (Kildal's) feed (Olsson et al., 2004). The radiation pattern of these feeds have a typical average power level of −10 dB, with respect to the peak, at 53° (half width).

We obtained the full antenna radiation patterns using a commercial electromagnetic simulator software (CST Microwave Studio, 2003) that uses finite differences in the time domain method (FDTD) in conjunction with a *perfect boundary approximation technique* (Krietenstein et al., 1998), which allows for accurate modeling of curved structures.

Due to computational constraints, we designed two very simple linear polarized feeds, at around 200 and 100 MHz, respectively, to provide the desired reflector illumination to the simulator. The first feed, is a circular TE_{11} mode horn with an

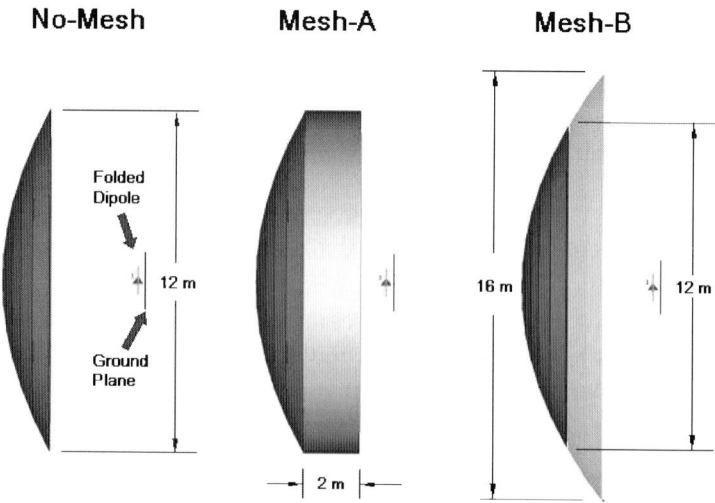

Figure 1. Prime focus symmetric antenna configurations analyzed: No-Mesh, 12 m⌀ Paraboloid (left). Mesh-A, 12 m⌀ paraboloid with a 2 m axial-mesh (center). Mesh-B, a 12 m⌀ paraboloidal reflector with a parabolic-mesh extension from 12–16 m⌀.

aperture of 1.8 m, is optimized to give the appropriate radiation pattern at 200 MHz. The second feed is a dual folded dipole over a circular ground screen of 2 m diameter, optimized to provide the appropriate radiation pattern at 100 MHz. The simulation includes the blockage effects of the Gregorian sub-reflector of the US-SKA symmetric antenna concept.

We calculated full antenna radiation patterns for each of the reflectors geometries illuminated by the folded dipole feed from 60–140 MHz, and by the circular TE_{11} mode feed from 160–220 MHz, in 20 MHz steps for each case.

The antenna patterns were sampled in steps of one degree, both in θ and ϕ. Once the pattern was obtained, we export the data to our own software and proceed to calculate the antenna noise temperature as a function of elevation from $0°$–$90°$. Our in-house software includes models for atmospheric and ground emission, as well as Galactic noise contributions.

3. Analysis and results

3.1. RADIATION PATTERNS

Figure 2 shows a sample set of the antenna for field pattern cuts at 100 MHz, on the left, and 200 MHz, on the right, respectively, and from top to bottom are presented: the reference *No-Mesh* geometry antenna pattern, the *Mesh-A* geometry antenna pattern, and the *Mesh-B* geometry antenna pattern, respectively. For each case, we show co-polar cuts at $\phi = 0°$, $45°$ and $90°$, as well as a cross-polar cut at $\phi = 45°$.

Initial assessment of the patterns shows a reduction in the far side-lobe levels, with both *Mesh-A* and *Mesh-B* shield types, compared with the reference case. One trade off of using the noise shield is an increase in the cross-polarization level, which is more noticeable at 100 MHz, with the *Mesh-A* shield geometry (i.e., −24 dB), than at 200 MHz (i.e., −27 dB). In fact, at 200 MHz, the calculated peak cross-polarization level with the *Mesh-B* shield geometry is very similar to the reference case, or about −30 dB.

The value of the near side-lobe level of the reference case is about −18 dB at 100 MHz and less than −20 dB at 200 MHz. The patterns corresponding to the *Mesh-A* antenna geometry show an increase of the near side-lobe level of approximately 2 dB at both 100 and 200 MHz. In contrast, the *Mesh-B* shield geometry yields lower side lobe levels when compared with the other two geometries, with a near side-lobe level below −24 dB at 200 MHz.

3.2. ANTENNA NOISE TEMPERATURE

Figure 3 shows a sample of the calculated antenna noise temperature as a function of elevation from $0°$–$90°$, at 100 and 200 MHz, respectively. The top graphs correspond to the total antenna noise temperature (T_A), the middle graphs to the atmospheric

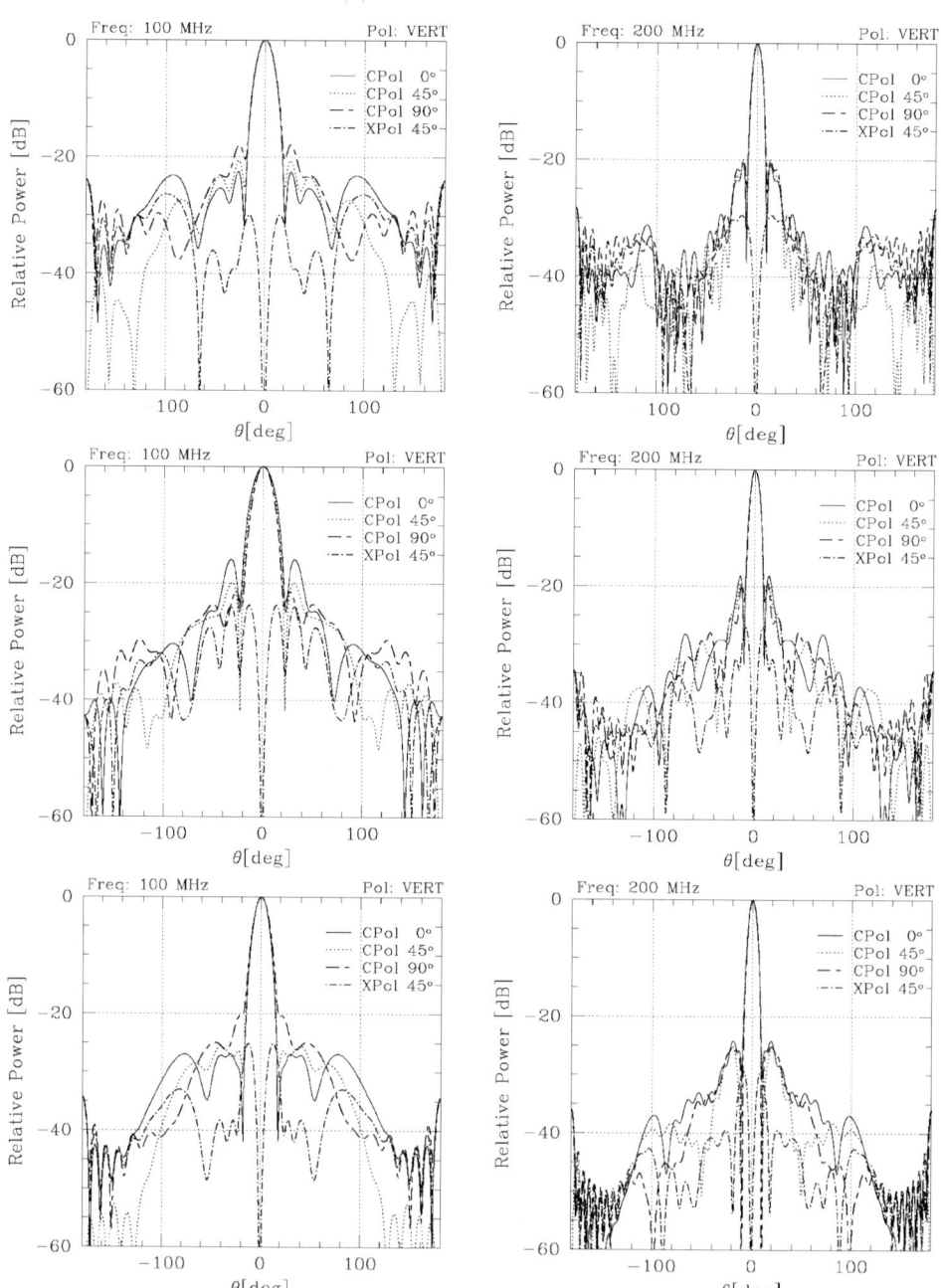

Figure 2. Radiation patterns cuts for the three antenna configurations. Left is at 100 MHz and right at 200 MHz. From top to bottom: No-Mesh case, Mesh-A and Mesh-B cases, respectively. The figure shows co-polar cuts at $\phi = 0°$, $45°$ and $90°$ and a $45°$ cross-polar pattern cut. (Reference polarization is vertical).

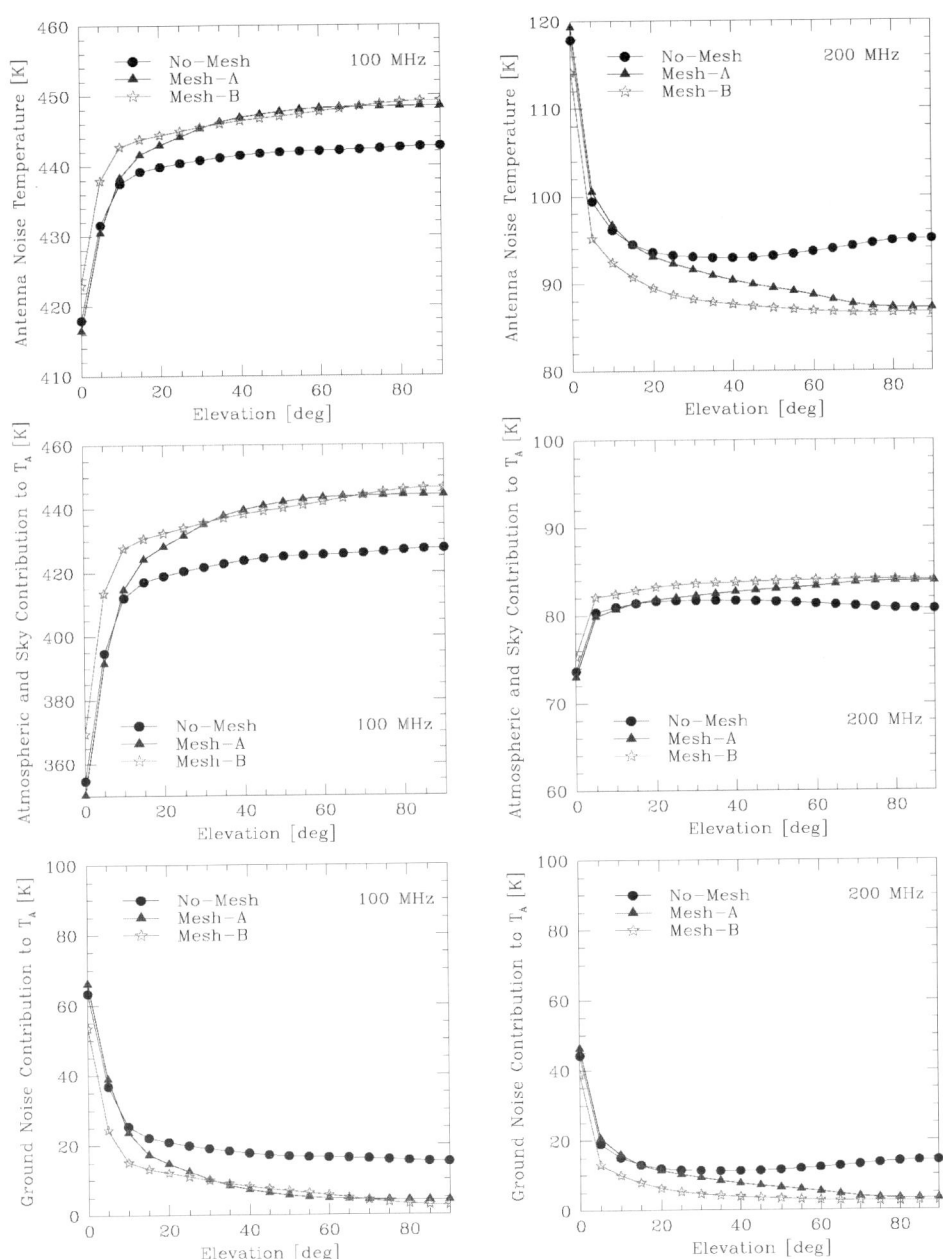

Figure 3. Calculated antenna noise temperature as a function of elevation from 0°–90°, for the three antenna configurations. Left is at 100 MHz and right is 200 MHz. From top to bottom: Antenna noise temperature T_A, atmospheric and sky contribution to T_A, and bottom, ground noise contribution to T_A.

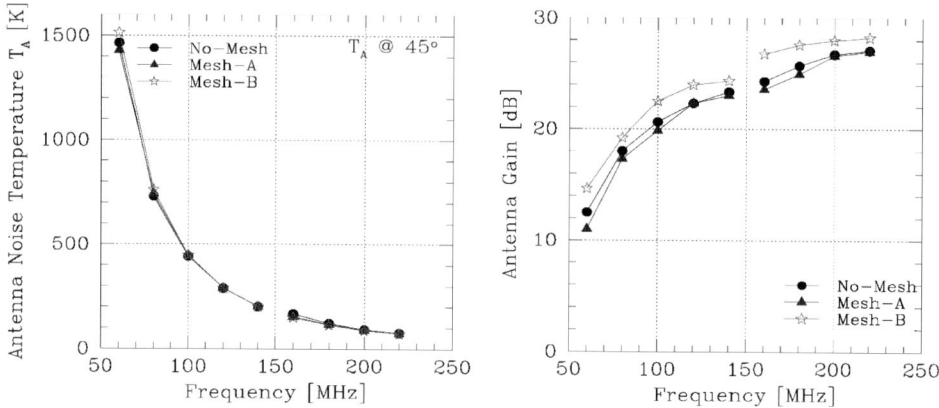

Figure 4. Antenna noise temperature T_A at 45° elevation as a function of frequency (Left), antenna gain as a function of frequency (Right).

and Galactic noise (T_{sky}) contribution to T_A, and the bottom graphs, to the ground noise (T_{gnd}) contribution to T_A, with $T_A = T_{sky} + T_{gnd}$.

At low frequencies, from 100 to 200 MHz, the antenna noise temperature is dominated by Galactic noise which has an inverse frequency power spectrum (van Cappellen, 2002) with an spectral index[1] $\beta \approx 2.5$ between 100 and 200 MHz. This is reflected in the graphs as T_{sky} is more than four times larger at 100 MHz than at 200 MHz. In fact, the mesh shielding (A or B) effectively increases T_{sky}. On the other hand, the bottom graphs show that T_{gnd} decreases with the use of the mesh, and *Mesh-B* is more effective at reducing the antenna noise pick-up from the ground.

Finally, Figure 4 (left), shows the antenna noise temperature at 45° elevation of all three cases as a function of frequency.

The conclusion is that at these low frequencies, due to the Galactic noise emission, the overall antenna noise temperature actually increases with the use of the mesh shielding. However, since we are interested in the ratio of A_{eff}/T_{Sys}, we still need to assess A_{eff}, which we do in the next section.

3.3. Antenna Gain and A_{eff}/T_{Sys}

Figure 4 (right), shows the antenna gain as a function of frequency. The figure shows that the *Mesh-A* shield configuration reduces the antenna gain slightly, in contrast with the *Mesh-B* shield which improves the antenna gain by at least 2.0 dB when compared with the reference case. This improvement is more noticeable at the high end of the band.

Figure 5 (left) shows the antenna sensitivity A_{eff}/T_{Sys} as a function of frequency for each of the three antenna configurations, with T_A calculated at 45° elevation,

---

[1] We use the value of Galactic emission in the direction of the Galactic poles.

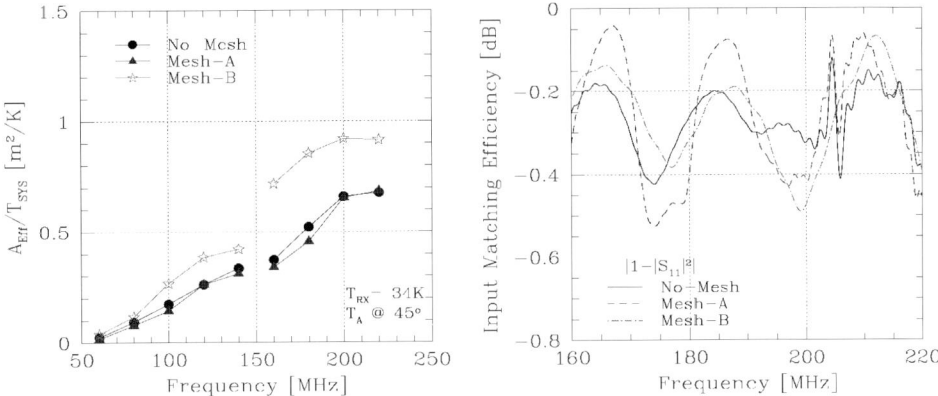

Figure 5. (Left) Antenna sensitivity $A_{\text{eff}}/T_{\text{Sys}}$, for $T_{\text{RX}} = 34$ K (Right), input matching efficiency $|1 - |S_{11}|^2|$ for the three different antenna configurations.

and a receiver noise temperature of $T_{\text{RX}} = 34$ K. The graph shows a marked improvement in sensitivity in the case of the *Mesh-B* at 200 MHz; unfortunately it still gives only 83% of the 5000 m^2/K specified for the SKA at 200 MHz (Jones, 2004).

It must be noted that the increase in antenna sensitivity comes from an increase in antenna gain since for all effects the antenna noise temperature at 45° elevation is not very sensitive to the particulars of the shield configuration (see Figure 4, left).

At 100 MHz, the improvement is a factor of 1.5 with respect to the reference. Although there is not an official antenna sensitivity specification at 100 MHz, for the SKA, at this frequency, the best shield configuration (*Mesh-B*) with a 12 m base reflector will provide only about 1200 m^2/K with 4500 antenna elements, (with $T_{\text{RX}} = 34$ K), which is already less than the specified goal at 60 MHz of 2500 m^2/K.

3.4. INPUT MATCHING EFFICIENCY

As a final consideration, prime focus symmetric antennas have phenomenon that reduces the antenna efficiency due to the presence of standing waves between the prime focus feed and the main reflector that result in sinusoidal variations of the input matching efficiency as a function of frequency. This is shown in Figure 5 (right) for the three cases under consideration; without any correction this could result in peak-to-peak gain variations of the order of ±0.2 dB.

For the US-SKA symmetric antenna concept there will be a pedestal in the form of a truncated cone holding the second high frequency feed for the upper frequency band of operations at the Gregorian focus. This pedestal will minimize the effects of the standing waves.

4. Conclusions

We have characterized a 12 m parabolic reflector antenna with two types of noise shielding attached to the periphery of the dish, around 100 and 200 MHz.

The use of this type of shielding, either *Mesh-A* or *Mesh-B* effectively reduces the coupling of ground noise emission to the antenna. Nevertheless, at these low frequencies Galactic noise dominates the antenna noise temperature making it somewhat insensitive to the particular type of shield configuration. On the other hand, there is an increase in antenna sensitivity produced by an improvement in antenna gain, for which, *Mesh-B*, i.e., a parabolic shielding extension to 16 m is more effective. This particular shielding extension brings the antenna only marginally within the specifications for the SKA at 200 MHz.

References

Computer Simulation Technology: 2003, 'CST Microwave Studio 5, HF Design and Analysis', *CST-Computer Simulation Technology*, Dec. 11, 2003.

Jones, D. L.: 2004, 'SKA Science Requirements: Version 2', *Square kilometre Array US-SKA Memo 45*, Feb. 24, 2004.

Krietenstein, B., Schumann, R., Thoma, P. and Weiland, T.: 1998, 'The perfect boundary approximation technique facing the challenge of high precision field computation', *Proceedings of the XIX International Linear Accelerator Conference LINAC'98*, Chicago, USA, pp. 860–862.

Olsson, R., Kildal, P.-S. and Weinreb, S.: 2004, 'A Novel Low-Profile Log-Periodic Ultra Wideband Feed for the Dual-Reflector Antenna of US-SKA', *IEEE 2004 AP International Symposium*, Monterey, CA. June, 2004.

Van Cappellen, W.: 2002, *LOFAR-ASTRON-ADD-009 report*, Oct 10, 2002.

RADIO ASTRONOMY ANTENNAS BY THE THOUSANDS

ROGER SCHULTZ
BSME, Consultant to Cal Tech, RAL, UC Berkeley & JPL, U.S.A.
(e-mail: schultz_assoc@pipeline.com)

(Received 18 August 2004; accepted 8 December 2004)

Abstract. Large number of microwave antennas of size and surface accuracy appropriate for the Square Kilometre Array (SKA) have not been manufactured previously. To minimize total cost, the design needs to be much more carefully considered and optimized than would be affordable for a small number of antennas. The required surface area requires new methods of manufacture and production-line type assembly to be considered. A blend of past antenna construction technology, creativity, and new technology is needed to provide the best possible telescope for the proposed SKA science goals. The following key concepts will be discussed with respect to reflector antennas and many supporting photographs, figures and drawings will be included.

- Surface and supporting structure – comparison of panels with a one-piece shell as produced by hydroforming.
- Combined reflector and mount geometry – performance/cost materially governed by this geometry which must be optimized for SKA requirements which are significantly different from typical communications antennas
- Types of fully steerable mounts – king post, turntable bearing and wheel and track
- Pointing accuracy – factors effecting cost, non-repeatable and repeatable errors
- Axis drive concepts – traction devices, gears, screws, etc.
- Life cycle costs – maintenance and power costs must be considered
- Synergistic design – all of the above factors must be considered together with the wideband feed and receiver system to optimize the whole system

Keywords: antenna, hydroform, life cycle cost, pointing accuracy, resonant frequency, SKA, surface error, throated reflector, turntable bearing

1. Introduction

Large number of microwave antennas of size and surface accuracy appropriate for the Square kilometre Array (SKA) have not been manufactured previously. To minimize total cost, the design needs to be much more carefully considered and optimized than would be affordable for a small number of antennas. A blend of past antenna construction technology, creativity, and new technology is needed to provide the best possible telescope for the proposed SKA science goals.

Automobiles built around the year 1900 were hand made and very expensive but not so efficient from a performance standpoint compared to what we all expect today. Today we all expect excellent performance automobiles all the way from very low price to extravagant. For radio astronomy to flourish, a similar evolution of antennas needs to advance. SKA construction will be the first time thousands of antennas are built. The evolution must progress to meet cost goals.

2. How many pieces?

We begin by examining some of the features of existing antennas that lead to new economies. Dishes larger that about 4 m have for many years been built of many pie shaped panels supported by frame structures. The panel structure held the panel surface shape but made no contribution to the frame support that sustained the overall parabolic shape (Figure 1).

The microwave reflective surface is a paraboloid of revolution that produces the antenna energy collective surface. Early antennas were built with stretch-formed sheet metal panels where the double curvature was inelastically imposed by aircraft stretch forming.

Figure 1. Typical 15.5 m pie panel antenna.

Figure 2. Flat sheet onto pie panel tooling.

Figure 3. Panel tool being shaped by sweep boom template.

In this process the edges of the sheet are pulled over a curved mold. This proved too expensive in the competitive ground antenna market.

It was found that flat sheet could be elastically curved if the pie panel was made of enough smaller pieces. These pieces were pulled down to shaped tooling (see Figures 2 and 3) where hat and Z section frame work was glued and riveted to the curved sheet metal (see Figure 4). When the assembled panel was released from the tooling it had a parabolic shape. The tooling was over-curved to compensate for spring back when the panel was released (Figures 5 and 6).

Pie panel size is limited by the ability to mount it upon reflector back structure in light winds (see Figure 7). There must be a reflector frame support point at the corner of each panel. This causes unnecessary structural complexity in the back structure.

Figure 4. Hat and Z section framework holding back side of pie panel made with slat sheetmetal.

Figure 5. Finished "bonded" pie panel made from flat sheet metal pieces (photo courtesy Antedo, Inc.).

Figure 6. Stretch formed pie panel (photo courtesy Patriot Antenna Systems).

Figure 7. Panel size kept small for handling in light wind.

The panel frame structure only supports the panel (see Figure 8). The reflective sheet and framework make no contribution to the overall reflector back structure rigidity and are dead weight.

Alignment of panels into a paraboloid is tedious. It is accomplished by first rough aligning the support clips of the back structure with a tape measure stretched

Figure 8. Dark members support panel surface; white members are reflector back structure frame.

from a center theodolite station and setting them to an elevation angle. After the panels are installed, holes are drilled by a chord linkage into which targets are inserted (Figure 9). Final alignment is done by a worker who climbs through the back structure to each panel mounting clip and adjusts it to the instructions of the theodolite operator atop the dish (Figure 10). No surprise to anyone that this climber works at a very high pay rate owing to the extreme risk involved. Best fit data reduction of the theodolite angles yields angular changes needed to finish setting reflector surface to the desired surface accuracy. Even with fully automated total station instruments that eliminate the need for uniform radius target drilling, this process will continue to be very costly at the site installation.

This pie panel process based large reflector proved to be competitive for years for small quantities of antennas. But now it is time, in the face of the need for such enormous quantity of surface as is required for the SKA project, that consideration be given to methods that require expensive development and tooling in order to drastically reduce the cost of the reflective surface.

3. Fewer pieces

The complexity seen thus far is greatly reduced for the 350 antennas forming the Allen telescope array (ATA). Figure 12 shows a one piece 6 m dish hydroformed

Figure 9. Linkage sets alignment hole drilling for targets from theodolite at dish center.

Figure 10. Theodolite used to set elevation angles to panel mounting clips and final alignment targets. Tape stretched to panel clip shown here.

Figure 11. Sometimes reflectors were aligned by a sweep boom in the factory.

Figure 12. Dave DeBoer, SETI Institute, and John Anderson, Anderson Manufacturing, examine first ATA 6 m dish.

to curvature. This process is now producing 0.2 mm RMS manufacturing surface accuracy.

Figure 13 shows a very simple reflector structure and mount – minimum complexity compared to pie panel antennas shown earlier.

4. Innovative antenna geometry

This 6 m antenna is an offset Gregorian geometry. It is an extremely low weight/low cost antenna at about $1500 per sq. meter. It has a "throated" reflector with its mounting gimbal mechanism within the reflector structure. This allows the elevation axis to be close to the dish center. Less wind torque is thereby produced allowing use of a less costly mount to achieve pointing.

Figure 13. First ATA 6 m antenna assembly.

5. Process challenges

In order to scale this 6 m success to 12 m, the hydroforming process needs development. Spring back and thinning in the hydroforming process are reasonably repeatable and can be modeled with nonlinear FEA. Figure 14 shows calculated spring back for a 12 m dish. Spring back and thinning calculations are used to adjust the mold to produce the desired dish shape. This dish is made by a proprietary hydroforming process belonging to Anderson Manufacturing.

Ohio State University will continue spring back and thinning simulations for SKA. All previous size increases at Anderson (and there have been many) have been successful!

Another challenge is to split the 12 m into two halves that will be allowed down the road with a permit. Figure 15 shows two types of split detail for a hydroformed shell. For the USSKA initiative, a strawman antenna is being developed to incorporate cost savings concepts. The 12 m shell reflector splits into two pieces. For surface accuracy the back structure is integrated onto the shell before the splitting. The 2 m extension of the shaped surface that yields a 16 m dish for low frequency operation is 10% projected solids mesh. Dark members define split plane; center hub and dark members are split and flanged for shipping.

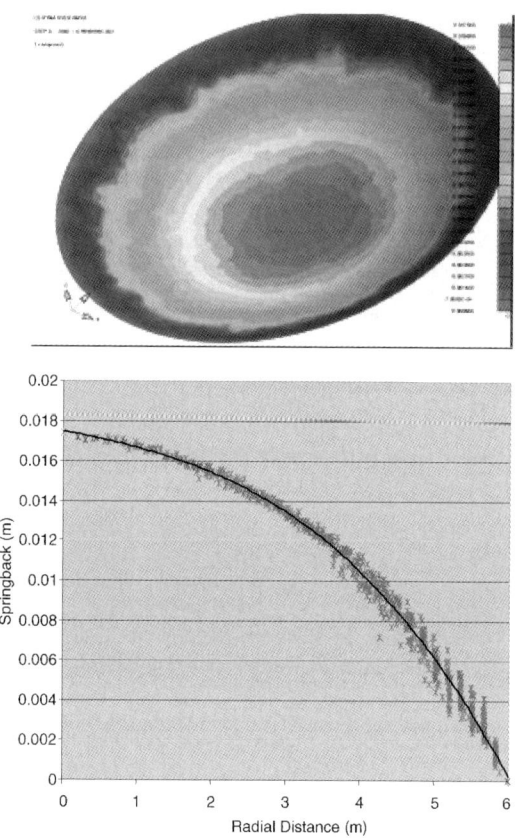

Figure 14. Figures show calculated spring back for 12 m dish (figures courtesy Dimitri Antos, JPL).

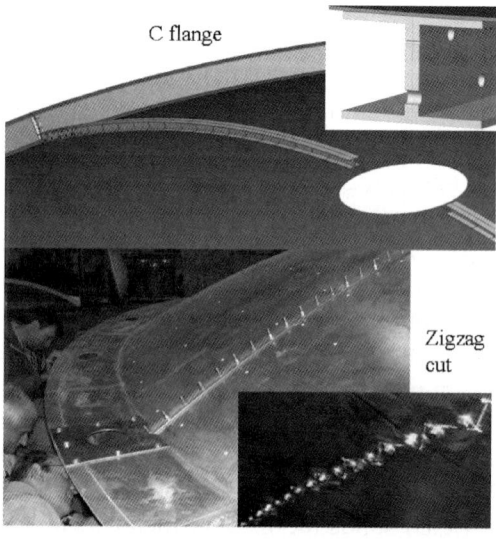

Figure 15. C flange and zigzag splitting methods (photo courtesy Rotek, Inc.).

Figure 16. USSKA 12–16 m strawman antenna.

The reflector 12 m shell IS VERY STIFF and forms the front half of the "backstructure". This shell replaces all front radials and hoops seen in previous designs, eliminates most diagonals and decreases the number of radial trusses. Sub reflector, tripod legs and 2 m skirt are removed for shipping. The reflector structure is "throated" for low cost high performance (Figure 16).

Mass produced components such as castings, stampings, CNC machinings and mature industrial components (turntable bearings, gear boxes, etc.) will be used in the mount.

Minimized site assembly and alignment will result from there being many fewer pieces of metal in the design.

There will need to be site handling equipment that is able to unload antenna halves and mounts from trucks that will permit direct assembly and erection in light winds.

The reflector half is a 6 m wide load that can pass down major highways with a permit in many parts of the world. Figure 17 shows a truck trailer passing under a 4.6 m (15 feet) high underpass with standard trailer. If the underpass is lower at 4.0 m (13 feet) passage is possible using a "low boy" trailer.

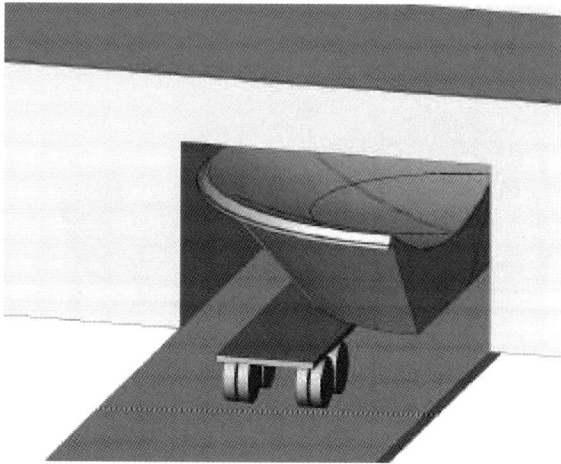

Figure 17. Normal height trailer with dish half passes under 4.6 m underpass.

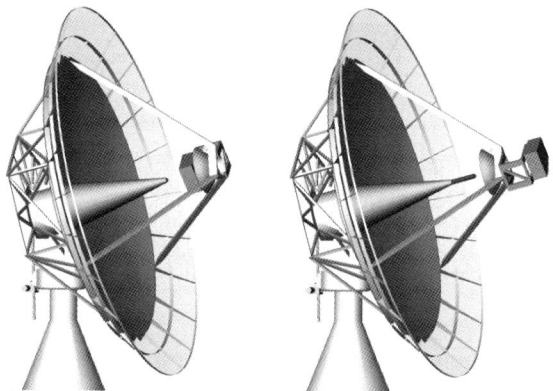

Figure 18. The 12–16 m operating configurations for high and low frequency.

The left view of Figure 18 shows 16 m prime focus operation down to 100 MHz. The right view shows 12 m Gregorian 1.2–34 GHz operations plus flipped feed, wide field, 0.1–1.5 GHz operations.

6. Geometry comparisons

Figure 19 shows typical 12 m antenna geometry composite of several current antenna products. Figure 20 shows the incorporation of reflector shell integration into reflector structure combined with "throated" reflector design. We compare features.

A shell reflector with throat is 30–50% lighter than for a typical pie panel reflector. The reflector cost will be significantly less and the mount will also be less (less reflector inertia, less mount needed for pointing dynamics).

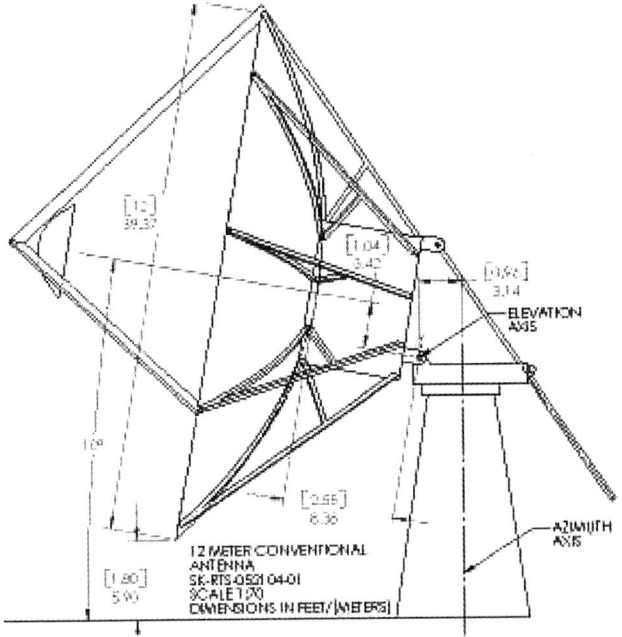

Figure 19. Composite of typical commercial antenna geometry.

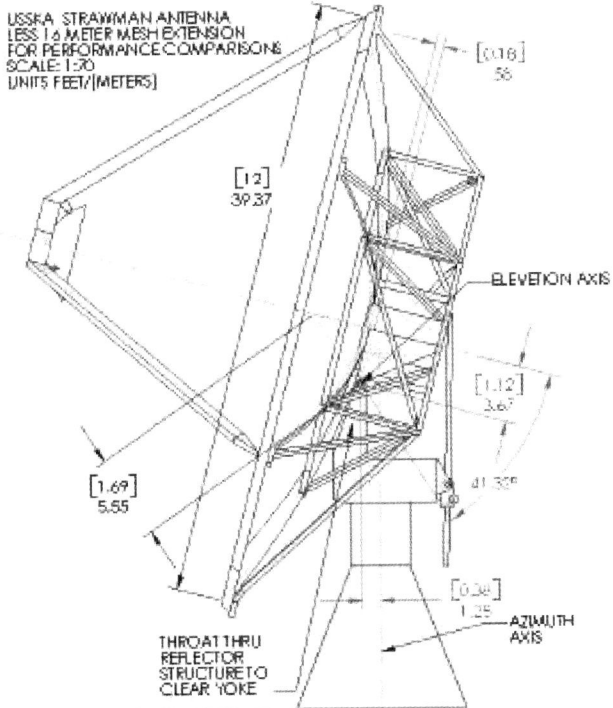

Figure 20. USSKA 12–16 m strawman antenna less 2 m extension.

The conventional antenna azimuth peak wind torque is 50% higher than for the "throated" antenna because the thrust term moment arm is much shorter. Conventional antenna elevation peak torque is 100% higher than for the "throated" antenna. The conventional antenna has the elevation axis unbalance all to one side at a long moment arm where the "throated" antenna has an even "split" of unbalance to both sides. Again, the thrust term moment arm is much shorter.

Wind "gust gain" is lower in proportion to the steady wind comparisons made above. Lowest resonant frequency (LRF) requirements go down in proportion to about the square root of the decrease in wind gust gain. This in turn reduces LRF requirement to meet absolute pointing requirement. Antenna costs are significantly reduced by reductions of reflector weight, drive torque and LRF requirements.

Why do we care about LRF?

- Wind gusts deflect antenna pointing.
- Servo system reacts to restore pointing.
- RMS pointing error controlled by higher servo band pass response.
- Band pass of servo system must be lower than the lowest resonant frequency of the antenna to *prevent violent instability*.

Figure 21 shows a twisting of the tripod at 5.7 Hz for tripod leg sideways bending. The tripod legs may need to be thickened to stiffen their bending to raise their resonant frequency.

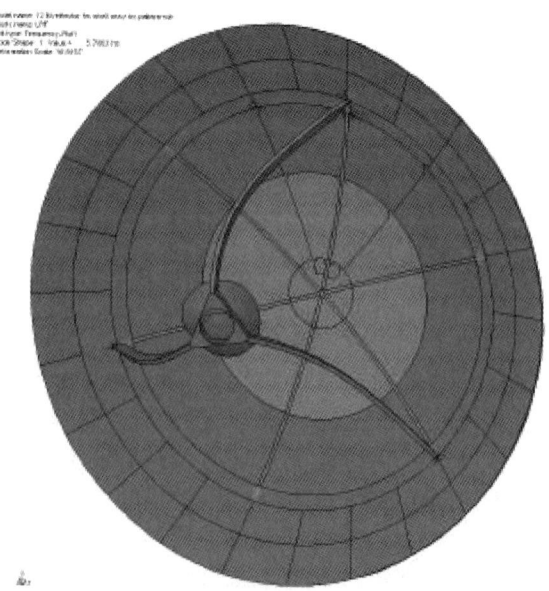

Figure 21. FEA analysis of lowest vibration mode of 12–16 m reflector.

7. Mount type

Fully steerable antennas have often been designed in the Elevation-over-Azimuth axis configuration which is now thought to be the most economical. Even optical telescopes are now built this way for economy. For microwave mounts, large diameter bearings have often been used to form the axis about which the azimuth axis rotates. The ATA antenna pictured previously is this type.

For larger antennas (30 m and up) azimuth rotation was provided by a circular track with wheels at the base of a large frame called an alidade. The antenna was mounted on this frame. Such an antenna is pictured in Figure 22.

The low costs of the railroad type wheels and track by themselves seem enticing but the entire structural cost of a wheel and track antenna must be considered to derive an even handed comparison with an antenna mounted on a single turntable bearing (Figure 23).

Wheel and track antennas are typically very rigid mounts since track diameters can be made a large fraction of the reflector diameter and since support frame

Figure 22. A 30 m communication wheel and track mounted antenna (TelesatCanada).

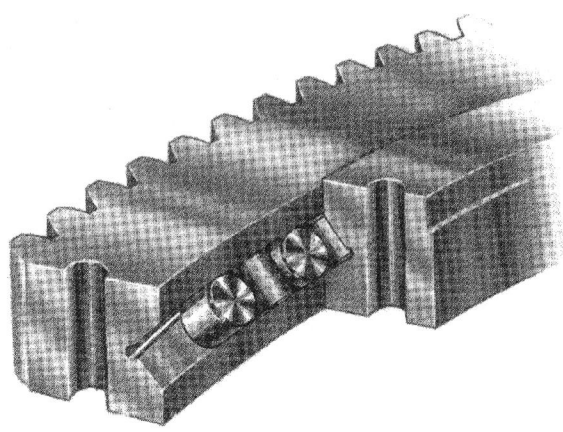

Figure 23. Cross section of geared turntable bearing (photo courtesy Rotek, Inc.).

structures that rise from the track to the elevation axis also can have matching base lines. This large baseline is the reason for the high stiffness. This innate stiffness not only makes resonant frequency high for excellent wind gust servo response but wind deflections are kept low.

The track requires a stiff and strong foundation under the entire periphery of the track. At any one time much of this strength and stiffness is not used since the rail only supports load where the wheels are. Also, there is a high cost of leveling the rail, or mapping the axis wobble to be used to calculate pointing to adjust for the irregularity in the rail height.

Traction is always an interesting design problem for wheel and track antennas. The up-wind wheels tend to slip as wind overturning moment shifts weight onto the wheels down-wind. The SKA requirement will be less of a problem since it is to operate only up to 50 MPH; communications antennas are usually expected to operate in higher winds.

Costs of turntable mounted antennas first scale as little more than the square of the bearing diameter in the region of sizes that would be used for a 12 m antenna. This relation does not hold up to 30 m antennas, particularly if they are expected to have reasonable pointing at centimeter wavelengths.

The elements of cost are the bearing itself and the material and machining costs of the housing. Machining form tolerance (housing cylindricity and flatness) is always an issue. Bearing manufacturers have kneeled to new equipment competitive cost considerations and recommended tolerances that encourage antenna designers to allow crude tolerances. However, the mounting costs to replace VLA azimuth bearings reminds the science community that carelessness at the point of machining new antenna housings can be very expensive later. (They are remachining the housings for the VLA as bearings are replaced.) The wasted money could have been spent on new science.

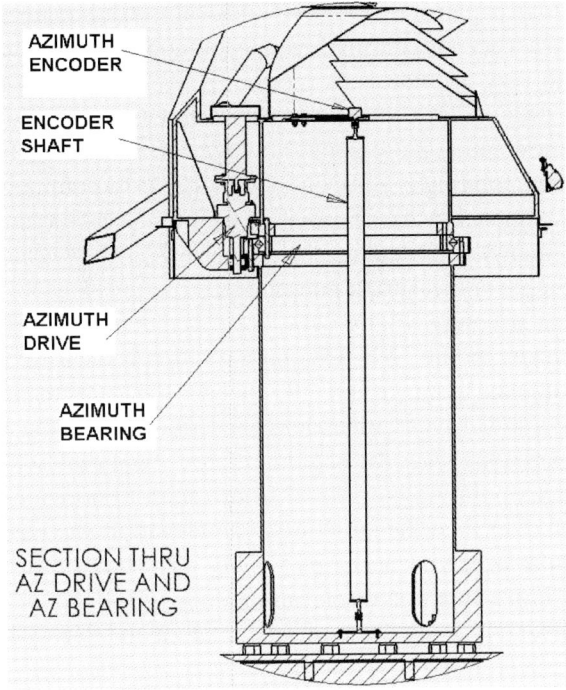

Figure 24. Section through azimuth encoder.

Turntable bearings can come with gearing cut in the same rings that are bearing races; this represents great price/performance design.

8. Pointing

Absolute pointing for radio astronomy arrays is more challenging than the relative pointing usually planned for communications antennas, where autotracking homes in on the most intense pointing to the discrete signal. In radio astronomy reference stars can be used to fine tune pointing but this process detracts from observing time.

In a perfect world, rotation of reflectors on mount structures is simple to measure. Angular encoders simply report the angle to the servo system which adjusts the pointing to the science requirements. In the real world wind and thermal deflection, initial misalignments, axis wobble and zero reference errors inevitably challenge the antenna designer to find ways to point accurately at the best cost.

For our 12 m strawman antenna we require a $0.005°$ RMS pointing error, a small fraction of the 3 db beam width for higher frequency work. Less pointing accuracy confuses the gain at various off axis locations within the field-of-view. This is much more stringent than pointing error considerations for a communication antenna.

There are two different strategies. First, build the antenna so rigid and thermally stable that adequate pointing is achieved. This turns out to be very expensive.

Massive, low thermal expansion, designs could do this. Carbon fiber resin composite structures can have very low thermal expansion and very low deflection in wind and be made with very accurate axis motion. However, there is no precedent for such an antenna, even in large quantity, to be manufactured for low costs.

Second, we know much more about how we can produce deforming antennas where the deformed pointing is compensated in the servo command. There are two compensation strategies. 1) Measure pointing by reference structures that are not deformed by the elements. 2) Calculate corrections based on measurable parameters.

Here is an example of measuring pointing by structures that are not deformed by wind on the mount structure (Figure 24). There is much "wind up" in the azimuth drive assembly on any mount. When azimuth rotation is measured by a shaft down the center of the antenna to the ground, the encoder measures the rotation of the antenna above the azimuth drives with no loss of accuracy due to the wind up in the azimuth drives and structure since the angular reference is a shaft with no torque on it.

As elevation axis pointing is lowered from zenith to horizon, the actual beam pointing droops farther than the structure at an elevation encoder would indicate. Most of this angular offset is caused by gravity distorting the structure. This distortion is then a simple function of elevation angle which can be compensated by calibrating the offset and using those offsets in a lookup table to digitally calculate the correction in the pointing command.

A more challenging compensation is to correct for the differential thermal distortion experienced in the antenna structure as the sun shines on it. Temperature sensors are placed in several locations in the structure. The heart of the challenge is to minimize the number of sensors and produce the equations that predict the required compensation.

Pointing considerations must be considered from the beginning of the design. It may be that the structure can be minimized to the dynamic wind gust requirements and all the deflections compensated by the above techniques.

9. Axis drives

Several generic mechanisms have been used to position antenna axes. Gearing is often used since it can produce a high gear ratio. High ratio is used primarily for torque amplification (used to allow low power motors to resist and compensate for gust deflection). The ratio is limited by motor speed and slew speed requirement. Additionally higher ratios emphasize the need for the motor's ability to accelerate itself. The relative effective rotational inertia of the motor and antenna are related by the square of the gear ratio. If the ratio is very high the preponderance of acceleration torque will be consumed accelerating the motor.

The focus for antenna designers is the drive where it connects to the antenna axis, the high torque end of a gear train. Here the stiffness considerations for lowest resonance frequency are most important. The structural baseline at the point of connection is the most important design parameter. For azimuth drives, the outer race of the azimuth bearing is often used also as the bull gear. So the torsional stiffness of the antenna is largely determined by the pitch radius of the bull gear. Higher radius, higher stiffness, higher resonant frequency. For the elevation axis driven by a jack screw the action radius is the important baseline dimension. The square of this baseline is proportional to the net overturning moment stiffness achieved by the jack.

Wheel and track antennas provide the most understood example of traction drives. Here the wheels are powered by gear boxes and motors. Since track diameters, mentioned above, are usually a high percentage of reflector diameter, the baseline or diameter is quite high, making the azimuth drive torsionally rigid. Again the concern is traction in high winds where wheels can slip.

Gear ratios added to wheel and track ratio or added to bull gear and pinion are best made up of commercially available speed reducers. These are, as a class, industrially mature commodity components. They are however used in uncommon circumstances in antenna drives. Few gear box manufacturers understand how to rate torsional stiffness, life and efficiency of their own product as it is used in antenna service. Experienced antenna designers must rely on antenna field experience and antenna related testing to determine design parameters. The most common generic types of reducers that have been used are planetary, cyclo and multiple stage helical (Figure 25). Dual differential planetary and worm drives have not done well.

High drive-to-stow torque usually determines the capacity of the drive needed.

10. Life cycle costs

Historically the three largest costs of maintaining the structural mechanical part of antennas have been lubrication, painting and replacement of failed components. A newer consideration is the operating power cost; how much money does it take to pay the power bill to run radio astronomy sky scans 24 h–7 days, year after year? The SKA power bill world wide could be 3–40 million dollars a year for the axis drives according to how well the antenna is designed.

Obviously initial design will be very influential in determining how costly maintenance will be.

For many high minded engineering reasons, circulating oil lubrication systems have been designed into many antennas. These have been first cost and maintenance cost disasters – mostly absorbed by governmental agencies that seem unphased by such outcomes.

The initial cost of circulatory systems often equals the cost of the gearing it is lubricating. The pumps that circulate the oil wear out and fail causing primary gearing failure, and so on. Greasing requires much less attention but must be engineered to ensure machinery life. Some gearing is intrinsically more suited to grease lubrication.

Figure 25. Planetary drives for 15.5 m antenna of Figure 1 being tested for smoothness.

Antenna surface finishes can have a major cost impact on the maintenance bill. White painted antennas, for thermal control, are well known for their high cost. The "telephone company" mentality has chosen galvanized surface finish as would utility industry for electric transmission towers. ATA has chosen bright zinc electrolytic finish for its steel pedestals and a natural aluminum finish for reflectors. The thermal impact on pointing needs to be quantified to see if surface finish reflectance alone can yield adequate pointing accuracy. If not, and active control is utilized, less reflective but permanent finishes such as bright zinc should receive serious consideration for SKA.

Some components will always be subject to wear such that replacement is required. Brakes have wear surfaces that will need repair. Brake wear adjustment is not taken care of automatically by most industrial brakes. The few dollars additional cost for brakes that do adjust themselves is wise for overall best cost.

Large active components such as azimuth bearings should be designed to fail infrequently. Failure costs should be balanced with the first costs. Full cost analysis will reveal that dollars spent on housing accuracy in the beginning will produce better failure cost ratios than are now being experienced on arrays such as the VLA.

11. Conclusion for SKA development today

The contrast between conventional and advanced antenna design has shown that, for cost reasons, the antenna design and manufacturing process must be advanced if we are to achieve a good cost for the SKA.

Developments, including those shown here, can be expected to produce $US100–200 M savings over conventional antenna concepts. Shell reflector reflective surface included in the structure (hydroforming or stretchforming), optimized antenna geometry (throated reflector structures), automated castings of complex geometries, and more, will be examined in the next few years of development. Other fruitful topics include axis bearing, axis drive, and instrumentation optimization. These are some of the topics that will be examined as we produce lowest cost concepts to meet SKA antenna performance.

In the examples above it is seen that there is tremendous interaction between the various parts in an antenna assembly. No one part can be optimized in a meaningful way without considering how it interacts with the balance of the system. This kind of design is not conquered by simply running some few parameter, optimizing, computer program.

An interactive blend of theoretical and practical experience is required to find the magic combinations of conceptual design detail and effective industrial process to achieve our cost/performance goals. Clearly an enthusiastic and spirited, but conceptually clear, dialog will speed this process along. We should all understand why the end result is optimum so we can all get behind this project's grand discovery capability.

Experimental Astronomy (2004) 17: 141–148

© Springer 2005

A SPHERICAL LENS FOR THE SKA

JOHN S. KOT[1,*], RICHARD DONELSON[2], NASIHA NIKOLIC[1], DOUG HAYMAN[1], MIKE O'SHEA[3] and GARY PEETERS[3]

[1]*CSIRO ICT Centre/Australia Telescope National Facility, Radiophysics Laboratory, P.O. Box 76, Epping, NSW 1710, Australia;* [2]*CSIRO Manufacturing and Infrastructure Technology, Clayton South, VIC 3169, Australia;* [3]*CSIRO Molecular Science, Clayton South, VIC 3169, Australia*
(*author for correspondence, e-mail: john.kot@csiro.au)

(Received 2 August 2004; accepted 30 December 2004)

Abstract. Spherical refracting lenses based upon the Luneburg lens offer unique capabilities for radioastronomy, but the large diameter of lens required for the Square Kilometre Array (SKA) means that traditional lens materials are either too dense or too lossy. We are investigating a composite dielectric that theoretically offers extremely low loss and low density, and is suitable for low-cost mass production. We describe our progress towards realising this material and demonstrating the manufacturing concept, via the manufacture and testing of a small (0.9 m) spherical lens.

Keywords: artificial dielectric, Luneburg lens, Square Kilometre Array (SKA)

1. Introduction

In 1944, R. K. Luneburg in his Brown University lecture notes on "Mathematical Theory of Optics" considered the general problem of ray paths in media in which the refractive index could be expressed as a continuous function $\eta(r)$ of the radial coordinate r, independent of the angular coordinates θ and ϕ, where (r, θ, ϕ) are the standard spherical polar coordinates. He used this result to obtain the solution for a graded-index spherical lens that imaged a point P on the surface of a sphere onto a point Q on the surface of a concentric sphere of a different radius, where both P and Q lay in a homogeneous medium exterior to the lens, thus generalizing a solution discovered by Maxwell, the "Maxwell fish-eye". In the case where the point P approached the surface of the unit sphere and Q receded to infinity, the solution had a simple closed form

$$\eta(r) = \sqrt{2 - r^2}$$

and this solution is often referred to in the literature as "The Luneburg Lens". The advantages of this optical system as a radiotelescope antenna are clear: by moving a small feed over the surface of a fixed Luneburg lens a beam can in principle be scanned over the full sky, and because of the symmetry of the lens the beam shape is independent of scan angle. For the Square Kilometre Array (SKA) this would offer significant advantages: operationally, the full aperture of the instrument would be available to multiple, independent users simultaneously, maximizing use of the infrastructure; scientifically, the wide field-of-view would offer a chance to explore new regions of phase space, such as the transient radio sky. A number of microwave antennas based upon the Luneburg lens have been made, and a large lens for radioastronomy has been proposed (Gerritsen and McKenna, 1975). The practical aspects of a microwave antenna based on a spherical lens have been considered by Cornbleet (1965) where he showed that a spherical lens made from a number of discrete shells, optimized for the operating frequency band and with the feed some distance from the lens surface, gave superior performance to a lens that attempted to approximate the continuous Luneburg solution.

For application to radioastronomy, and for a spherical lens-based SKA component in particular, the engineering problems come down primarily to the lens material. A practical lens requires materials with dielectric constants in the range of approximately 1.1 for the outer shell to 1.7 for the core. Because the lens is at ambient temperature and is placed in front of the feed and LNA, any loss in the lens contributes directly to the system temperature, so it is important to keep the dielectric loss as low as possible. The lens material also has to be cheap, and of low density, since to achieve a large collecting area with an array of identical spherical lenses, the total amount of material required is proportional to the diameter of an individual lens. For the SKA, lenses of several metres diameter are likely to be needed; this immediately rules out a number of traditional microwave lens materials on cost grounds alone.

While the SKA application is particularly challenging, because of the need for lenses with such large aperture, there are clear advantages to be gained by reducing the mass, loss, and manufacturing cost of spherical microwave lenses in a number of other areas where smaller lenses might be used, such as wide field-of-view imaging, and antennas for mobile satellite communications, especially for airborne platforms where mass is critical.

Whole system considerations such as the trade-offs involved in reducing lens size have already been treated extensively in an SKA white paper (Hall, 2002). In the present paper, we shall concentrate on the development of dielectric materials and the manufacture and testing of a small prototype spherical lens, and describe the progress made so far.

2. Low-loss, lightweight dielectric materials for a spherical microwave lens

Small spherical microwave lenses available commercially are typically made from foamed polymers of varying density. Low-loss polymers such as polyethylene, PTFE, and polystyrene have dielectric constants in the range 2–2.5 when solid. The dielectric constant of typical low-density foamed polymers, such as commonly found as packaging, or in disposable drinking cups, is around 1.02; so making a variable dielectric constant material by foaming a low-loss polymer produces a suitable range of dielectric constants for making a spherical lens. The "mixing rule" for these foams (that is, the relationship between averaged dielectric constant and the volume fraction of polymer) is linear to a very good approximation, allowing us to calculate the weight of such lenses, or we can scale from the manufacturers' published data. Either way, the practical limit for the diameter of this type of lens is not much greater than 1 m. The structural problems of large foam lenses can be addressed by using a half lens on a ground plane (the "virtual source" Luneburg lens). However, the mass of raw material needed leads to costs that are much too high for the SKA.

Two approaches to make lenses of larger diameter are to use artificial dielectrics made from highly conducting metal particles suspended in a low density foam matrix, or composite dielectrics or dielectric mixtures. Larger lenses have been made successfully using artificial dielectrics; however, the reported dielectric loss tangents are still too high (Baker, 1951), limiting their usefulness for radioastronomy over the SKA frequency band. Our initial experiments looked at both artificial dielectrics and dielectric mixtures and, while both approaches produced dielectric materials of reduced density, the initial results with dielectric mixtures looked more promising, so that is where we have focused our effort.

Our approach is based on composite dielectrics where there is a high contrast between the different component dielectrics forming the composite. In particular, we have focused on Titanium Dioxide (TiO_2) particles dispersed uniformly in low-density polymer foam. We chose TiO_2 because it is manufactured cheaply on an industrial scale for applications such as a pigment in paint, and in its rutile form, it is used as a dielectric at microwave frequencies. The rutile form is somewhat anisotropic, with a relatively high average dielectric constant ($\simeq 100$). It has moderately low loss, and is widely used in microwave applications such as dielectric resonators. For these high contrast mixtures, the mixing rules can be highly non-linear, and controlling the distribution of the different fractions of the mixture to optimize the mixing rule appears to be a key to obtaining low density materials.

3. The physics of high-contrast dielectric mixtures

The study of the electromagnetic properties of mixtures has a long history going back to the mid-1800s, and a very comprehensive overview of this area is given by Sihvola (1999). In our study of the lens material we have used mixing rules of the Maxwell Garnett type (Maxwell Garnett, 1904). Although more sophisticated models for mixtures exist, these simple rules are sufficiently accurate over the domain of parameters under consideration: particles size very much less than the operating wavelength, and volume fraction of the high-permittivity particles of a few percent.

For high-contrast mixtures, the distribution of the different fractions has a very strong effect on the resulting mixing rule. A good example of this is an opalized structure, which is a regular packing of spherical particles. Because of the regular

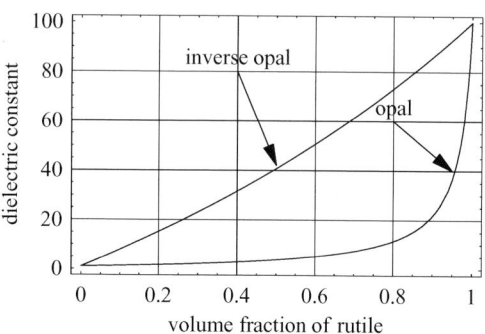

Figure 1. Effective dielectric constant of TiO_2/air opal and inverse opal.

structure, we can accurately calculate the effective dielectric constant for an opal formed by small dielectric spheres in air and, by duality, that of its inverse formed by spherical air bubbles in solid dielectric. The result is shown in Figure 1 for TiO_2 with a dielectric constant of 100. We can see that, for an opal, the mixing rule is highly non-linear. At the maximum theoretical packing density for spheres, the effective dielectric constant is around 10, despite 70% of the mixture being a material with dielectric constant 100. For the inverse structure, the mixing rule is close to an ideal, linear rule where effective dielectric constant is proportional to volume fraction.

An inverse opal is not a practical structure for a lens, but it gives us a clear indication of the type of distribution needed to approach an ideal, almost linear, mixing rule. To illustrate the importance of this for a Luneburg-type lens, the maximum dielectric constant needed is typically less than 2, so a mixture of shaped TiO_2 particles in a low-density polymer foam could achieve this dielectric constant with a volume fraction of TiO_2 of around 1%, if an ideal mixing rule could be obtained. The density of rutile is approximately 4000 kg m^{-3}, and typical low-density polymer foam around 20 kg m^{-3}, so the density of the lens core could be as low as 60 kg m^{-3}. In comparison, a polymer foam lens core might approach 1000 kg m^{-3}, so this type of mixture offers dielectric materials of extraordinarily low density; less than 1/10 the density of a foam lens.

4. Manufacture and testing of a TiO_2/polymer foam material

Various particle shapes were explored for their theoretical electromagnetic properties and manufacturability, and thin disks of TiO_2 were chosen as they gave a close to ideal mixing rule, and the shape could be well approximated by a practical manufacturing process. The manufacturing process produced flakes of rutile TiO_2 in the size range of 1 to 2 mm, with an aspect ratio approaching 100 : 1 (i.e., thickness 10–20 μm), using a standard tape-casting process, followed by sieving to separate out the smaller flakes (having lower aspect ratio). The mixing rule for 100 : 1 disks is plotted in Figure 2, and compared to the rule for "ideal" disks (i.e., in the limit of

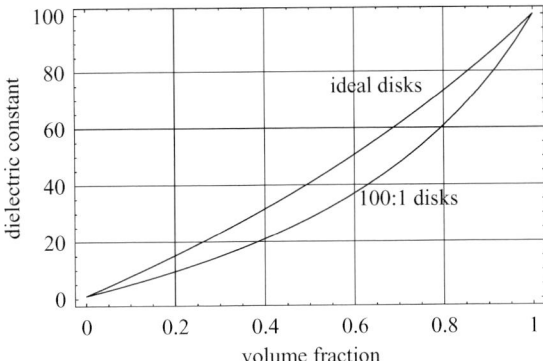

Figure 2. Dielectric mixing rules for TiO$_2$ disks, ideal and finite thickness.

infinite aspect ratio). For this type of mixture, as the dielectric contrast between the disks and the background material increases, so the aspect ratio needed to approach an ideal mixing rule also increases.

Samples were prepared by mixing the TiO$_2$ flakes with polymer and foaming the mixture using an extrusion process. The polymer (polypropylene), a gaseous foaming agent, and the rutile flakes were heated, mixed and extruded through an orifice using a twin-screw type extruder, producing a long 'sausage' of rutile-loaded polypropylene foam. The resulting loaded foam was chopped, then this material was molded into rectangular blocks, dimensioned to fill a WG-14 waveguide cavity. The dielectric properties of the blocks were measured by measuring the resonant frequencies and Q for the loaded and unloaded cavity, using an HP-8510 vector network analyzer, and from this the dielectric constant and loss tangent of the blocks were determined.

The measurement process was calibrated by measuring known samples, such as PTFE. In all cases, the loss tangent for the loaded foam samples was below 10^{-4}, with the mean measured value approximately 5×10^{-5}. The measured dielectric constant is plotted in Figure 3, where it is compared with the theoretical curves

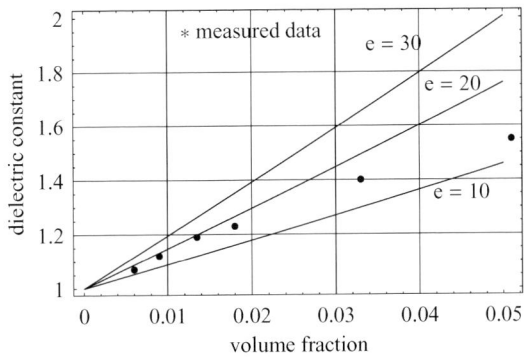

Figure 3. Dielectric mixing rules for TiO$_2$ disks, measured and theoretical.

for mixtures of disks with aspect ratios $e = 10:1$, $20:1$, and $30:1$. We can see immediately that the measured samples are not behaving like mixtures of $100:1$ particles. When the samples were investigated further it was found that the large aspect ratio particles were being broken up by the extrusion process, and that in the resulting material the aspect ratio of the particles was between $10:1$ and $30:1$. The results presented in Figure 3 are the result of a lot of optimization of the process, and we believe that this is about the best that can be achieved with an extrusion process and this size of particle.

5. Manufacture and testing of a small prototype spherical lens

In order to test the concept of manufacturing a spherical lens using molded components made from a loaded foam, we designed and manufactured a small (0.9 m) spherical lens to operate up to a maximum frequency of 5 GHz. The range of dielectric constants used was 1.1–1.55 and the target f/D was 0.75. A five-layer lens comprising a thin outer protective layer of fiberglass, three inner layers, and a spherical core was designed. The design method used an exact spherical-wave expansion method to solve the problem of the spherical lens illuminated by an elementary dipole source, and this exact forward solver was placed within an optimization loop. The dipole source was placed at the desired focus, and the antenna gain was optimized by varying the dielectric constant and thickness of each layer. For this operating frequency, a four layer lens was found to be adequate, but a five layer design was chosen for ease of manufacture. Restricting the maximum dielectric constant to 1.55 caused a small reduction in the achievable gain, but we were not confident of reliably molding parts with a higher dielectric constant at the time. The parameters of the final design are shown in Table I.

A trial assembly of the lower hemisphere of the lens in its protective fiberglass shell, but without the core present, is shown in Figure 4. This shows how the spherical shells are constructed from molded pieces using a spherical tiling. The final assembly of the lens was done using a very small amount of hot-melt

TABLE I
Parameters of the 5-layer, 0.9 m prototype lens

Layer number	Outer radius of layer (mm)	Dielectric constant
1 (core)	191	1.55
2	297	1.50
3	388	1.36
4	450	1.22
5 (shell)	451	3.40

Figure 4. Trial assembly of prototype lens components.

polyethylene adhesive to secure the layers in place, previous measurements having verified that this adhesive contributed negligible loss. Two types of test were done on the completed lens: antenna radiation pattern measurements in an anechoic chamber, and outdoor radiometric measurements with the lens placed inside a metal radiation shield and looking at the cold sky. While the radiometric measurements confirmed the low dielectric loss, the antenna radiation pattern measurements showed considerable deviation from the predicted patterns:

1. The experimentally-determined f/D ratio was considerably smaller than the predicted value of 0.75, and it was difficult to determine an accurate focus experimentally.
2. The aperture efficiency was close to the predicted efficiency (around 80%) at 2 GHz, but fell sharply with frequency, reaching approximately 50% at 4.2 GHz, whereas the theoretical model predicted high efficiency to over 5 GHz.

Near-field measurements confirmed that the loss of efficiency was due to systematic phase error over the lens aperture. The phase error showed a characteristic elliptic contour that rotated with the feed polarization. This suggests that the material of the lens has significant radial anisotropy. We are presently manufacturing new test samples, and a new test cavity, in order to verify this by direct measurement of the isotropy.

6. Conclusions

Theoretically, an optimized rutile TiO_2/polymer foam mixture offers a dielectric material with extraordinarily low density for the range of dielectric constants from 1.1 to 2. Our measurements have confirmed that such a mixture can have very low dielectric loss. If both the measured loss and the theoretical lower bound for the density could be closely approached in practice, then an SKA component based around refracting spherical lenses, with its tremendous advantages operationally and scientifically, would become feasible. To realize a material that has both low density and low loss requires distributing optimally-shaped particles uniformly in a low-density matrix. We have attempted this with an extrusion process, but the particles are damaged by the process and so we did not achieve a mixing rule for our material that was close to optimal. To proceed further we need to develop a new process. While we plan to continue this work, we recognize that on the current SKA project timescale we are unlikely to demonstrate the very challenging specifications needed for the SKA, so we shall not be actively pursuing the Luneburg lens further as an SKA concept.

References

Baker, E. B.: 1951, 'Dielectric Material', US Patent **2,716,190**.
Cornbleet, S.: 1965, *Microwave J.* 65–68.
Gerritsen, H. J. and McKenna, S. J.: 1975, *Icarus* **26**, 250–256.
Hall, P. J. (ed): 2002, http://www.skatelescope.org/documents/SKA_AUS_CONCEPT_Luneburg_17072002.pdf.
Maxwell Garnett, J. C.: 1904, 'Colours in Metal Glasses and Metal Flms', Trans. of the Royal Society, London, CCIII, pp. 385–420.
Sihvola, A.: 1999, 'Electromagnetic Mixing Formulas and Applications', *IEE Electromagnetic Waves Series* **47**, IEE.

EFFICIENCY ANALYSIS OF FOCAL PLANE ARRAYS IN DEEP DISHES

MARIANNA V. IVASHINA,* JAN SIMONS and JAN GERALT BIJ DE VAATE
ASTRON, P.O. Box 2, 7990 AA Dwingeloo, The Netherlands
(*author for correspondence, e-mail: ivashina@astron.nl)

(Received 19 August 2004; accepted 28 February 2005)

Abstract. In this paper we demonstrate the practical analysis of a dense Focal Plane Array (FPA) for a deep dish radio telescope. The analytical model is used to optimize efficiency and bandwidth of deep dish system by varying design parameters such as feed size and FPA aperture field. A prototype FPA was evaluated on the Westerbork Synthesis Radio Telescope (WSRT) and the resulting efficiency is discussed.

Keywords: reflector antennas, dense focal plane arrays, aperture efficiency, radio telescope

1. Introduction

The main figure of merit of a radio telescope is the Aeff/Tsys ratio, where Aeff is the effective area of the reflector and Tsys is the noise temperature of the system. Successful hybrid reflector design requires separate knowledge of Aeff and Tsys. This information is especially important when developing a new feed system where characteristics such as aperture efficiency and system loss can be, or should be, optimized. Additional system level challenges in a FPA-based hybrid include developing techniques and design procedures for the dense FPA. This work provides a detailed analysis of the Aeff of reflector systems and relates this to design parameters of FPAs. Maaskant (2004) introduces a model for the noise temperature for these systems. A more extensive study of noise in FPAs is underway.

The aperture efficiency of a hybrid reflector antenna is described by the properties of the feeds' far-field patterns and the geometry of the reflector (Balanis, 1982). However, precise modeling of far-field patterns for dense arrays can be difficult due to strong mutual coupling between array elements. In addition, analytical models for these structures do not provide sufficient accuracy for practical application (www.astron.nl, R&D project THEA; Boryssenko and Schaubert, May 2004; Craye, 2004).

The question arises of how to estimate the efficiency of a hybrid reflector antenna with a small F/D, and how to relate efficiency to the design parameters of the FPAs.

This paper outlines a simple FPA model, starting with a given feed size and a specified cosine function aperture field distribution. The model has been applied to reflector antennas with small F/D ratios varying from 0.3 to 0.6. In this model we address finite-size focal plane arrays with 7–20 elements in a row, with element spacing of 0.2–0.5λ. This number of elements was chosen to ensure efficient reflector illumination and to keep the number of RF signals to be combined within reasonable limits. For such systems infinite array approximations, as well as large finite array simulations break down. For an infinite array approximation there are too few array elements and the element patterns will vary significantly due to edge effects. The infinite array approximation is valid when the distance between the centre and edge elements is larger than 2 wavelengths. Therefore, this approximation would require a minimum of 23 elements (3 centre elements + 10 elements from both sides) for an array with elements spaced 0.2λ. On the other hand, current finite array simulator such as HFSS, CST, and IE3D do not have the capacity to analyze arrays larger than 5 × 5.

This paper also demonstrates the first FPA prototype made of a dense array of Vivaldi elements. The prototype was designed for the Westerbork Synthesis Radio Telescope (WSRT) reflector antenna, which has an F/D ratio of 0.35. The prototype was designed using a conjugate field matching method and near field measurements. Results comparing modeled and measured reflector aperture efficiencies from the FPA model and prototype, respectively, are presented. While our approach was validated for the WSRT with an F/D ratio of 0.35, the results are not limited to the WSRT application.

2. Modeling the focal plane array

2.1. FPA AS A FEED FOR REFLECTOR ANTENNAS

In practical analysis of reflector antennas, the array feeds are often modeled with a given element far-field pattern defined by a cosine function to the n-th power (Galindo-Israel et al., 1978; Balanis, 1982; Rahmat-Samii and Lee, 1983; Lam et al., 1985). In many practical antennas including horns and coaxial waveguides, the pattern represents the major part of the main lobe. Horn and waveguide arrays are usually sparse and the "active" element patterns are similar to the "isolated" element patterns. However, for the dense arrays with the element spacing less than 0.5λ, this condition is not valid due to strong mutual coupling. Furthermore, the patterns are different from element to element, depending on the location of the element in the array (Boryssenko and Schaubert, 2004; Craye, 2004).

Bird (1982) successfully approximates the behavior of a circular waveguide array using a different feed model excited with the TE_{11} mode. But we do not have *a priori* known information about the operation modes of the elements in dense arrays.

Here, instead of modeling each element separately, the field distribution of a dense array feed is specified on the feed aperture in cosine-form. For a circular

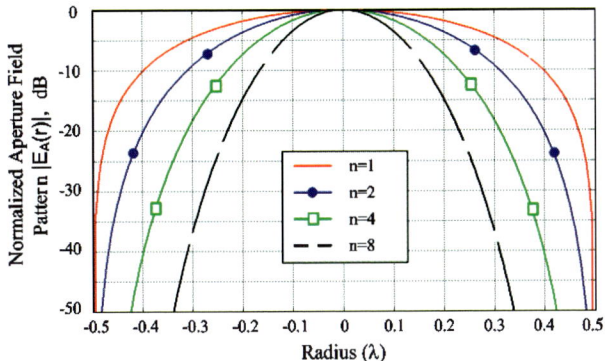

Figure 1. Aperture-field pattern cuts for the FPA model with $r = 0.5\lambda$. $E_T^{(n)} = 0$ (no edge effects).

aperture of the feed the field distribution is defined as:

$$|E_A(r)| = E_T^{(n)} + E_0^{(n)} \cos^n(\pi r/2r_{\max}), \tag{1}$$

where $0 \leq r \leq r_{\max}$, $r_{\max}$ is the radius of the feed; $E_0^{(n)}$(V/m) is a normalization constant; and $E_T^{(n)}$(V/m) is the illumination taper at the feed edges. This model has three design parameters including $r_{\max}$, n, and $E_T^{(n)}$. Figure 1 shows the field magnitude $E_A^{(r)}$ for a circular feed distribution with a 0.5λ radius and constant phase over the aperture. A practical range of aperture field distributions is covered by setting the cos-power value n to 1, 2, 4, and 8. The constant $E_0^{(n)}$ is normalized in such a way that the feed radiates the same amount of power for all n values. In this article we will not use $E_T^{(n)}$, effectively setting it to zero. The phase is assumed to be constant across the aperture and the cross-polarization component is negligibly small.

The main beam and side lobes could also be described by continuous fields, such as $|\sin(r)/r|$, allowing more advanced field distributions to be analyzed. However, such a model would only be possible with large arrays in order to approximate the required distribution with sufficient accuracy.

Given a specific efficiency, the corresponding feed size can be determined using conjugate field matching for the main beam and its first sidelobe (Born and Wolf, 1975; Clarricoats and Poulton, 1977). Operating efficiencies of 80–90% can be obtained for various *F/D* ratios and an ideal feed with a radius close to $r = \lambda$. A 5% increase in efficiency can be obtained by matching the second side lobe in the focal field using a feed radius of $r = 1.5\lambda$. In order to realize this high efficiency with a FPA, the aperture field must be sampled with a large number of elements. A practical field synthesis requires a minimum of about five sample points per wavelength (Boryssenko and Schaubert, 2004; Ivashina et al., 2004). For example, a 0.2λ-cell rectangular array will need at least 15×15 elements for sampling the main beam and its first sidelobe in the aperture field. Inclusion of the second sidelobe will increase the required number of elements to 25×25, capturing (only) 5% more of the total power. This approach only applies to arrays with large numbers of elements.

Lam et al. (1985) have shown that the superdirectivity phenomenon would appear for a dense array feed whose elements produce an amplitude distribution with a number of sidelobes on the array aperture, and have a phase either 0 or π. This would make the system highly sensitive to variations in the excitation coefficients, and, therefore, in practice very difficult to realize. Brisken et al. (2004) found a similar excitation form for an optimized Vivaldi array feed of the VLA. The last feed consists of a $9 \times 10 \times 2$ element array and is excited with an amplitude distribution which has a large number of sidelobes and phase changes between adjacent elements. The predicted efficiency of this system is 60–80% over a 25% bandwidth. It remains to be studied, however, if such an aperture field is realizable with Vivaldi arrays.

In this paper we will only model the FPA whose aperture field distribution is represented by a cosine function, according to Equation (1).

2.2. Efficiency of arbitrary reflector antennas with a simplified FPA model

The aperture efficiency of reflectors with arbitrary F/D ratios was simulated for different design parameters, r and n, of a circular feed array. The following definition of the aperture efficiency for reflector antennas in the transmit situation was used for these simulations (Balanis, 1982):

$$\eta_A = \eta_{sp}\eta_T\eta_{Ph}\eta_{Pol}\eta_b, \qquad (2)$$

where η_{sp} is the spillover efficiency, determined by the fraction of the total power that is radiated by the feed, intercepted, and collimated by the reflecting surface; η_T is the taper efficiency, related to the uniformity of the amplitude distribution of the feed pattern over the surface of the reflector; η_{Ph} is the phase efficiency, related to the uniformity of the phase distribution of the feed pattern over the surface of the reflector; η_{Pol} is the polarization efficiency; and η_b is the blockage efficiency. The efficiencies are related to the reflector geometry and the far-field pattern properties of the feed.

The FPA model definition can be simplified to $\eta_A = \eta_{sp}\eta_T$ using the following assumptions: (i) the spillover and taper efficiencies dominate and (ii) the phase, polarization, and blockage efficiencies are ideal (equal to 1). In these simulations far-field patterns were recalculated from the aperture fields of the feed specified in Equation (1). Using the GRASP8 software package (http://www.ticra.com) the fields specified in the aperture plane were transformed into equivalent currents. The radiated pattern in the far field was then determined using these equivalent currents. The GRASP model was an unblocked parabolic reflector with a feed located in the focus.

TABLE I
Optimal radius

Aperture field distribution	Circular feed radius	Equivalent rectangular feed radius ($r_{eq} = r\sqrt{\pi/2}$)
$\cos(r)$	0.54λ	0.67λ
$\cos^2(r)$	0.69λ	0.87λ
$\cos^4(r)$	0.85λ	1.07λ
$\cos^8(r)$	1.15λ	1.44λ

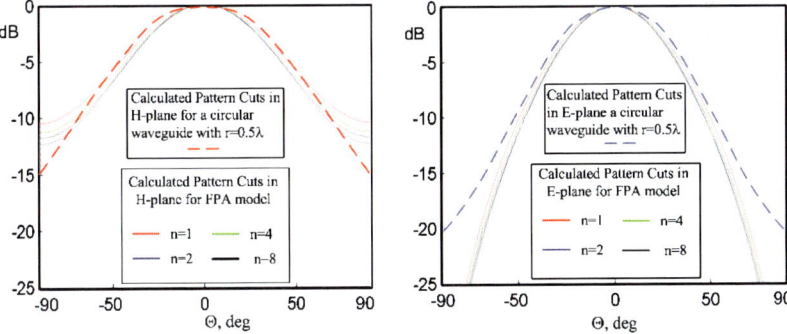

Figure 2. Far-field pattern cuts for the FPA model with the optimal radius and for the circular waveguide with $r = 0.5\lambda$.

2.3. NUMERICAL RESULTS

The calculated far-field pattern cuts in E- and H-plane for the FPA model with the optimal radius (see conclusions and Table I) are illustrated in Figure 2. Patterns of a conventional waveguide feed with the radius of 0.5λ supporting the dominant mode of propagation were derived using the same software for reference. Given a practical reflector with a half-opening angle $<70°$, this FPA model is in good agreement with a circular waveguide.

Figure 3 shows the simulation results of the aperture efficiency and efficiency components obtained for the 0.35 F/D reflector with the FPA model. The results indicate that a small feed provides a high η_T and a lower η_{sp} due to the broad far-field pattern. Conversely, a large feed provides high η_{sp} and a low η_T due to the narrow far-field pattern. From Figure 3 we draw the following conclusions with regards to the aperture efficiency:

1. The circular feed model demonstrates a maximum aperture efficiency of 72% for a reflector F/D ratio of 0.35. This value represents the trade-off between the taper and spillover efficiencies.

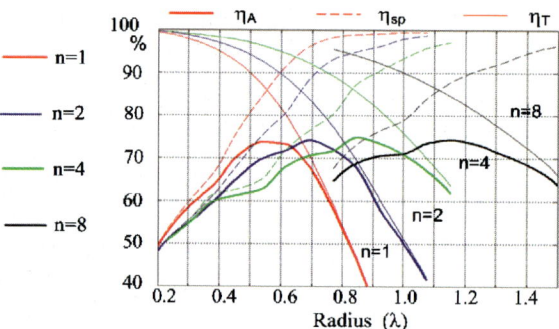

Figure 3. Aperture efficiency η_A, spillover efficiency η_{sp}, and taper efficiency η_t of the 0.35 F/D reflector fed with the FPA model.

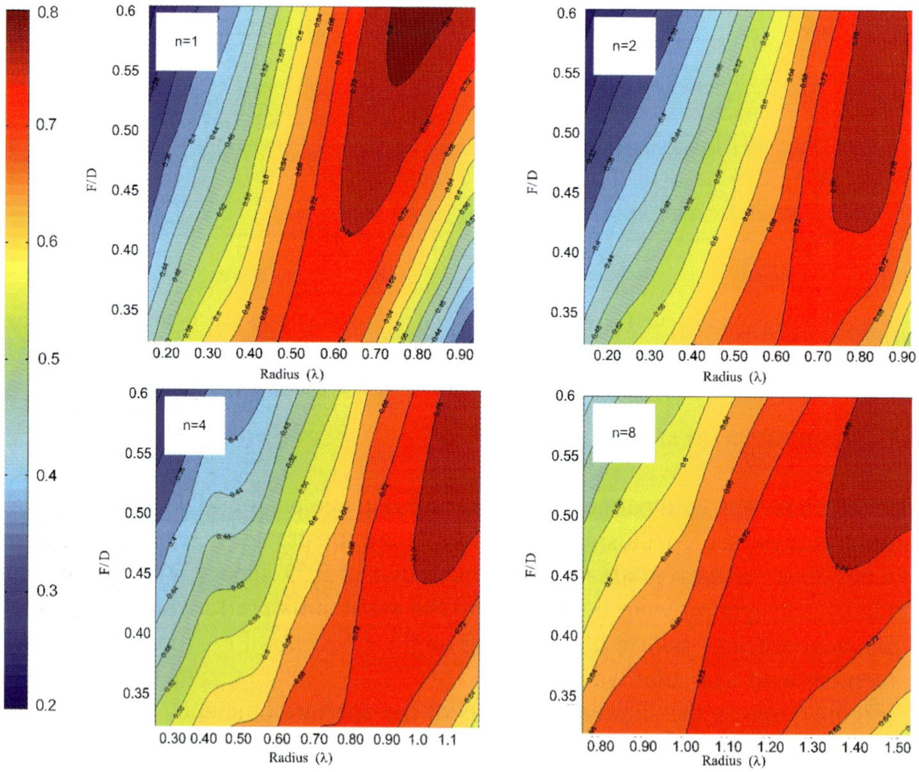

Figure 4. Aperture efficiency of the arbitrary F/D reflectors fed with model FPAs of the radius r and $\cos^n(r)$ aperture distribution.

2. The optimal radius of the feed, providing maximum reflector efficiency, doubles when n is varied from 1 to 8 (see the Table I). This implies that an electrically large feed needs to be excited with an aperture distribution with fast and strong tapering at the edges. (See an example for $n = 8$ in Figure 1). For dense arrays, a

strong tapering can only be realized by adding several rings of dummy elements around the feed, which significantly increases the total array size.
3. The optimal radius values in Table I correspond to approximately the half power radius in the aperture field distributions for the different *n* values.

Comparable simulations were performed for the reflectors with *F/D* ratio of 0.3–0.6. These results are presented in Figure 4 with the aperture efficiency as a function of the *F/D* ratio and the radius of the circular feed. We then find:

4. For any aperture field (any design parameter *n*), the maximum efficiency η_{max} is between 72 and 80%. Efficiency varies as a function of *F/D*; a larger *F/D* results in a higher efficiency. The numbers are comparable with the values of efficiency obtained with a different approach (Clarricoats and Poulton, 1977).
5. For the reflectors under study the optimum radius is $r_{opt} = 0.5$–1.55λ for the circular feed, and $r_{opt} = 0.63$–1.94λ for the equivalent rectangular feed.
6. A 60–80% aperture efficiency is possible for the FPA model over a wide frequency band (3:1).

3. FPA prototype for the WSRT 25 m reflector antenna

3.1. SYNTHESIS PROCEDURE

Based on the results obtained for the FPA model in Section 2 the question arises: Is the modeled aperture field distribution of the array feed appropriate for Vivaldi arrays? Is it possible to attain the predicted efficiency with a practical feed that does not have theoretical attributes assumed in the FPA model, i.e. phase uniformity, ideal polarization purity, and small edge effects?

To demonstrate the implementation of the approach an example is shown for the actual design of the FPA developed for the WSRT 25 m reflector antenna with an *F/D* ratio of 0.35 (http://www.astron.nl/ Westerbork Observatory). A wide-band Vivaldi array with a rectangular grid of 8 × 9 elements spaced 2.7 cm apart (Chio and Schaubert, 1999) was used as the prototype FPA.

The reflector system was modeled in a receive situation using GRASP8 and the conjugate field method as described in Ivashina and van 't Klooster (2002). The focal field distribution of the incident field was computed using both the modeled and simulated results. Later, these results were used as a reference for the aperture field synthesis of the receiving feed design. The aperture-field pattern was sampled with the array elements, which were arranged into rings and excited according to the focal field structure. In the initial FPA design the positions of sample rings were fixed and the physical and effective areas of the array elements were assumed to be equal. In reality, the strong mutual coupling in the Vivaldi array causes the effective area of the sample rings to appear larger, requiring excitation coefficient correction. The correction factor was determined by comparing the aperture patterns of the fixed grid sample rings to direct near-field measurements. More information about the synthesis approach can be found in Ivashina et al. (2004).

3.2. RESULTS FOR DESIGN PARAMETERS AND PERFORMANCE

Low and high frequency sub-bands are represented by 5×5 and $5 \times 5/3 \times 3$ array feeds, respectively. The elements of the 5×5 and 3×3 feeds were combined into 6 and 3 rings, respectively. For verification of the FPA synthesis, the efficiency of the reflector system with the FPA prototype was analyzed. First, the far-field patterns of the array feed were determined using direct near-field measurements. Then, using Equation (2), the aperture efficiency of the WSRT reflector was calculated and the feed far-field patterns obtained.

Figure 5 illustrates experimental results for the 5×5 element FPA prototype at 2.3 GHz. The aperture-field pattern cuts of this feed and the aperture-field patterns modeled with $n = 8$ on the equivalent rectangular aperture of the same area are illustrated in Figures 5a and 5b, respectively. In this example the model of the field distribution agrees fairly well with aperture field of the Vivaldi array. The major part of the beam pattern follows the modeled cosine distribution and the phase is nearly uniform. The magnitude and phase variations close to the edges of the feed ($r \to r_{max}$) show some deviations from the ideal model. Figure 5c presents

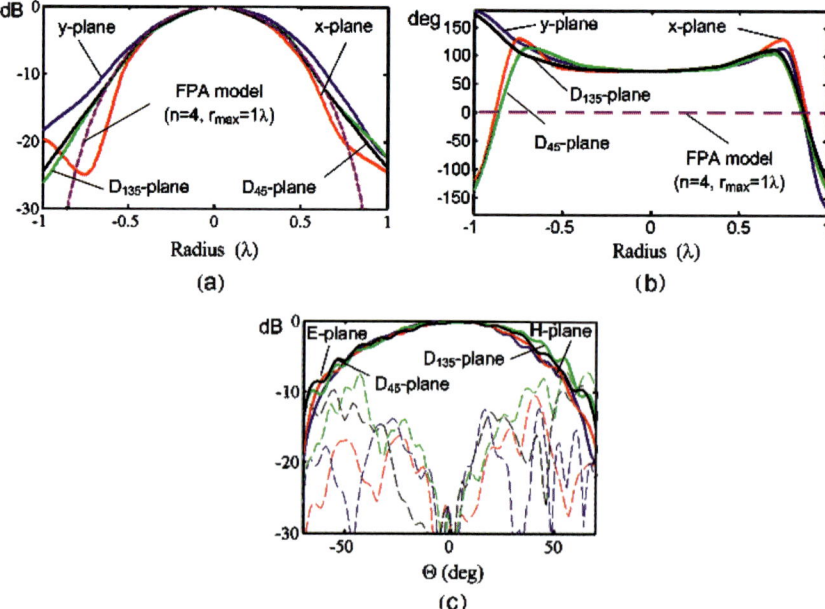

Figure 5. Amplitude distribution cuts (a) and phase distribution cuts (b) of the aperture-field patterns for the FPA prototype (bold lines) and the FPA model (dashed lines); Far-field pattern cuts (c) for co-polar components (bold lines) and cross-polar components (dashed lines) of the FPA prototype.

TABLE II

Design parameters and performance characteristics of FPA developed for the WSRT single dish of 0.35 F/D

Frequency	Element spacing	# elements	Efficiency components and aperture efficiency of WSRT with FPA, %					Total aperture efficiency %
			η_{sp}	η_T	η_{Ph}	η_{Pol}	$\eta_A = \eta_{sp}\eta_T$	
2.3 GHz	0.20λ	5 × 5	99	61	96	99	60	58
4.8 GHz	0.432λ	3 × 3	98	63	85	99	62	52
5.5 GHz	0.5λ	5 × 5	98	60	78	99	59	45

measured far-field pattern cuts (magnitude) for the FPA prototype and shows the phase distribution of the far-field pattern to be nearly uniform (Ivashina et al., 2004). The prototype's measured cross-polar component is low and was not considered in the model. Measured and modeled cross-polar components are illustrated in Figure 5c. A similar far-field amplitude distribution, with a stronger phase variation, has also been experimentally observed at 5.5 GHz (Ivashina et al., 2004).

Table II presents the design parameters and performance characteristics of the 3 × 3 and 5 × 5 FPAs optimized for the WSRT. The optimized results were obtained for the array feeds operating from 2.3 to 5.5 GHz with a passive beamformer. The passive beamformer provides optimal excitation coefficients of the array elements and consists of several combiners, phase shifters, and attenuators. The actual beamformer was developed at the high frequency sub-band around 5 GHz and connected with a 3 × 3 array feed. This system was evaluated by mounting it onto one of the dishes of the WSRT (number RT6) and comparing the outcome to that of standard reflectors (number RT5).

The FPA prototype has an efficiency of 60% at 2.3 GHz (see column $\eta_A = \eta_{sp}\eta_T$). The modeled efficiency for the equivalent circular aperture with the radius of 1λ is 70% (see Figure 3 for $n = 4$). Considering the model's simplicity, this efficiency is in good agreement with that of the prototype. The 10% discrepancy is primarily caused by the model's constraints on zero-illumination taper of the edges of the feed. As Figure 5a shows, the field levels close to the edges are higher for the 8×9 array with only 5 × 5 elements excited and the remaining elements dummy loaded.

4. Comparing efficiency of the FPA model with the prototype for WSRT

Summarizing the results obtained from the FPA model and prototype analysis we can draw the following conclusions:

1. The maximum aperture efficiency ($\eta_{max} = 72\%$) simulated with the FPA model is 14–27% higher than of the aperture efficiency obtained for the FPA prototype. In both cases the element spacing varies from 0.2λ to 0.5λ.

The total error in definition of the efficiency with the FPA model is related to the following effects:

2. The phase efficiency is dismissed in the analysis approach, (setting $\eta_{Ph} = 100\%$), which introduces sampling errors. At the low frequency of the band this error is about 4% (5 samples/λ), while at the high frequency the resulting error can be as high as 22% (3 samples/λ).
3. Another assumption was zero taper in illuminating the FPA model edges. The error in the definition of $\eta_{sp}\eta_T$ remained stable over the frequency band and was equal to 10–13%. In practice, complete loading of dummy elements can improve the edge taper.

Conclusions 1, 2, and 3 can be generalized for a practical dense array with an element spacing ranging $d = 0.2 - 0.5\lambda$ feeding a reflector system with an F/D ranging from 0.3 to 0.6. The following statements can be made by combining these conclusions with 4 and 5 of Section 2.2:

4. The maximum aperture efficiency of reflectors with an FPA element spacing of $d = 0.2$–0.5λ is $\eta_{max} = 45$–66%. The size of the FPA without dummy radiators is at least 5×5 and 3×3 elements for $d = 0.2\lambda$ and $d = 0.5\lambda$, respectively. Two rows of dummy elements surrounding the feed are loaded.

   Over a frequency band the highest aperture efficiency ($\eta_{max} = 58$–66%) occurs at the lowest frequency, when the elements spacing of the FPA is $d = 0.2\lambda$. Similarly, the lowest aperture efficiency ($\eta_{max} = 45$–53%) is at the highest frequency, with an element spacing of $d = 0.5\lambda$.

   The lower efficiency at the higher frequencies is primarily due to the effects of increased element spacing. This makes it difficult to illuminate the reflector with uniform phase across the band. This effect was observed also in Ivashina et al. (2004).
5. The efficiencies which can be attained with FPAs (Table II) are very close to what can be realized with existing horn antennas (Ivashina et al., 2004). Thus, the FPAs enable one to construct a closely interspaced multi-beam system without degrading total efficiency relative to existing horn antennas.

5. Demonstration of the FPA at WSRT

Tests with the FPA were performed the Westerbork Synthesis Radio Telescope (WSRT) with one polarization in the 4590–4750 MHz and 4850–5010 MHz frequency bands (http://www.astron.nl, Westerbork Observatory). Figure 6 shows a photograph of the 3×3 FPA with the beamformer on the back of the Vivaldi array feed. The nine central elements are connected with LNAs and then combined in rings. The rings are fed to vector modulators which perform amplitude and phase control of the first and second ring, after which they are added to the central element

Figure 6. 3 × 3 FPA synthesized with the Vivaldi antenna.

Figure 7. The FPA on the test MFFE going up to the WSRT.

signal. The vector modulators for this test (on the upper board) were designed and developed at ASTRON using off-the-shelf LNAs. Figure 7 shows a photograph of the array feed installed on a Multi Frequency Front End (MFFE), ready for mounting in the radio telescope (RT6). In tests, the prototype is compared to one of the standard reflectors (RT5), which has a conventional 6 cm horn feed in its primary focus.

The test consisted of drift scan and holographic measurements. Holographic measurements were done in order to observe the illumination patterns of the dish with the FPA. For these types of observations a strong point source is needed, so the source 3C84 (of about 30 Jy) was used. In general, to achieve high accuracy for the holography a long observation time is required (37 × 37 points in 12 h). However, taking into account the basic exploration purposes of the tests with the FPA, a shorter time of 3 h (17 × 17 points) was chosen. The amplitude and phase distribution results of the holographic tests with the FPA are presented in Figure 8. Efficiencies

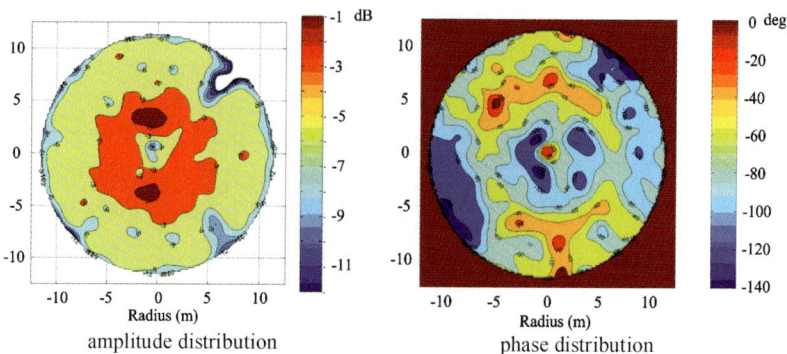

Figure 8. Measured aperture field for the WSRT reflector of 0.35 *F/D* ratio with the receiving Focal Plane Array. Holographic tests at 4.86 GHz.

characterizing the amplitude taper and the phase uniformity were calculated directly from the holographic patterns using Equation (2) and found to be 95% and 82%, respectively.

The first-generation beamformer designed for these tests was inaccurate and it introduced small errors in the beamforming coefficients, as well as some overall loss. Consequently, the illumination pattern of the FPA with the active beamformer was broader than the optimal pattern of the FPA with the passive beamformer (see Table II) resulting in a lower spillover and higher taper efficiencies. The blockage effect from the struts and the focus box of the reflector system also produced illumination pattern differences. In fact, the calculated taper efficiency of 95% and phase efficiency of 81% include these effects.

For beam scans a bright radio source was used (Cassiopeia A). Scans in both north–south and east–west directions were obtained. Figure 9 shows solid lines for the FPA prototype and dotted lines for the standard WSRT dish (RT5). Gaussian

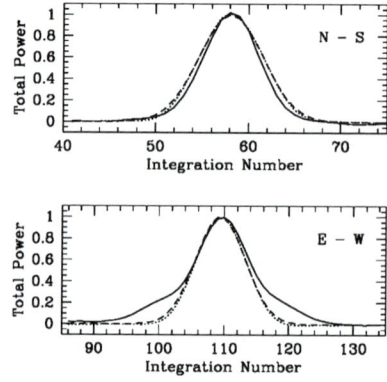

Figure 9. Beam scans with the FPA prototype and a standard WSRT dish. All curves were rescaled to the same peak position and height.

TABLE III

Deconvolved beamwidths in arcsec

	FPA	WSRT-horn
N–S	565	650
E–W	670	650

curves were drawn in dashed lines for comparison purposes. The beam widths analyzed in Table III show that for the N–S scanning direction the FPA realized a narrower beam corresponding to a more uniformly illuminated 25 m dish. In the E–W scan we see equal beam widths for both feeds and some pattern broadening. (This effect was also described in Ivashina et al. (2004)). As a first result this is very promising.

6. Conclusions

In this article we have demonstrated a practical beamformer design for a reflector antenna feed using a dense phased-array with strong mutual coupling. A circular aperture feed model was introduced with specified parameters such as a radius and an aperture-field pattern of the $\cos^n$-form. This model allowed us to relate the maximum efficiency of the reflector system with arbitrary F/D to the required radius and aperture distribution of the array feed. The simulations were done for $F/D = 0.3$–0.6 and $n = 1$–4.

The theoretical FPA model suggests that aperture efficiency between 60% and 80% can be reached over a wide frequency band (3:1). The model based on FPA near-field measurements demonstrates that in practice efficiencies are about 10% lower than those theoretically predicted. This result is primarily due to FPA model simplifications such as zero illumination taper at the edges in finite arrays. From this practical model the attainable efficiency values were recalculated and these were experimentally verified with a Vivaldi FPA for the WSRT reflector ($F/D = 0.35$).

The results of our FPA model and its application to the design of a wide-band FPA have brought us to conclude that over a frequency range of minimum 2:1, the maximum aperture efficiency of deep-dish reflectors is $\eta_{max} = 45$–66%. The size of the FPA must be at least 5×5 with an element spacing of $d = 0.2\lambda$ (3×3 at $d = 0.5\lambda$). Due to edge effects, the active elements must be surrounded at least by 2–3 rings of dummy elements, further increasing the minimum array size to 7×7. Our model predicts that the highest aperture efficiencies occur at the smallest element spacing. At the larger element spacing the aperture efficiency will be reduced, which is mostly due to the increasing error in phase uniformity of the reflector illumination.

The FPA design parameters suggested in the paper ensure efficient reflector illumination over a wide frequency band and that the number of RF signals combined and processed in the beamformer remains reasonable. These parameters provide system simplicity ensuring good noise-performance required for radio astronomy.

Acknowledgements

This publication arose out of the FARADAY Project supported by the European Community–Access to Research Infrastructure action of the Improving Human Potential Programme, and the Commonwealth Scientific and Industrial Research Organisation of Australia.

References

'Astron Project Documentation for THEA and SKA' , (see www.astron.nl, R&D project THEA).
Balanis, C.: 1982, *Antenna Theory, Analysis and Design*, Wiley, New York.
Bird, T. S.: 1982, *Proc. IEEE (H)* **129**, 293–298.
Born, M. and Wolf, E.: 1975, *Principles of Optics*, Pergamon, Oxford, UK.
Boryssenko, A. O. and Schaubert, D. H.: 2004, 'UMass Method of Moment Electromagnetic Solver', in *Focal Plane Array Workshop*, Dwingeloo, The Netherlands (see www.astron.nl, R&D project FARADAY).
Brisken, W., Craeye, C., Veidt, B. and Napier, P.: 2004, 'Focal Plane Arrays for VLA?', in *Focal Plane Array Workshop*, Dwingeloo, The Netherlands.
Chio, T.-H. and Schaubert, D. H.: 1999, 'Design of Dual-Polarized Tapered Slot Antenna Arrays', in *Technical Report*, Antenna Laboratory, ECE Department, University of Massachusetts, Amherst, MA 01003, USA.
Clarricoats, P. and Poulton, G. T.: 1977, *Proc. IEEE* **65**, 1470–1504.
Craye, C.: 2004, 'Efficient Simulation of Finite Wideband Arrays for Focal Plane Applications', in *Focal Plane Array Workshop*, Dwingeloo, The Netherlands.
Galindo-Israel, V., Lee, S-W. and Mittra, R.: 1978, *IEEE Trans. AP* **AP-26**, 220–228.
Ivashina, M. V., Bij de Vaate, J. G., Braun, R. and Bregman, J. D.: June 2004, 'Focal Plane Arrays for Large Reflector Antennas: First Results of a Demonstrator Project', in *Astronomical Telescopes and Instrumentation, SPIE Conference*, Glasgow, UK.
Ivashina, M. V. and van 't Klooster, C. G. M.: 2002, 'Focal Fields in Reflector Antennas and Associated Array Feed Synthesis for High Efficiency Multi-Beam Performances', in *25th ESA Antenna Workshop on Satellite Antenna Technology*, Noordwijk, The Netherlands.
Lam, P., Lee, S-W., Chang, D. C. D. and Lang, K. C.: 1985, *IEEE Trans. AP* **AP-33**, 1163–1174.
Maaskant, R.: 2004, 'On the Overall Signal to Noise Ratio of Focal Plane Array Receiving Systems', in *Focal Plane Array Workshop*, Dwingeloo, The Netherlands.
Rahmat-Samii, Y. and Lee, S.-W.: 1983, *IEEE Trans. AP* **AP-31**, 463–470.
http://www.ticra.com.
http://www.astron.nl, Westerbork Observatory.

MODEL VALIDATION AND PERFORMANCE EVALUATION FOR THE MULTI-TETHERED AEROSTAT SUBSYSTEM OF THE LARGE ADAPTIVE REFLECTOR

MEYER NAHON*, CASEY LAMBERT, DEAN CHALMERS and WEN BO
Department of Mechanical Engineering, McGill University, Montreal, Quebec, Canada H3A 2K6
(*author for correspondence, e-mail: meyer.nahon@mcgill.ca*)

(Received 16 August 2004; accepted 28 February 2005)

Abstract. The Canadian design for the Square Kilometre Array radio telescope includes a large multi-tethered aerostat to support the telescope's receiver. To validate this design concept, two parallel tracks have been undertaken: a numerical simulation of the multi-tethered aerostat system has been assembled, and a one-third scale prototype of the system has been constructed. This paper describes the experimental facility, presents results from initial tests of the uncontrolled system and compares these results to the predictions of the computer model of the system. Generally, the results compare very favourably. Using the simulation, we arrive at two important design philosophies to be used in the design of the full-scale system: (a) perturbations on the confluence point should be minimized, and (b) the system stiffness should be maximized to ensure minimum response to disturbances.

Keywords: SKA, radio telescope, tethered aerostat, feed positioning

1. Introduction

Astronomers and engineers at the National Research Council of Canada's Herzberg Institute of Astrophysics have proposed a conceptual design for the SKA that consists of an array of about 50 very large antennas. The novel antenna design, depicted in Figure 1, is called the Large Adaptive Reflector (LAR) (Fitzsimmons et al., 2000). The LAR design includes two central components. The first is a 150–200 m diameter parabolic reflector, with a focal ratio of 2.5, composed of actuated panels, mounted on the ground. The second is a focal package held aloft at the focal point by a large helium balloon (aerostat) and a system of three or more taut tethers. The telescope is steered by modifying the shape of the reflector, and simultaneously changing the lengths of the tethers with winches so that the receiver is positioned on the surface of a hemisphere of radius 375–500 m, centered at the reflector. The variable-length tethers also allow a measure of control of the receiver position in response to disturbances such as wind gusts (Fitzsimmons et al., 2000; Nahon et al., 2002). The aerostat's limited size relative to the reflector, and the system geometry imply that there would be minimal shadowing of the reflector ($<5\%$), and that only in the zenith configuration.

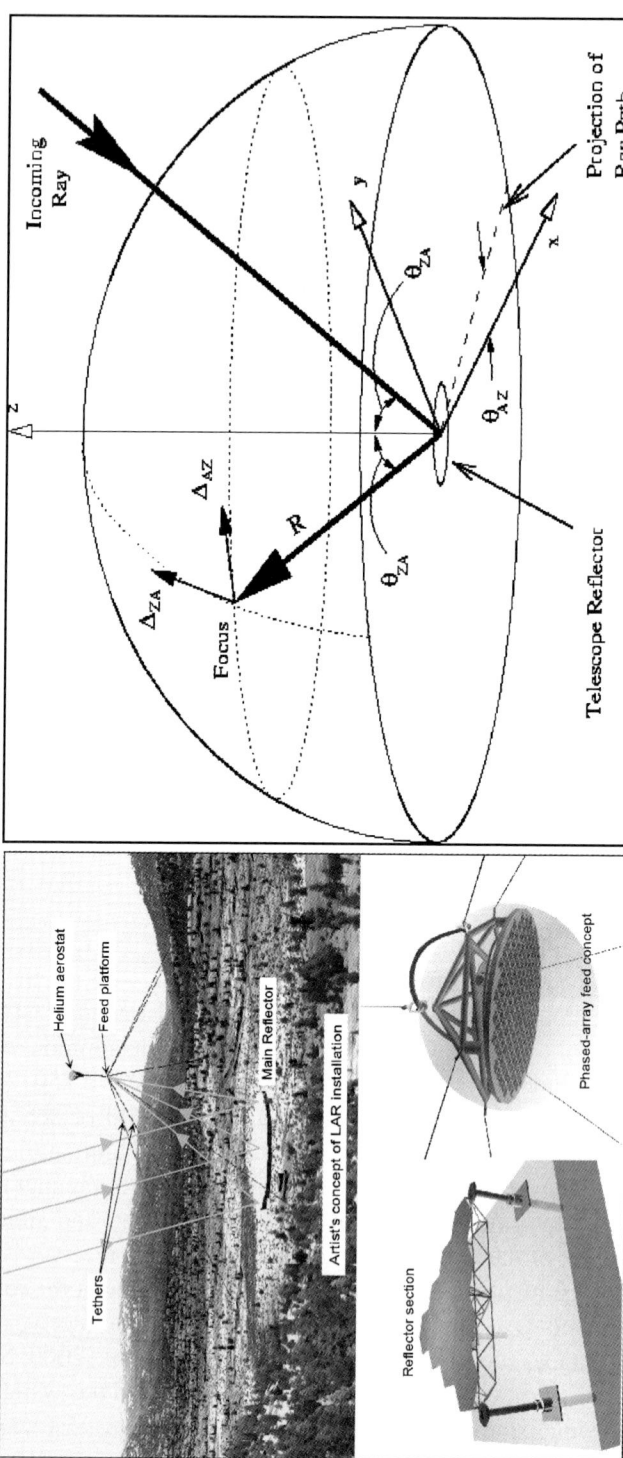

Figure 1. LAR design concept.

To further develop the LAR multi-tethered aerostat concept, a computer simulation has been developed to predict the motion of the system in response to disturbances and control inputs. It can be used to weigh alternative design options, determine how typical gusts are likely to affect the receiver positioning accuracy, and perform preliminary evaluation of the winch control algorithms. In parallel, a one-third scale prototype of the multi-tethered aerostat subsystem for LAR has been constructed as a proof of concept for this design (Lambert et al., 2003).

Validation of the numerical model was considered crucial to allow confidence in using the model for full-scale design studies. This paper first reviews the model, then presents a description of the experimental setup, followed by a comparison of results from the model and the experiments. Based on the model results, guidelines are then drawn for an optimal design of the system.

2. System model

The dynamics model of this multi-tethered aerostat system has been discussed extensively elsewhere, but is briefly reviewed here. Figure 2 shows a layout of the system, which includes an aerostat, the *confluence point* where all tethers meet and the payload (receiver) is located, a *leash* (the single cable connecting the aerostat to the confluence point), three to six tethers connecting the confluence point to the ground, and winches at the base of each tether.

The aerostat may be spherical or streamlined. The spherical aerostat is modeled as a sphere whose drag varies with Reynolds number (Nahon et al., 2002). Modeling the streamlined aerostat is more involved and is discussed in Lambert and Nahon (2003) – a component breakdown approach is used to model the aerodynamic behavior of the hull and each fin. These are then summed to find the overall behavior of the aerostat. The confluence point is modeled as a sphere, similar to the spherical aerostat. Each cable is discretized into elements and the mass of each element

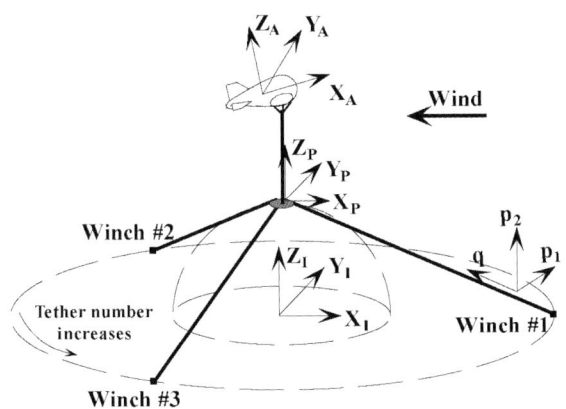

Figure 2. System layout.

is lumped at its endpoints (Lambert and Nahon, 2003). The equations of motion for each lumped mass are then formulated, including the internal forces (stiffness, damping) and external forces (aerodynamic, weight). The disturbances to the system are due to the turbulent wind. This is modeled using a mean wind varying with height, due to the earth's planetary boundary layer. Superimposed on this, are turbulent gusts with a von Karman spectrum. Further details are given in Nahon et al. (2002). The entire system of second-order nonlinear dynamics equations is assembled, put into first- order form and solved using a fourth-order Runge–Kutta numerical integration routine (Press et al., 1992).

3. Experimental system

The numerical model should be able to predict the behavior of the multi-tethered aerostat system in response to typical winds and gusts. However, a parallel track was undertaken in which a scaled-down version of the system was constructed in Penticton, B.C. to gain operational experience with a large tethered aerostat, and to validate the results of the numerical model.

A systematic scaling analysis was performed to design the experimental system (Lambert et al., 2003). Buckingham's Pi theorem (Massey, 1989) was used to determine the relevant non-dimensional parameters that had to be maintained in the scaled system. It was not possible to ensure exact duplication of all these parameters in the scaled system and a choice was made to ensure Froude scaling over Reynolds number scaling. Simulations were used to confirm that this choice would yield a closer correspondence between the full-scale system and the scaled prototype. A scale factor of 1/3 was chosen as a reasonable compromise between the high cost/ideal accuracy of a system with unity scale factor and the much lower cost and ease of experimentation provided by a lower scale factor.

A streamlined aerostat was chosen over a spherical one since its drag was expected to be 3–4 times lower, and it was expected that this would reduce the disturbance on the confluence point. The Helium aerostat shown in Figure 3, built by Worldwide Aeros, was selected for the prototype system. It has a length of 18.3 m, and a maximum net lift of 240 kg. In operation, the lift decreases over time as the Helium becomes contaminated with air. Once the lift drops below about 180 kg, the Helium can be purged and replenished.

The main tethers are made by Cortland Cable. They are of braided construction, 6 mm in diameter, and made of *Plasma*, an advanced high-strength, high-stiffness synthetic material similar to Kevlar. A high strength-to-weight ratio is critically important as it allows a reduction of the sag of the tethers for a particular tension. This then allows the main tether structure to remain stiff in the presence of gust disturbances. At present, there are three main tethers, though we expect to increase this to six in the near future to further stiffen the system. The upper cable connecting the confluence point to the aerostat, referred to as the *leash*, is of similar construction

Figure 3. Worldwide Aeros 18.3 m aerostat.

to the main tethers, but contains conductors that are used to transmit data from the aerostat. An additional cable (not shown in Figure 2), which remains slack during all tests, connects the confluence point to the center of the winch circle. This cable allows system deployment and retrieval and permits high-speed data transfer from the airborne instrument platform. The full-scale system would use optical fibers for data, and DC for any power transmission, thus avoiding the possibility of radio interference.

At the bottom end of each cable, a winch allows active control of the cable length. As mentioned earlier, adjusting the cable lengths allows the receiver to be repositioned on the focal hemisphere, thereby pointing the telescope to a new astronomical position. The receiver can be positioned at any azimuth angle, down to $60°$ zenith from the vertical. In addition, the winches can be controlled at higher frequency to compensate for disturbances due to wind gusts. Practically, the main limitation of this control is the flexibility and length of the tether, which creates a time delay from actuation at the bottom end to motion at the top end. For example, a longitudinal wave would take 50 ms to travel from one end to another of a highly tensioned 300 m Plasma cable.

The winch drum pivots relative to its frame on a vertical hinge to allow the drum to face toward the confluence point. A fairlead and levelwind ensure that the tether wraps onto the drum in an orderly manner. The computer-controlled winch motor, made by Bosch Rexroth, drives the drum through a 100:1 gearbox. All three motors are controlled centrally by a PC-based system through a high-speed fiber-optic line, using the control architecture shown in Figure 4.

Rather than a receiver at the confluence point, the prototype system includes an instrument platform to allow a detailed sensing of the motion at that point. The

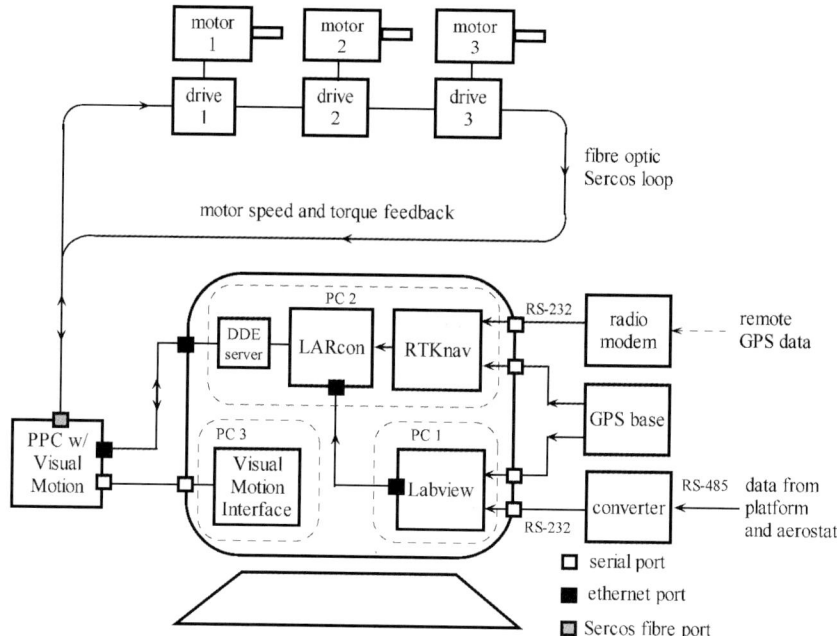

Figure 4. Schematic of PC-based motion control system.

0.8 m diameter platform is attached to the main tethers using Teflon-coated hangers. Upon this platform are mounted a GPS antenna and receiver, a 2-axis tilt sensor, a digital compass and a 2-axis wind sensor. An inertial measurement unit, with three accelerometers and three rate gyros is also included on the platform for future motion measurements at higher frequencies. A load cell is incorporated in each of the main tethers and the leash to measure cable tensions. This suite of sensors thus allows a complete and precise knowledge of the motion of the confluence point.

In addition to the above, the aerostat has its own suite of sensors, including a GPS antenna and receiver, a 2-axis tilt sensor and a digital compass. These are used to allow independent validation of the aerostat model. Finally, mounted at reference ground locations, we have a third GPS antenna and receiver, as well as an additional wind measuring station.

All sensors are analog, and their data is digitized on the platform at a rate of 10 Hz. The GPS data is sent down via an onboard radio modem, while the remaining data is transmitted to the ground through conductors in the central tether, using an RS-485 protocol. All data is sampled using Labview, time stamped and stored in files for later analysis. For the open-loop tests, the data is not fed back for closed-loop control of the winches.

As shown in Figure 4, a total of three PCs are used to collect data and control the system. PC 1 is host to a Labview program that collects instrument platform data and stores it to file. PC2 is host to RTKnav, which processes the GPS data from the three receivers, as well as LARcon, the control program that uses the GPS data to calculate the desired winch motions to compensate for the position error. Those commands are sent to the winches by the DDE server, a Bosch–Rexroth proprietary application. The third PC is host to the Visual Motion interface program, which allows tuning of the Bosch–Rexroth motor controller. As noted above, the results shown in this paper are only for the uncontrolled system, and so all the winch control elements are inactive.

4. Comparison of results

Three experimental flight tests were conducted at the Dominion Radio Astrophysical Observatory (DRAO) in Penticton, B.C. during the Spring of 2004. The purpose of the flights was to study the passive (i.e. no control) performance of the system and to provide a means for assessing the validity of our dynamics model of the system. Tests were performed at two different winch base radii and two different zenith angles to investigate how changes in tether geometry affect the behaviour of the system. Table I gives a summary of the tether configuration for the three flights and Figure 5 shows an overhead view of the experimental layout. The focal length (distance from center of the base circle to the confluence point) for the flights was 175 m and the aerostat leash length (from confluence point to aerostat) was 150 m.

TABLE I
Summary of tether configuration for test flights

	Flight 1	Flight 2	Flight 3
Base radius, R_B (m)	250	400	400
Zenith, θ_{za} (deg)	2	29	2
Azimuth, θ_{az} (deg)	−101	41	−101

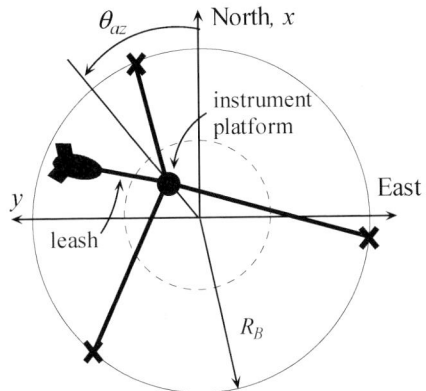

Figure 5. Overhead view of experimental system.

TABLE II
Experimental and simulation results for Spring 2004 test flights

		Flight 1 Exp	Flight 1 Sim	Flight 2 Exp	Flight 2 Sim	Flight 3 Exp	Flight 3 Sim
Mean wind, U (m/s)		4.94	4.95	3.01	3.08	3.79	3.81
Turb. intensity, σ_U/U		0.14	0.12	0.11	0.13	0.13	0.14
Platform position, mean and standard deviation (m)	x	−1.22	−1.25	65.79	65.72	−1.42	−1.46
	σ_x	0.022	0.029	0.041	0.070	0.027	0.016
	y	−6.12	−6.13	−57.60	−57.54	−7.00	−7.10
	σ_y	0.031	0.065	0.051	0.065	0.036	0.035
	z	175.44	175.40	150.15	150.10	172.85	172.54
	σ_z	0.133	0.113	0.173	0.169	0.268	0.196
Leash tension, T_L (N)		2473	2101	1886	1908	2623	2575
σ_{T_L} (N)		182	219	37	84	69.6	137.7

The experimental and simulation results for the three flights are given in Table II. The performance of the system is assessed by the motion of the instrument platform. The standard deviation of the platform position gives a good indication of the platform's motion. From Table II, it is observed that all tests show relatively small (a few cm) deviation in the horizontal x and y directions and slightly larger deviation in the vertical z direction. It is worth mentioning that during the flights, the horizontal deviation of the aerostat position ranged from 2 to 5 m. Considering the size of the overall system and the size of the aerostat motion, the platform deviation of a few cm demonstrates the remarkable stiffness of the tether tension structure.

To compare simulation results to the experimental measurements, the test conditions for each flight were duplicated in the dynamics model. The measured wind speed at the confluence point is used as a basis for modeling the entire wind field while incorporating boundary layer and turbulence effects as mentioned in Section 2. It is acknowledged that our limited information of the actual wind field, resulting from only one airborne sensor, restricts our ability to achieve strict coherence between the simulation and the experimental results. Moreover, since the disturbances to our model are provided by a statistical approximation of the wind, we can only hope to get a statistically similar match to reality. The results for the simulation comparison are given in Table II for the wind speed, turbulence intensity, platform position and leash tension. Turbulence intensity is the standard deviation of the wind speed divided by its mean. The wind conditions at the platform are quite similar for the simulation and the experiments, but the similarity of the wind at all other locations is unknown.

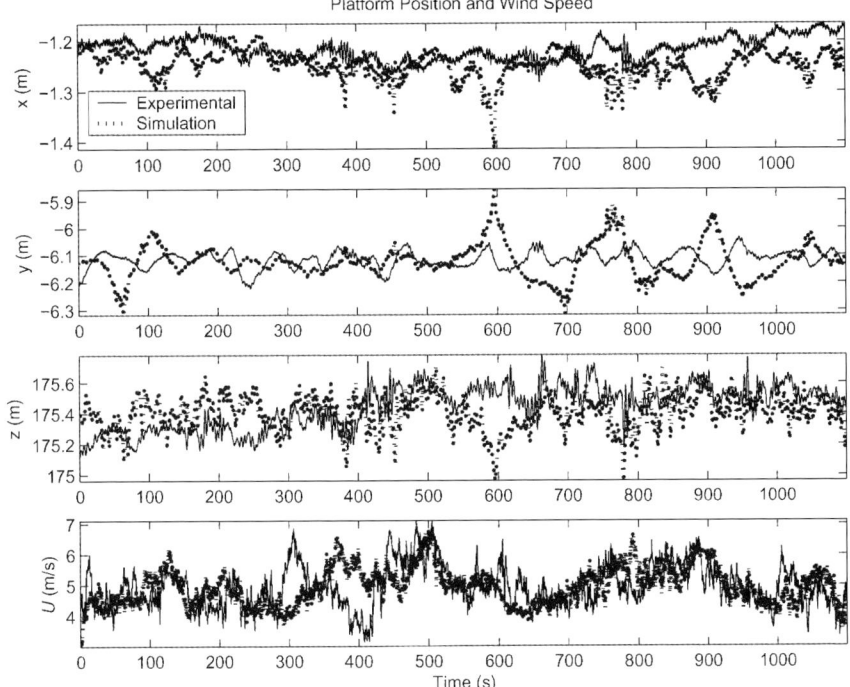

Figure 6. Comparison of experimental and simulation results for Flight 1.

Overall, the simulation results for the platform motion appear to be quite good. A plot of the experimental and simulation platform position and wind speed for Flight 1 is given in Figure 6. A comparison was also performed in the frequency domain, and showed good correspondence between simulation and experimental results. It is noted that the simulation tends to slightly overestimate the platform motion, which is likely due to larger amplitude perturbations from the aerostat leash. By far the main source of disturbance to the platform is from the leash loading and since the standard deviation of the leash tension tends to be significantly higher in the simulation, the overestimated platform motion is expected. There are two possible causes for this discrepancy: (1) the simulated wind field provides a greater disturbance than in reality or (2) the aerodynamics model of the aerostat results in greater leash disturbances than in reality. Without more comprehensive wind measurements, it is not possible to gain a better understanding of the actual wind field. Considering these limitations, the simulation produces results that agree quite well with the experimental data.

5. System design

With confidence that the simulation produces realistic results, we can use it to evaluate the ability of the proposed full-scale design to maintain a stable feed position during an observation. With the present design, we expect that, *without control*, the confluence point motions of a 6-tether system in its worst-case configuration would reach peaks of 5 m in a 10 m s^{-1} wind with typical turbulence. Active control of the winches is expected to reduce these peak excursions to 0.5–1 m (Nahon et al., 2002). These excursions would be corrected using a separate mechanism located at the confluence point. The mechanism would also orient the feed toward the center of the ground-based reflector. Commercially-available Stewart platforms are available with the desired range of translational motion (Bonev, 2004), but they are relatively heavy and do not have the angular travel required for our application. Therefore, a special-purpose parallel mechanism is being designed at Laval University to provide the needed positioning and angular travel in a lightweight package. Thus, we are confident that the overall system, including the feed mechanism, can stabilize the feed motion to cm-level accuracy. Assuming pointing accuracy of one-tenth of a beamwidth, and a focal ratio of 2.5, if the feed can be stabilized to 1 cm accuracy, and other telescope components have performance to match, the telescope could be used to observe up to 7.5 GHz.

The simulation can also be used to improve the design of the system. These improvements could lower system cost, ease the burden on the winch control system and the feed mechanism and allow observation at higher frequencies. Design studies were undertaken to determine the design features that would minimize the feed motion, for a given level of turbulence. Two avenues are available to minimize the confluence point motion: minimizing the disturbances and minimizing the response to those disturbances. Physically, these correspond to system changes *above* the confluence point, and below it.

5.1. MINIMIZING DISTURBANCES

The ambient wind is essentially the only disturbance to the multi-tethered aerostat system that prevents it from staying immobile. It is important to note that a *constant* wind is of no consequence to the confluence point motion because it can be perfectly compensated by using the winches to adjust the main tether tensions. Of greater concern are the wind gusts which cause transient motions of the confluence point that are too quick to be compensated for by the winches. We expect these to be gusts at a frequency greater than 0.25 Hz.

The bulk of the disturbance on the confluence point is due to the wind gusts acting on the aerostat. The resulting aerodynamic force on the aerostat is transmitted through the leash. To determine the proportion of the disturbance due to this mechanism, we forced the leash tension to remain constant at its mean value in the simulation, for a number of cases. The confluence point motion was reduced to 10% or less of its original value. Thus, if the leash tension could be made constant, we could expect a dramatic reduction in the confluence point motion. To move in this direction, we could (a) choose an aerostat with low response to wind gusts, and (b) select a leash that will filter out the disturbances before they reach the confluence point.

The response of an aerostat to wind disturbances is determined by its aerodynamic shape. Off-the-shelf aerostats come in a limited range of shapes: spherical, streamlined and hybrid lifting bodies. Our choice of a streamlined aerostat in Penticton was motivated by its aerodynamic efficiency and the fact that it is operationally well understood. The aerodynamic efficiency means that drag forces are relatively low, which is desirable. However the streamlined aerostat is also relatively efficient at producing lift, which is undesirable at high frequencies. Simulations show that the transient lift generated by a streamlined aerostat in wind gusts may needlessly disturb the system (Nahon et al., 2002), and that the rms components of the leash tension could be reduced by 80–90% with a spherical rather than a streamlined aerostat. It should be noted that our simulation does not include the lateral oscillations typical of spherical aerostats due to vortex shedding (Govardhan and Williamson, 1997), and so may underestimate those rms tensions. A 'small' 3.5 m spherical aerostat is presently being tested at McGill University to investigate its behaviour in typical winds. If those results confirm our simulation results, a larger spherical aerostat will be developed and tested at the Penticton facility.

Lowering the stiffness of the leash is another important design modification shown to be effective in the simulation. The stiffness of two tethers of the same length but different material depends on the product EA, where E is Young's modulus and A is the cross sectional area – the latter being selected to obtain sufficient strength. Although there exist materials with much lower Young's modulus than the Plasma leash presently being used, they also have lower strength. Realistically, we can only reasonably expect to get two orders of magnitude reduction in tether stiffness, for example by using a natural rubber bungee cord instead of Plasma. The reduced stiffness also comes at a cost of additional weight, which itself tends to reduce the system performance. Using simulations, we find that if we reduce the leash stiffness by two orders of magnitude, we can reduce the rms of the leash tension by a factor of two. However, the more important benefit of the flexible leash is that it reduces the frequency of the tensions transmitted to the confluence point, thus making it easier for the winch controller to control those perturbations. We are presently testing a partial leash made of bungee cord in Penticton to evaluate its benefits in practice, as well as its operational characteristics.

5.2. Minimizing System Response

Once a particular disturbance is applied by the leash on the confluence point, the resulting confluence point motion will be determined by (a) the stiffness of the multi-tether system below the confluence point, and (b) the ability of the control system to provide counteracting forces.

Maximizing the system stiffness is accomplished only partly by proper selection of the material properties. A tether of length L has three tension regimes: (a) slack, where its stiffness is zero (b) completely taut, where its stiffness is EA/L and (c) in between the first two. Only in the second regime will the material properties completely determine the tether behaviour. This implies that the aerostat should be sized to ensure that the main tethers should remain in the second regime, even at high zenith angles. Thus, increasing our aerostat's net lift to about 82 kN (from a presently presumed 35 kN) should reduce confluence point motion by a factor of 2–3.

The winch radius also has an important effect on the stiffness of the multi-tether system. Our studies show that vertical stiffness increases monotonically with decreasing winch radius, but that there exists a winch radius, which maximizes the system's horizontal stiffness. Also, the number of tethers has a strong impact on system stiffness. To date, we have focussed primarily on 3- and 6-tether systems, and the latter appears superior in all respects; typically having confluence point motions about three times lower in the worst-cases configurations than a 3-tether system. It is expected that adding further tethers could further reduce the confluence point motion, but at the penalty of added system costs and control complexity.

Once the design exercise has exploited all possible advantages, proper control can then yield significant further benefits. With our present system design, simulations show that the maximum confluence point motion can typically be reduced by a factor of about 5 using optimally-tuned PID winch controllers, relative to the uncontrolled system. When the controller gains are increased too much, system instabilities develop due to the time lag for a winch action to be felt at the confluence point. We are presently investigating the use of more advanced controllers, which would take this delay into account as a means to further improve the benefits of control.

We expect that a formal design optimization exercise, improved controllers and improved confluence point motion sensing should allow us to develop a multi-tethered aerostat system that can position the confluence point within approximately 25 cm of its desired location. If the feed mechanism were able to reduce this error down to a few mm, this would allow telescope observation at up to 22 GHz.

6. Conclusions

This paper discussed the validation of the simulation of the multi-tethered aerostat subsystem of the LAR concept for the Square Kilometre Array. A comparison of the numerical and experimental results showed remarkably good correspondence, given the limitations of incomplete knowledge of the wind field existing in the experiments.

Once the simulation was validated, it was used in a design study to find the system design that would minimize the motion of the confluence point. Two design approaches were used: minimizing the disturbances on the confluence point, and minimizing the system's response to these disturbances. It was found beneficial to have a relatively flexible leash; main tethers with a high stiffness and low weight; an aerostat with large enough buoyancy to ensure the main tethers remain in a high stiffness regime and possibly, a spherical rather than a streamlined aerostat.

Acknowledgements

The authors would like to thank the Canadian Foundation for Innovation, the British Columbia Knowledge Development Foundation, the Natural Sciences and Engineering Research Council and the National Research Council for their financial support to this project. Also we would like to thank our component suppliers, including AGO Environmental, AMEC Dynamic Structures, Bosch Rexroth, Cortland Cable, Novatel, Waypoint Consulting and Worldwide Aeros for their assistance in construction and development of the facility.

References

Bonev, I.: 2004, 'Parallemic – The Parallel Mechanisms Information Center web site', http://www.parallemic.org/WhosWho/CompSims.html.
Fitzsimmons, J.T., Veidt, B. and Dewdney, P.: 2000, *Proc. SPIE* **4015**, 476–487.
Govardhan, R. and Williamson, C.: 1997, *J. Wind Eng. Ind. Aerodyn.* **69–71**, 375–385.
Lambert, C. and Nahon, M.: 2003, *AIAA J. Aircraft* **40**, 705–715.
Lambert, C., Saunders, A., Crawford, C. and Nahon, M.: 2003, 'Design of a One-Third Scale Multi-Tethered Aerostat System for Precise Positioning of a Radio Telescope Receiver', in *CASI Flight Mechanics and Operations Symposium*, Montreal, Canada.
Massey, B.: 1989, *Mechanics of Fluids*, 6th Edn., Van Nostrand Reinhold, London.
Nahon, M., Gilardi, G. and Lambert, C.: 2002, *AIAA J. Guid. Control Dyn.* **25**, 1107–1115.
Press, W.H., Teukolsky, S.A., Vetterling, W.T. and Flannery, B.P.: 1992, *Numerical Recipes in C: The Art of Scientific Computing*, Cambridge University Press, New York.

MODELING OF A FEED SUPPORT SYSTEM FOR FAST

WENBAI ZHU[1,*], RENDONG NAN[1] and GEXUE REN[2]

[1]*National Astronomical Observatories, CAS, Beijing, 100012, P.R. China;* [2]*Department of Engineering Mechanics, Tsinghua University, Beijing, 100084, P.R. China*
(*author for correspondence, e-mail: wbzhu@bao.ac.cn)

(Received 19 July 2004; accepted 28 February 2005)

Abstract. As an engineering demonstrator for SKA, the Five-hundred-meter Aperture Spherical Telescope (FAST) is proposed in China. This paper is focused on one of the most critical components of FAST, the feed support system. The engineering concept, the configuration and results from model experiments are presented. The mechanical characteristics of the structure are analyzed. The performance of the feedback control system of the model is described. The feasibility of the design is tentatively confirmed by the experiments described at the end of the report.

Keywords: cable, feed support, large telescope, SKA, Stewart stabilizer

1. Introduction

Chinese astronomers have been involved since 1993 in a world wide long-term cooperation in building the SKA, the Square Kilometre Array. As soon as the SKA was proposed, a Chinese engineering concept (Figure 1) to realize an individual element of the array began to be investigated. FAST, a Five-hundred-meter Aperture Spherical Telescope, is an Arecibo-type antenna with a number of innovations. The reflector of FAST is active, the illuminated area being deformed to a paraboloid during observation, which enables spherical aberration to be corrected on the ground (Qiu, 1998; Nan et al., 2000, 2003). The feed support system of FAST is "platformless", the focus cabin being driven by a cable system to realize the pointing and tracking of the telescope without having a heavy platform in the air above the reflector like Arecibo (Duan et al., 1996). Large numbers of depressions have been surveyed in southern GuiZhou province for the future antenna locations (Nan et al., 1996; Peng and Nan, 1997).

Some basic parameters of FAST are illustrated in Figure 2. It has a main spherical reflector with radius of $R \sim 300$ m and a projected diameter of 500 m. The illuminated aperture is about 300 m, and the focal length is about 0.467R. This geometrical configuration enables FAST to achieve larger sky coverage, up to a zenith angle of $60°$, compared with $20°$ for the Arecibo telescope. Conventional feeds and receivers cover continuously the frequency range 150–2000 MHz, with possible capability up to 5 GHz. During astronomical observations, the feeds

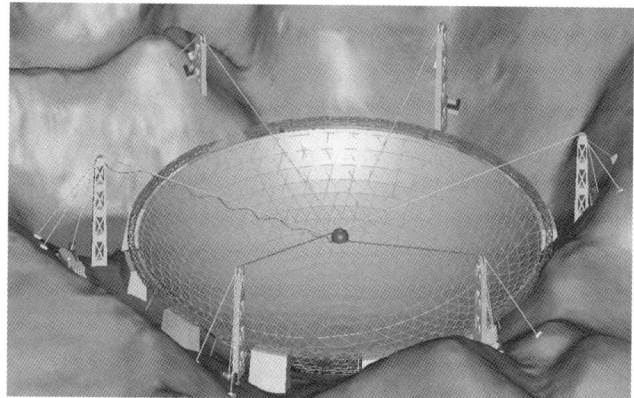

Figure 1. 3-D image of FAST configuration (Courtesy of Mrs. Cao Yang).

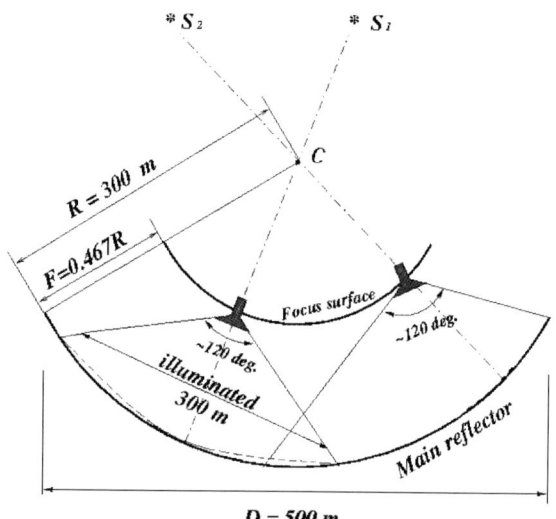

Figure 2. FAST optic geometry.

are required to track with a positioning precision of 4 mm at a maximum speed up to 11 mm/s on a spherical cap of 250 m diameter above the reflector (Zhu, 1999).

FAST is "pointed" by moving the focus cabin, while the illuminated surface on the main reflector is deformed in coordination. One of the pivotal technologies is the cable support system for the feed, which has the potential to reduce construction expenditure. This report addresses work on the test model for the feed support system of FAST.

2. Concept of platformless feed support system

Two different concept designs have been proposed for the FAST platformless feed support system. One is Duan's six-tower design (Duan et al., 1996). The other is a four-tower cable-car feed support configuration put forward by Ren in 1998 (Nan et al., 2000; Ren et al., 2001). In both concept designs, the dynamical movement of the focus cabin along the required locus is realized by adjusting lengths of the suspension cables, and a secondary feed-stabilizer is employed in order to suppress the vibration induced by the wind and other disturbances. However, six towers and six suspension cables are used in Duan's design, while in Ren's design, four towers, twelve cables, a six-wheel trolley and an orienting mechanism are used (see next section for details).

The configuration of the platformless feed support system consists of three parts:

1. The cable-driven system (Figure 1) is controlled by a central computer. Given the difference between the observed and required positions, the central computer drives each servo-mechanism to adjust the lengths of the cables, so that the movement of the focus cabin along its locus can be realized.
2. A Stewart platform in the focus cabin serves as a stabilizer, to provide a fine adjustment from some tens of centimeters achieved by the cable-driven system to several millimeters. A typical Stewart platform (Stewart, 1965) is composed of six length-adjustable actuators connecting a moveable part to a base. As the lengths of the actuators change, the moveable part is able to move in all six degrees of freedom with respect to the base. In the case of the Stewart platform in the focus cabin of FAST, a group of feeds/receivers are mounted on the moveable plate.
3. The third part is the measurement system by which the positions of the feeds and tension of the cables are read and recorded in real time. The information then is fed back to the central computer to realize the global loop control.

3. Experiments and results

3.1. REVIEW OF THE PREVIOUS MODEL EXPERIMENTS

Based on Duan's design, a system for testing platformless feed support system was built in Xidian University in 2002. The system includes six concrete towers and six cables connecting to the focus cabin. In 2002, an experiment with this 50 m cable driven model was carried out. The required position of the focus cabin was successfully achieved. At the same time, the nonlinear dynamical process of the cable system was simulated, and the position precision of the moveable plate was investigated by using kinematics, inverse position computation and singularity analysis of the stabilizer (Duan, 1999; Su, 2000; Su et al., 2001, 2003).

According to Ren's design, a smaller 20 m scale model for a cable-car feed support system was built in Tsinghua university in 2001. It consists of four towers supporting orthogonal pairs of cable tracks. A trolley that carries the stabilizer rolls along the tracks drawn by four driving cables from the tower tops. Four tie-down cables attached to the trolley are used to increase the stiffness of the cable network and to reduce its vibration. The well-separated parallel cables making up each track mitigate the rotational motion of the trolley. There is an orienting mechanism between trolley and focus cabin to control their relative azimuth and elevation. This mechanism helps to control the orientation of the focus cabin when observing at large zenith angle. The results of experiments on this 20 m model were encouraging (Yu et al., 2004). The mechanical characteristics and vibration control of the design were studied (Cheng et al., 2003; Ren et al., 2004).

3.2. Recent experiments

In Tsinghua university, a more complete 50 m field model of the cable car feed design was built in 2003. The experiments and results are presented below.

3.2.1. *Stabilizer experiment*

In order to test the dynamic characteristics of the stabilizer whose workspace is about $10 \times 10 \times 10$ cm, experiments were first made in the laboratory (Figure 3). To simulate working conditions, the stabilizer was hung from the ceiling with rubber ropes, and the base plate of the stabilizer was shaken by an oscillator with stimulations composed from a standard wind spectrum. Position and orientation were observed and fed back to the controller. Laser ranging was applied to read the coordinates of the moveable plate and evaluate the effectiveness

Figure 3. Stabilizer experiment in the laboratory.

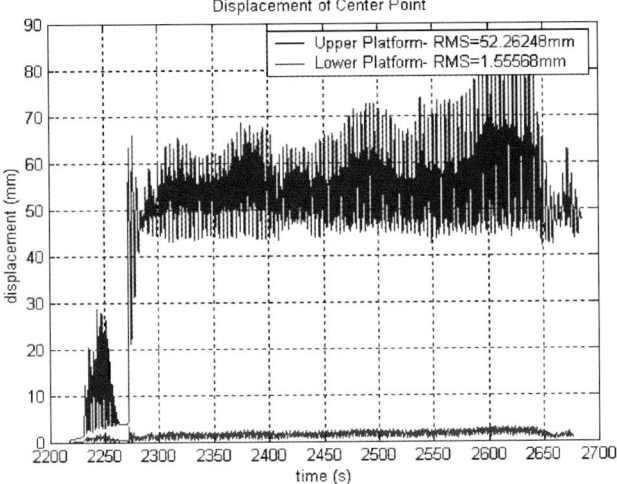

Figure 4. Results of stabilizer experiment in the laboratory.

of the control. As shown in Figure 4, the average displacement of the base plate is about 50 mm and the moveable plate could be stabilized to within r.m.s errors below 1.56 mm, which indicates the remarkable capability of the system adjustment.

3.2.2. *Combined model experiment and result*

A 50 m down-scaled (1:12) model of the cable-car support system, combining four steel towers, cable network, trolley, orientation mechanism and Stewart stabilizer, has been built in Tsinghua University in 2003 (Figure 5). The trolley is driven by simultaneously adjusting the lengths of the twelve cables (Figure 6). Cables 1–4, referred to as the driving cables, are anchored to the trolley. Cable pairs 5a,b and

Figure 5. Field model of 50 m, including four steel towers, cable network, trolley, orientation mechanism and stabilizer.

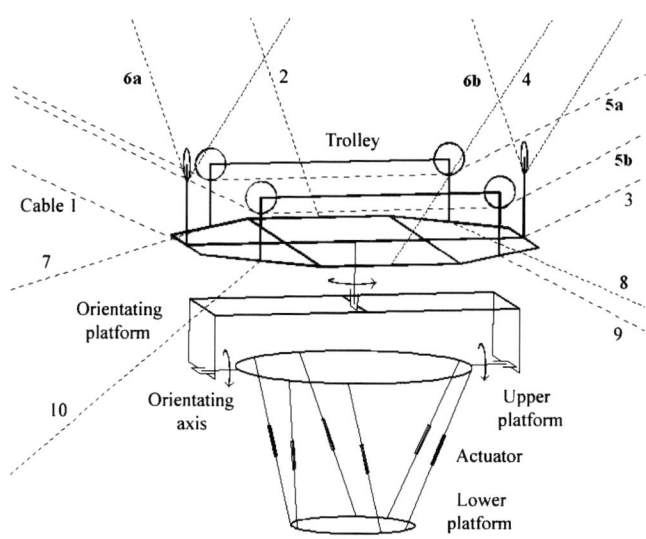

Figure 6. Assembly of the trolley, the orientation mechanism and the Stewart platform.

6a,b are suspension cables. They form an orthogonal flexible rail for the trolley by passing through both the fixed pulleys on the towers and those on the trolley. Cables 7–10, the pre-tension tie-down cables, are used to adjust the stiffness of the architecture. The orientation mechanism (Figure 6) is used to orient the upper platform relative to the trolley. It has two orthogonal axes: azimuth ($-270°$–$270°$) and elevation ($-20°$–$20°$). In this application the cable-driven architecture controls the position, while the orientation mechanism controls the orientation of the upper plate. The Stewart platform (Figure 6) is used to further reduce position and orientation error of the lower platform.

While the trolley traces the required locus at a velocity of 3 mm/s within the workspace of $10 \times 10 \times 5$ m around the center of the field, laser ranging is employed to measure the position of the upper and lower plates. Tension feedback is applied to avoid the relaxation of the cables. A wind meter at height of 2.0 m is used to record the wind speed throughout the experiment. To test the controllability of the mechanism and the reliability of the control laws, various experiments have been carried out. Dynamical coupling of the two plates of the Stewart stabilizer appears to be prominent only when PID coefficients and/or control time intervals are improperly selected.

Position and orientation errors for a typical experiment are shown in Figures 7 and 8. As a summary of the experiments, in the wind of about 2 m/s, the precision of the upper platform is about 0.9–1.6 mm in position and 1.9–2.6 $\times$ 10^{-3} radian in orientation and the lower platform is about 0.4–0.6 mm in position and 1.0–2.0 $\times$ 10^{-3} radian in orientation. Assuming linear scaling, positional accuracy of 4–7 mm would be expected for focus tracking of the future FAST.

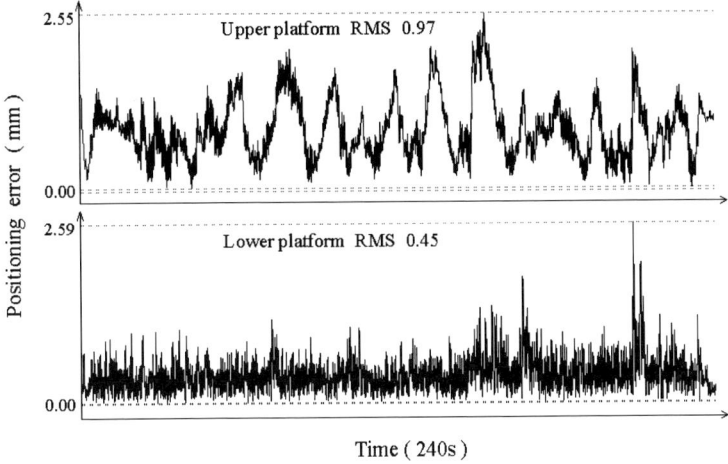

Figure 7. Position error of the base (upper) plate and moveable (lower) plate.

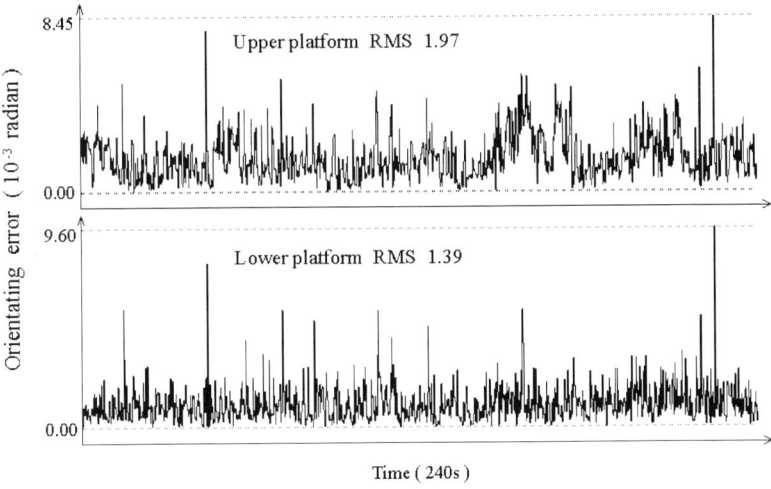

Figure 8. Orientation error of the base (upper) plate and moveable (lower) plate.

4. Concluding remarks

This light cable structure for feed support can avoid the heavy and expensive platform at the focus of an Arecibo type antenna. The Stewart stabilizer is able to compensate the displacements of the cabin and cables caused by wind impacts so that the feed is precisely located. The monitoring of tension is critical for the safety of suspension cables and high accuracy control. The key technology in the feedback loop control is remote position measurement of high accuracy (~several millimeters) with a rather fast sampling rate up to 10 Hz for the future FAST.

Acknowledgements

We thank Lu Yingjie, Zhu Lichun, Wu Shaobo, Zhou Qian and Qi Lin who have been participants in the experiments. We also thank Dr. Jin Chengjin and Prof. Peng Bo for their good comments to the report. This work is supported by the research fund for large radio telescope, National Astronomical Observatories of CAS, the fundamental research fund (No. JC1999031) of Tsinghua University and NNSF (No.10172049 & No.1017305).

References

Cheng, Y., et al.: 2003, *IEEE Trans. Robot. Autom.* **19**(3), 489–493.
Duan, B., et al.: 1996, 'Study of The Feed System For A Large Radio Telescope From The Viewpoint of Mechanical and Structural Engineering', in R. G. Strom, B. Peng and R. Nan (eds.), *Proceedings of the LTWG-3 &W-SRT, IAP*, pp. 85–102.
Duan, B.: 1999, *Mechatronics* **9**(1), 53–64.
Nan, R., et al.: 1996, 'Site Surveying for The LT in Guizhou Province of China', in R. G. Strom, B. Peng and R. Nan (eds.), *Proceedings of the LTWG-3 &W-SRT*, IAP, pp. 59.
Nan, R., et al.: 2000, *ASP* **213**, 523.
Nan, R., et al.: 2003, *Acta Astron. Sin.* **44**(Suppl.), 13–18.
Peng, B. and Nan, R.: 1997, *IAU Symp.* **179**, 93–94.
Qiu,Y.: 1998, *MNRAS* **301**, 827–830.
Ren, G., et al.: 2001, *Astrophys. Space Sci.* **278**(1), 243.
Ren, G., et al.: 2004, *Mechatronics* **14**, 1–13.
Stewart, D.: 1965, *Proc. Inst. Mech. Eng.* **180**(5), 371–386.
Su, Y. X. and Duan, B.: 2000, *Mechatronics* **10**(7) 819.
Su, Y. X. et al.: 2001, *Robot. Syst.* **18**(9), 507.
Su, Y. X., et al.: 2003, *Mechatronics* **13**(2), 95–110.
Yu, L., et al.: 2004, *EXPA* **15**(2), 113–122.
Zhu, W.: 1999, Description on Tracking Traces of FAST Feed, *MEMO to the FAST*.

CYLINDRICAL REFLECTORS

JOHN D. BUNTON
CSIRO ICT Centre, P.O. Box 76, Epping, NSW 1710, Australia
(e-mail: john.bunton@csiro.au)

(Received 3 August 2004; accepted 4 October 2004)

Abstract. In the early days of radioastronomy cylindrical parabolic reflectors were a popular technology. This paper traces some of the factors that have hindered their use in recent times. These factors no longer apply and as the cost of electronics decreases their applicability to the SKA becomes more and more favourable.

Keywords: cylindrical parabolic reflectors, radioastronomy, SKA

1. Introduction

In this paper we look at some of the historical development of cylindrical parabolic reflectors and show that the time has come to investigate their use as solution to requirements of the SKA. In particular they provide one of the lowest cost option for a wide field-of-view instrument.

2. A natural hybrid

At low frequencies the natural choice for a low cost radio telescope is a phased array of feed elements. Each feed element has a large effective area and is simple and cheap to construct. This has led to their use in telescopes such as of the 85 MHz Mills Cross (Mills et al., 1958) and LOFAR (Bregman, 2000), 30–220 MHz. At high frequencies the natural choice is a parabolic reflector as is used in ALMA [http://www.alma.nrao.edu] for frequencies above 40 GHz. The frequencies covered by the SKA span the range between LOFAR and ALMA. Thus parabolic dishes are an expensive solution at the low frequency end of the SKA range and even current proposals do not contemplate phased arrays beyond 2 GHz (European SKA Consortium, 2002).

The lack of an obvious technology has led to many options being considered as is exemplified by the seven major design concept white papers presented in 2002 (SKA design concept white papers 1–7) as well as many other less detailed proposals. Here we consider a natural hybrid of reflector and phased array technology: cylindrical parabolic reflectors or, for simplicity, cylindrical reflectors. A low cost cylindrical reflector is used to focus the incoming electromagnetic energy onto a line. A phased array then provides focussing in the orthogonal direction.

As the cylindrical reflector has only one axis of rotation and as the surface is curved in only one direction, it is cheaper to build than an equivalent dual axis fully steerable dish. This is exemplified by the Parkes and Molonglo telescopes which where built in the 1960s, for similar maximum frequencies (the Molonglo reflector was designed to work up to 1.4 GHz and Parkes to 3 GHz). Both telescopes cost about the same but Molonglo had a collecting area six times that of Parkes. The large collecting area allowed Molonglo to initially outstrip Parkes in pulsar discoveries.

3. Historical perspective

Because of their low cost collecting area cylindrical reflectors were popular: Ooty (Swarup et al., 1971), Northern Cross, DKR1000 (Artyukh et al., 1968) and Molonglo (Mills et al., 1963). However, none have been built since 1967. This has happened because they were unable to take full advantage of advances in receiver technology; it is the single focus parabolic dish that did this. Even the most expensive parametric or cooled receiver is a fraction of the cost of a fully steerable dish. Thus the major single dish antennas have seen a continual series of receivers with ever decreasing Tsys, increasing bandwidth and frequency range. On telescope arrays such as the VLA (Napier et al., 1983) and Westerbork (Baars et al., 1973) the pace at which improvement has occurred has been slower, but still significant, because each upgrade requires tens of receivers. On cylindrical reflector telescopes, there has been little improvement because of the high cost of upgrades, and many consider them as a technology that has had its day.

However, major single dish and synthesis telescopes are nearing the limit of possible receiver improvement and it will be the complex systems such as cylindrical reflectors that will benefit from the ever decreasing Tsys of low noise amplifiers (LNAs) and cost of electronics. In the mid SKA frequency range around 1.4 GHz, cylinders are set to overtake single dish antennas in terms of cost per unit of sensitivity and field-of-view, which makes them a prime candidate for use as an SKA technology.

4. Low-cost mechanical structure

Historically the cost of the mechanical structure of a cylindrical reflector has been one of its strengths. The Molonglo telescope has a collecting area of 40 000 m^2 and cost US $ 600 K in 1967. Adjusting for inflation this gives a cost of US $ 133 M per square kilometre (Bunton et al., 2002). The SKA specifications call for a sensitivity of 5000 m^{2}K^{-1} at 200 MHz (Jones, 2004). Assuming a sky temperature of 170 K and aperture efficiency of 0.7 then the total physical area needed is about 1.5 km^2, with either cooled or uncooled LNAs. Even if the current estimates are low by a factor of two the reflector cost is still well within the SKA budget of one billion dollars.

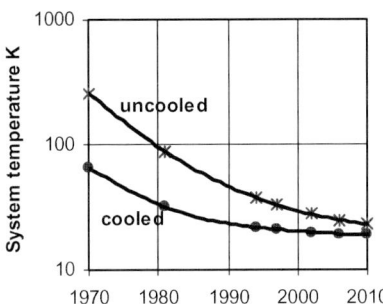

Figure 1. Estimated system temperature for narrow band 1GHz ratio telescope with cooled and uncooled LNAs

5. System temperature

The number of LNA needed for cylindrical reflector systems precludes the use of cooled amplifiers. Compared to parabolic dish systems this has resulted in a system temperature that could be up to ten times higher. This has negated the advantage due to the large apertures available to cylinders. However, other contributions to Tsys such as sky noise, spillover, and feed loss add about 18 K to the system temperature. As LNA noise temperature continues to fall this is leading to a convergence between the system temperature of cooled and uncooled systems. This is shown in Figure 1.

The data for uncooled LNAs is based on narrowband LNAs used at WRST (1970, 240 K) (Baars et al., 1973), Fleurs (1984, 60 K) (Jones et al., 1984) and Molonglo (1994, 20 K) (Large et al., 1994). The data shows a ten times reduction in 24 years. Adding the 18 K of other noise contributions gives the Tsys curve shown in Figure 1. Assuming that cooled systems have a five times lower LNA noise temperature give the curve shown for cooled amplifiers; extrapolating these results into the future shows that the system temperature of cooled and uncooled systems are now converging.

6. Line feeds

Many of the early cylindrical reflectors used stretched wire for a surface and could only receive one linear polarisation. The exception is Molonglo and work is now proceeding to demonstrate a dual-polarisation wide-band line feed. Currently a Foursquare feed (Buxton and Stutzman, 1999) is being considered which would give at least a 2:1 frequency coverage but uses only a single row of feed elements at focus. Use of Vivaldi arrays (Schaubert and Choi, 1999) would give increased frequency coverage at the expense of needing an array that is at least three elements wide. For the range of frequencies covered by the SKA multiple line feeds will be needed. To accommodate these a wide focal plane is needed. Initial work (James and Parfitt, 1999) has shown that an offset fed reflector with tilted focal plane can provide a solution.

7. Beamforming

Beamforming has been one of the weaknesses of cylinders. In the 1960s, the required delay or phase change was mechanical; for example, coupled rotating helical antennas at Molonglo, trombones at Ooty, and kerosene filled co-axial cables at the Northern Cross. The resulting systems were expensive and inflexible. In the 1980s, the shift to electronic systems started; for example Molonglo using 384 electronic phase rotators in the second stage of beamforming. This trend is continuing and now it is feasible to build a fully electronic beamformer at frequencies of 1 GHz. Each feed element would have an RF-CMOS downconversion system and digitiser. The data generated would then be transferred over fibre to a digital beamformer. With modern Field Programmable Gate Arrays (FPGA) providing processing power at a cost of about $10 per Giga operations per second, it is now economic to digitally beamform the individual elements of a 1.4 GHz linefeed: 1000 dual polarisation feeds on a 110 × 15 m reflector. The data from each feed is first analysed into a set of narrow band signals and then summed with an appropriate phase rotation to form beams. The current cost of a 200-beam 20 MHz beamformer is estimated at $180 per feed or $360,000 for an SKA antenna, which is comparable to the cost of the mechanical structure. Each beam is $2 \times 0.25°$ at 0.7 GHz and with the beams arranged contiguously the area of sky accessed is $2 \times 50°$. In 5 years time the bandwidth would increase to 200 MHz assuming costs continue to decrease according to Moore's Law. At this point the specifications match those needed to meet the 0.7 GHz survey speed requirements of the SKA. Further cost reductions due to Moore's law holding beyond 2009, the use of application specific integrated circuits (ASICs) or more efficient beamforming algorithms will make the beamforming cost much less than the mechanical structure by the time the SKA is built.

8. Cylindrical reflectors for the SKA

The SKA specifications call for a wide field of view at approximately 1 GHz (Jones, 2004). A cylindrical reflector with each feed having a downconverter and digitiser coupled into a digital beamformer matches the required specification. In this mode the signals from the line feed are optimally used, minimising the cost of feeds, LNAs, downconverters, digitisers and antenna signal transport. This coupled with the low cost mechanical structure provides a low cost solution to meeting the SKA specifications at frequencies of a couple of GHz. Above this a line feed with RF beamforming is needed to minimise cost and extrapolations with this scenario show that the cylinder is competitive up to about 20 GHz (Bunton, 2003). An alternative is a hybrid (Bunton and Lazio, this issue) with parabolic dish antenna, where the cylinders provide the wide field-of-view at low frequencies and the dishes cover the frequency range to 25 GHz and beyond.

9. Conclusion

Historically, the development of cylindrical reflectors has been hampered by the high system temperatures and the cost of electronics. Advances in electronics are rectifying both these problems making cylinders a viable option for the SKA.

References

Artyukh, V. S., Vitkevich, V. V. and Dagkesamanskii, R. D.: 1968, *Soviet Astron. – AJ* **11**, 792–799.
Baars, J. W. M., Van Der Brugge, J. F., Casse, J. L., Hamaker, J. P., Sondaar, L. H., Visser, J. J. and Wellington, K. J.: 1973, *Proc. IEE* **61**(9), 1258–1266.
Bregman, J. D.: 2000, 'Concept Design for a Low Frequency Array', *SPIE Proceedings Astronomical Telescopes and Instrumentation, Radio Telescopes*, Munich, Vol. 4015, pp. 19–32, SPIE Washington.
Bunton, J. D., Sadler, E. M. and Jackson, C. A.: 2002, 'Cylindrical Reflector SKA' *SKA Design Concept White Paper 7*, July 2002, http://www.skatelescope.org/pages/p_docsandpres.htm
Bunton, J. D. (ed.): 2003, 'Panorama of the Universe: A Cylindrical Reflector SKA', *SKA Design Concept White Paper 13*, May 2003.
Buxton, C. G. and Stutzman, W. L.: 1999, 'Implementation of the Foursquare antenna in broadband arrays', URSI National Radio Science Meeting, Orlando, FL, July 1999.
European SKA Consortium: 2002, 'European Concept for the SKA: Aperture Array Tiles', *SKA Design Concept White Paper 3*, July 2002, http://www.skatelescope.org/pages/p_docsandpres.htm
James, G. L. and Parfitt, A. J.: 1999, 'A Low-Cost Cylindrical Reflector for the Square Kilometre Array', in Smolders and van Haarlem (eds.), *Perspectives on Radio Astronomy: Technologies for Large Antenna Arrays*, ASTRON, Dwingeloo.
Jones, D. L.: 2004, 'SKA Science Requirements', *SKA Memo 45*, February 2004, http://www.skatelescope.org/pages/p_docsandpres.htm
Jones, I. G., Watkinson, A., Egau, P. C., Percival, T. M., Skellern, D. J. and Graves, G. R.: 1984, *Proc. ASA* **5**(4), 574–578.
Large, M. I., Campbell-Wilson, D., Cram, L. E., Davidson and Robertson, J. G.: 1994, *Proc. ASA* **11**(1), 44–49.
Napier, P. J., Thompson, A. R. and Ekers, R. D.: 1983, *Proc. IEEE* **71**(11), 1295–1320.
Mills, B. Y., Little, A. G., Sheridan, K. V. and Slee, O. B.: 1958, *Proc. IRE* **46**, 1958, 65–84.
Mills, B. Y., Aitchison, R. E., Little, A. G. and McAdam, W. B.: 1963, *Proc. IRE Aust.* **24**, 156–164.
SKA Design Concept White Papers 1–7: 2002, http://www.skatelescope.org/pages/p_docsandpres.htm
Schaubert, D. H. and Choi, T. H.: 1999, 'Wide Band Vivaldi Arrays for Large Aperture Antenna', in Smolders and van Haarlem (eds.), *Perspectives on Radio Astronomy: Technologies for Large Antenna Arrays*, ASTRON, Dwingeloo.
Swarup, G., Sarma, V. G., Joshi, M. N., Kapahi, V. K., Bagri, D. S., Damle, S. H. Ananthakrishnan, S., Balasubramanian, V., Bhave, S. S. and Sinha, R. P.: 1971, *Nat. Phys. Sci.* **230**, 185–188.

RF SYSTEMS

LOCAL OSCILLATOR DISTRIBUTION USING A GEOSTATIONARY SATELLITE

J. BARDIN*, S. WEINREB and D. BAGRI
Jet Propulsion Laboratory, California Institute of Technology, U.S.A.
(*author for correspondence, e-mail: joseph.c.bardin@jpl.nasa.gov)

(Received 2 August 2004; accepted 30 December 2004)

Abstract. A satellite communication system suitable for distribution of local oscillator reference signals for a widely spaced microwave array has been developed and tested experimentally. The system uses a round-trip correction method to remove effects of atmospheric fluctuations and radial motion of the satellite. This experiment was carried out using Telstar-5, a commercial Ku-band geostationary satellite. A typical Ku-band satellite has uplink and downlink capacity at 14–14.5 GHz and 11.7–12.2 GHz, respectively. For this initial experiment, both earth stations were located at the same site to facilitate direct comparison of the received signals. The local oscillator reference frequency was chosen to be 300 MHz and was sent as the difference between two Ku-band tones. The residual error after applying the round trip correction has been measured to be better than 3 ps for integration times ranging from 1 to 2000 s. For integration times greater than 500 s, the system outperforms a pair of hydrogen masers with the limitation believed to be ground-based equipment phase stability. The idea of distributing local oscillators using a geostationary satellite is not new; several researchers experimented with this technique in the eighties, but the achieved accuracy was 3 to 100 times worse than the present results. Since then, the cost of both leased satellite bandwidth and the Ku-band ground equipment has dropped substantially and the performance of various components has improved. An important factor is the availability of narrow bands which can be leased on a communications satellite. We lease three 100 kHz bands at approximately one hundredth the cost of a full 36 MHz-wide transponder. Further tests of the system using terminals separated by large distances and comparison tests with two hydrogen masers and radio interferometry of astronomical objects are needed.

Keywords: hydrogen maser, LO distribution, local oscillator, two-way satellite time transfer, TWTT, TWSTT

1. Introduction

In a synthesis array system, the angular resolution of the image is directly proportional to the longest baseline in the array. Therefore, an array such as the SKA must have several long baselines if high angular resolution is a design parameter. In an array in which all antennas are at a single site, all local oscillators can be derived from a single hydrogen maser frequency standard. However, when the antennas are placed large distances from one another, it becomes much more difficult to keep the local oscillators coherent. Unfortunately, any phase instability in the local oscillators translates directly into the received data. For instance, if there is a 10° phase drift in a local oscillator at a given site, then the down-converted data will have a 10° phase drift as well. In addition, there is a phase shift caused by turbulence in the atmosphere (Thompson et al., 2001). If atmospheric phase instabilities are the dominant source of phase noise in the measurements, then this disturbance can be calibrated out by the use of one or more phase calibration sources (Walker, 1999). What this means is that the local oscillators used in Very Long Baseline Interferometry must have very high stability requirements between calibrations.

The Very Long Baseline Array (VLBA) is equipped with a separate hydrogen maser at each antenna. In the case of the VLBA, where there are only eleven antennas, the cost of the hydrogen masers is not great enough to be a limiting factor. However, for a larger array, such as the SKA, equipping hundreds of antennas (or clusters of antennas) with their own hydrogen maser would impose too large of a cost burden. Therefore, it makes sense to evaluate alternative ways in which local oscillators can be distributed to a sparse array of antennas. In this paper an experiment in which a commercial geostationary satellite was used to distribute a LO signal will be described.

2. Two-way time transfer (TWTT)

Sending a LO signal through a satellite has several advantages over the use of hydrogen masers. First of all, the system cost is significantly lower than the hydrogen maser alternative. This is in part due to the satellite television industry. It is now possible to buy "off the shelf" low noise block downconverters, block upconverters, and antennas. These components are high quality and cost a fraction of what they would have cost prior to the satellite television boom. Perhaps a more important factor is the fact that the satellite companies have started leasing small amounts of bandwidth inside certain transponders. What this means is that one can now lease 100 kHz of a 36 MHz transponder for about 1/360th the price of the full transponder. Since the LO signal is a very narrow band signal by nature, this new leasing policy is quite an important breakthrough. There is also a significant performance advantage of the satellite distribution system over the classical hydrogen maser approach in that an absolute frequency reference is transferred. Therefore, unlike a hydrogen maser, the frequency of the transferred reference will not drift.

However, despite the advantages inherent in the use of a satellite LO distribution, there are several significant challenges which must be dealt with in order to use such a system. When sending a local oscillator signal through a satellite to a remote basestation, a phase shift will occur due to the electrical path length, the earth station hardware, and the satellite hardware. This phase shift will have time dependence and cannot be computed accurately enough to permit interferometry. To combat this

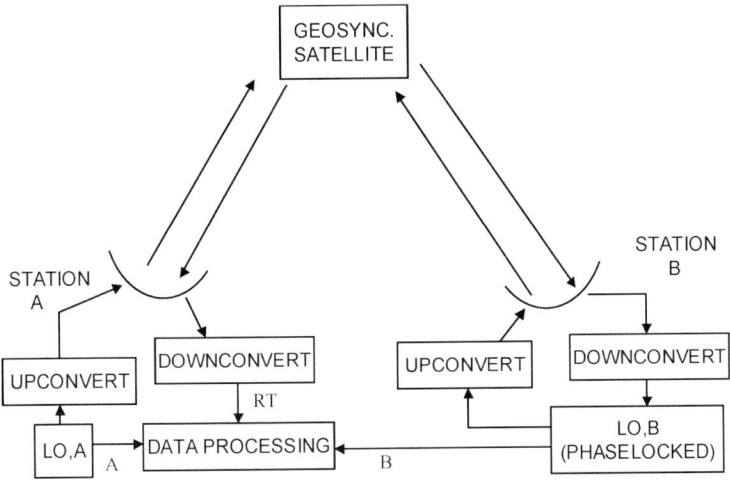

Figure 1. Two-way time transfer block diagram.

effect, a two-way time transfer technique was originally suggested by Yen et al. (1977).

A block diagram of a two-way time transfer system is presented in Figure 1 and photographs are shown in Figures 2 and 3. This experiment was conducted using the Telstar-5 Satellite. The oscillator located at station-A serves as the master LO and is a 5 MHz voltage controlled crystal oscillator that is multiplied to 300 MHz. The master local oscillator will be referred to as LO-A. The 300 MHz frequency standard is transmitted from station-A to station-B as the difference of two Ku band CW tones. Differential transmission of the LO is necessary to remove the effects of any phase noise which would otherwise be introduced by the down-conversion oscillator onboard the satellite. At station-B, the 300 MHz signal is recovered using a narrow-band phase lock loop (PLL) which serves as a tracking filter. The tracking filter has an equivalent double sideband noise bandwidth of approximately 1 Hz. The output of this PLL serves as LO-B, the transferred local oscillator. The phase error in a PLL is dependent on both the SNR of the input signal and the loop bandwidth of the PLL. This quantity can be written explicitly (Gardner, 1979)

$$\Delta\phi_{\text{RMS}} = \sqrt{\frac{BL}{SNR}}, \qquad (1)$$

where BL is the equivalent single sideband loop noise bandwidth and SNR is the ratio of input signal to the noise in a 1 Hz bandwidth.

In order to measure the phase shift that is encountered during the transmission path, LO-B is sent back through the satellite to station A. We can call this signal LO-RT, where RT stands for "round trip." Now, the phase difference between LO-A

Figure 2. Photograph of transmit/receive setup.

Figure 3. Station A Electronics.

and LO-B can be written as

$$\phi_B = \phi_t + \phi_{E,B}, \tag{2}$$

where ϕ_t is a phase shift due to the satellite link and $\phi_{E,B}$ is a phase shift due to the station B receiver electronics. The phase difference between LO-RT and LO-A can be written as

$$\phi_{RT} = 2\phi_t + \Delta\phi_t + \phi_{E,B} + \phi_{E,A}, \tag{3}$$

where $\phi_{E,A}$ is a phase shift due to the station-A receiver electronics and $\Delta\phi_t$ is an error term due to non-reciprocity of the link. One source of non-reciprocity is the difference in uplink (14.2 GHz) and downlink (11.9 GHz) frequencies coupled with a difference in dispersion due to the ionosphere at sites A and B; this has not been evaluated in the present experiment.

If the ground based hardware has been designed well, $\phi_{E,A}$ and $\phi_{E,B}$ will be constants which can be determined through calibration. However, ϕ_t will not be constant, due to radial motion of the satellite and atmospheric fluctuations. For this initial experiment, station-A and station-B were co-located in order to permit the direct measurement of ϕ_B. Thus, we can define a residual error as

$$\phi_{\text{residual}} = \frac{\phi_{RT}}{2} - \phi_B = \frac{\Delta\phi_t}{2} + \frac{\phi_{E,A}}{2} - \frac{\phi_{E,B}}{2} \qquad (4)$$

So, if the link is perfectly reciprocal and the earth station hardware does not drift, then the residual error will be a constant with noise determined by Equation (1).

3. Results

Using digital phase detectors, ϕ_B and ϕ_{RT} were sampled at 100 Hz. After the data were unwrapped, Equation (4) was calculated as a function of time. These data were then used to calculate the Allan standard deviation and RMS time error. For a description of the Allan standard deviation, see (Allan et al., 1991). Results appear in Figures 4 and 5.

Similar experiments were carried out in the eighties. Results from Cannon et al. (1982), Ables (1989), and Van Ardenne et al. (1983) also appear in the figures.

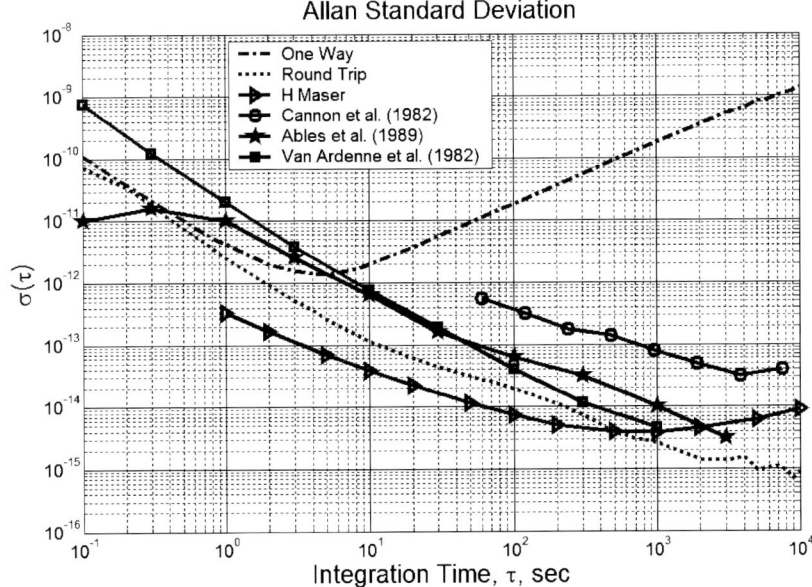

Figure 4. Measured Allan standard deviation (Cannon et al., 1982; Ables, 1989; Van Ardenne et al., 1983) one way refers to ϕ_B and round trip refers to ϕ_{residual}.

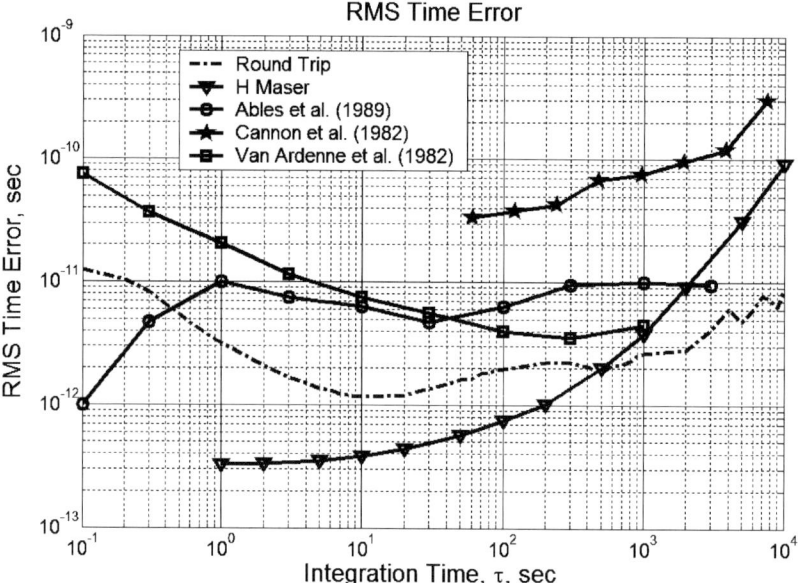

Figure 5. Measured RMS time error (Cannon et al., 1982; Ables, 1989; Van Ardenne et al., 1983).

4. Conclusions

A TWTT system has been demonstrated in which both terminals were located at the same location. An important conclusion which came out of this experiment is that the system performance is not hindered due to signals generated by other users. Further explanation of the effects of transponder occupancy is given in (Bardin, in preparation). While the results are better than previous results, there is still a need for improvement. There are two clear ways in which the system could be improved. The first improvement is to add a temperature stabilization system to the hardware. It seems that our largest source of error was temperature-related effects. In particular, we noticed a ripple in the data that correlated quite well with measured temperature changes in the room. The second way in which the system could be improved is to improve the signal to noise of the link. For this experiment, we transmitted less than 1/2 watt and used one-meter antennas. Although our contract permitted us to transmit four times more power, we simply were unable to increase our power levels without producing unreasonable intermodulation products. While the results obtained in this experiment look promising, more research is needed. A reasonable next step would be to add temperature control to the system and perform an experiment in which the terminals are significantly separated. Such a test must include hydrogen masers at the two sites and interferometry with astronomical sources to test the long-term stability.

Acknowledgements

This work was performed at the Jet Propulsion Laboratory, California Institute of Technology, under a contract with the National Aeronautics and Space Administration. The authors would like to thank Prof. P. Lubin at University of California, Santa Barbara, for supervision and laboratory space for a portion of the work.

References

Ables, S. T.: 1989, 'A Report on the Phase Stability of the AUSSAT Phase Transfer System', ATNF Internal Document.
Allan, D. W., Weiss, M. A. and Jespersen, J. L.: 1991, 'A Frequency-Domain View of Time-Domain Characterization of Clocks and Time and Frequency Distribution Systems', *Proceedings of the 45th Annual IEEE Symposium on Frequency Control*, pp. 667–678.
Bardin, J.: in preparation, 'Local Oscillator Distribution Using a Commercial Geostationary Satellite', M. S. Thesis, Electrical Engineering Department, University of California, Los Angeles.
Cannon, W. H. et al.: 1982, 'Phase Stable Long Baseline Interferometry Via a Satellite Link', *Very Long Baseline Interferometry Techniques*, Cepadues Editions, Toulouse, pp. 443–459.
Gardner, F.: 1979, *Phaselock Techniques*, John Wiley and Sons, New York.
Thompson, A. R., Moran, J. M. and Swenson, G. W. Jr.: 2001, *Interferometry and Synthesis in Radio Astronomy*, 2nd ed., Wiley Interscience, New York.
Van Ardenne, A. et al.: 1983, 'A High-Precision Phase-Comparison Experiment Using a Geostationary Satellite', *IEEE Trans. Instrum. Meas.*, **IM-32**, 370–376.
Walker, R. C.: 1999, 'Very Long Baseline Interferometry', *Synthesis Imaging in Radio Astronomy II*, Astronomical Society of the Pacific, San Fransisco, pp. 433–462.
Yen, J. L. et al.: 1977, 'Real-Time Very-Long Baseline Interferometry Based on the Use of a Communications Satellite', *Science*, 198.

RF DESIGN OF A WIDEBAND CMOS INTEGRATED RECEIVER FOR PHASED ARRAY APPLICATIONS

SUZY A. JACKSON

CSIRO Australia Telescope National Facility and Macquarie University, Sydney, Australia
(e-mail: suzy.jackson@csiro.au)

(Received 9 August 2004; accepted 9 December 2004)

Abstract. New silicon CMOS processes developed primarily for the burgeoning wireless networking market offer significant promise as a vehicle for the implementation of highly integrated receivers, especially at the lower end of the frequency range proposed for the Square Kilometre Array (SKA). An RF-CMOS 'Receiver-on-a-Chip' is being developed as part of an Australia Telescope program looking at technologies associated with the SKA. The receiver covers the frequency range 500–1700 MHz, with instantaneous IF bandwidth of 500 MHz and, on simulation, yields an input noise temperature of <50 K at mid-band. The receiver will contain all active circuitry (LNA, bandpass filter, quadrature mixer, anti-aliasing filter, digitiser and serialiser) on one 0.18 μm RF-CMOS integrated circuit. This paper outlines receiver front-end development work undertaken to date, including design and simulation of an LNA using noise cancelling techniques to achieve a wideband input-power-match with little noise penalty.

Keywords: astronomy, CMOS LNA, integrated receiver, radio-on-chip, RFIC, Square Kilometre Array, system-on-chip

1. Introduction

The scientific demands of large collecting area and large field-of-view (FOV) below 1 GHz mean that the SKA will require either a large number of elemental receptors (aperture phased array) or a smaller number of flux concentrators, each with some form of focal plane array (FPA) to extend the low-frequency FOV. In either case, a large number of inexpensive, high performance receivers are required.

One method of realising the cost reductions necessary to fabricate these systems is to design a single integrated circuit containing the entire receiver chain. The fabrication processes heretofore popular for amplifiers and mixers in radio telescope receivers, namely gallium-arsenide (GaAs) and indium-phosphide (InP), whilst offering high performance in terms of noise temperature and high transistor f_T (unity current gain frequency), are unsuited to the integration of complex digital logic and are comparatively expensive to manufacture.

Current trends in RF-CMOS performance indicate that a single-receiver IC implemented using advanced deep-submicron CMOS processes is likely to meet the receiver requirements of the SKA, at least by the time the array is constructed post-2010. In order to begin the development process and gain valuable design experience, a rather more modest receiver is being developed using current CMOS processes. This paper describes work to-date in development of this prototype receiver.

2. CMOS for RF

Radio-astronomy receivers generally use either GaAs or InP transistors and MMICs as active elements. GaAs and InP are applicable to low-noise, high frequency use, as the processes offer very high f_T, low-noise transistors, and perform well when cooled to cryogenic temperatures. However, InP and GaAs are not amenable to the implementation of complex digital circuits (such as multi-bit samplers) due to their relatively high power requirements, low integration levels, and prohibitive wafer costs.

Silicon CMOS is a mature process which is extremely popular for use with digital logic and low frequency analogue circuits. Enormous resources have been invested in CMOS over the last two decades in order to improve digital logic speed and levels of integration. More recently, the growth of the wireless networking and mobile telephone markets, with their requirement for low-power, inexpensive, highly integrated RF systems, has further driven developments in silicon CMOS, as well as creating a new Silicon Germanium (SiGe) process which adds hetero-junction bipolar transistors (HBTs) to a standard CMOS process (Subbana et al., 2000).

Notwithstanding promising developments with SiGe, modern deep-submicron CMOS processes are becoming more applicable for use in radio astronomy receivers as feature sizes decrease and transistor f_T increases. By extrapolating representative performance and feature sizes from the last five CMOS generations (IBM, 2004), as shown in Figure 1, we see that by the time the SKA is built, cutting edge silicon

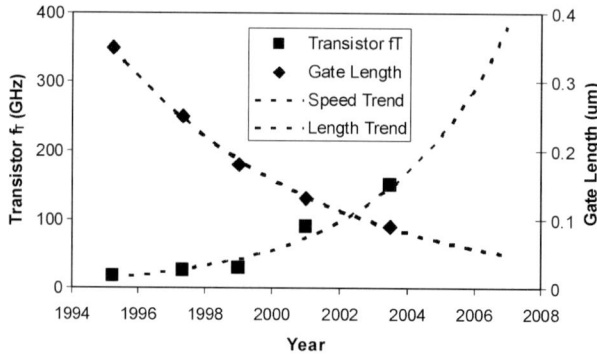

Figure 1. CMOS performance trends.

CMOS processes may offer transistor f_T of around 400–500 GHz, exceeding the performance of today's mainstream GaAs processes. As such, it is anticipated that CMOS could be the technology of choice for building large numbers of high performance, inexpensive receivers covering the frequency range of 1–10 GHz.

3. CMOS low noise amplifiers

One of the key challenges faced in implementing a radio astronomy receiver in RF-CMOS is that of designing a low noise amplifier (LNA) having adequate RF bandwidth and sufficiently low noise operation. Lee (1998) works through a derivation of MOSFET noise parameters. Whilst this approximation is accompanied by numerous caveats, most notably the validity of the three noise coefficients in the deep-submicron regime, it still provides some insight into the improvement in minimum noise temperature ($T_{N(\min)}$) with increased process f_T:

$$T_{N(\min)} = 290 \frac{2}{\sqrt{5}} \frac{f}{f_T} \sqrt{\gamma \delta (1 - |c|^2)} \quad (K)$$

where γ is the body effect coefficient, δ is the gate noise coefficient, and c is the gate/drain noise correlation coefficient. Use $\gamma = 2$, $\delta = 4$, $|c| = 0.395$ for typical short channel process.

Applying this equation to the published data for current processes allows us to plot the trend for minimum LNA noise temperature as shown in Figure 2, both for a nominal 10 GHz SKA RF frequency and a more modest 2 GHz. The latter figure is more applicable to a demonstrator project using current processes. In addition, noise performance of some representative CMOS LNAs is shown.

The vast majority of CMOS LNAs encountered in the literature (Leroux et al., 2002; Hayashi et al., 1998; Shaeffer and Lee, 1997) are designed for low power portable devices, such as mobile telephones, wireless network adapters, and global

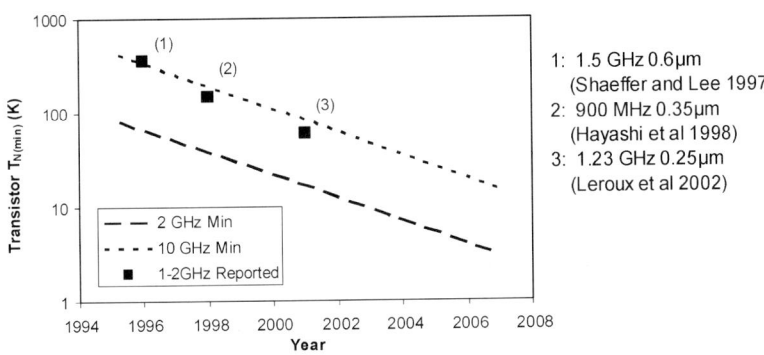

Figure 2. CMOS minimum noise trends.

positioning receivers. Power consumption in these typically battery powered devices is critical, and noise performance is traded in order to reduce power. The SKA application does not impose the strict operating power requirements of portable devices, so some design latitude is allowed.

Returning to the bandwidth–noise trade-off, we note that topologies providing a resistive input power match, such as the common-gate or shunt-series topology, add a significant noise penalty. In the case of the common-gate amplifier, this noise penalty is around 200 K (Lee, 1998).

In order to achieve noise temperatures close to T_{min} across the operating band, the usual practice is to employ some method of impedance transformation on the LNA input. Accepted approaches include one-fourth wave filter structures (large and noisy at low frequencies, and not amenable to integration on wafer) or, alternatively, using reactive components (inductive degeneration) to realise an impedance transformation for a common-source transistor. It is difficult to extend these topologies to broad band operation using integrated passive components, as the losses in the matching networks then dominate the LNA noise.

With the improvement in active device gain-bandwidth product it is possible, in principle, to depart from conventional RF topologies and to implement amplifiers using a variety of circuit designs hitherto restricted to the low frequency or video domain (Gray et al., 2001). One such approach is to utilise feedback or feedforward techniques to accomplish a broadband power match without sacrificing the amplifier noise match. The recent work of Bruccoleri (2004) is particularly relevant, as it establishes a means whereby the power matching and signal amplifying parts of the LNA may be separated and, by careful design, noise currents in the matching network can be cancelled in the amplifying section.

Figure 3 demonstrates the generalised noise cancelling principle. M1 and R_f form a shunt-series matching amplifier, with modest noise performance. Drain current noise in M1 is present as a voltage at the drain node and as an in-phase voltage at the gate node, whereas the input signal is in anti-phase between the drain and gate nodes. It is possible therefore to cancel out M1's drain current

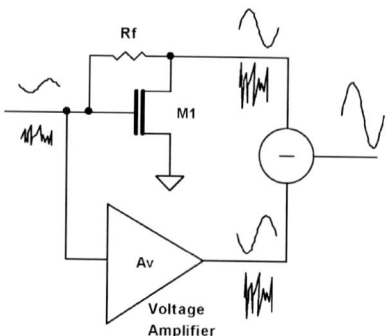

Figure 3. Noise cancelling principle (Bruccoleri, 2004).

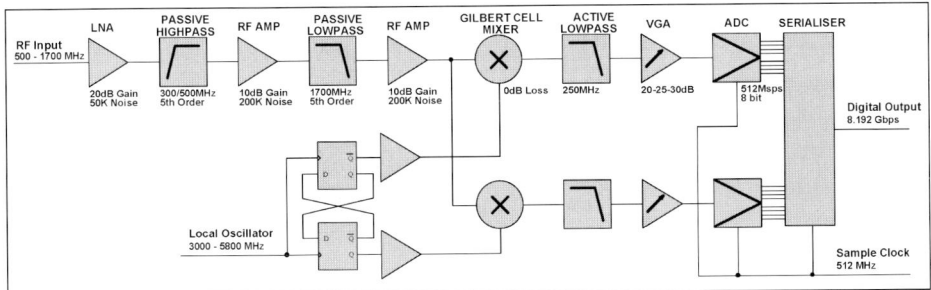

Figure 4. Receiver RFIC overall schematic.

noise using a voltage amplifier of the same gain as the matching network, and a difference network. In practice, the voltage amplifier adds its own noise, but the power matching constraints for this amplifier are greatly relaxed, and it can be matched instead for low noise.

4. A prototype integrated receiver

An overall schematic for the prototype receiver IC is shown in Figure 4. The receiver uses image-free direct conversion of a 500–1700 MHz RF band to baseband. The 500 MHz bandwidth complex baseband signals are digitised by two 512 Msps ADCs.

The RF bandwidth, IF bandwidth and noise specifications for the receiver have been optimised so that acceptable performance is possible with contemporary CMOS processes. Local oscillator (LO) and sample clock inputs, and digitised receiver output, connect directly to fibre interfaces, reducing leakage into the RF input. The conversion chain is designed to maximise dynamic range, so that it may be used in comparatively hostile RF environments.

5. Low noise amplifier

The topology chosen for the LNA (Figure 5) is based on the active feedforward noise-cancelling approach, using a common-gate matching stage with a common-source amplifying stage. In such an arrangement the noise cancelling criteria are met when the voltage gain of each signal path is equal.

The main amplifying transistor (M2) dominates the noise of the circuit, so the design process starts with the choice of an appropriately sized transistor and selection of a minimum noise bias point. Simulations on a simple test circuit showed that the minimum noise (and maximum transconductance) for the process of interest (IBM 0.18 μm) was achieved at around 20 $\mu A/\mu m$ drain current density, and

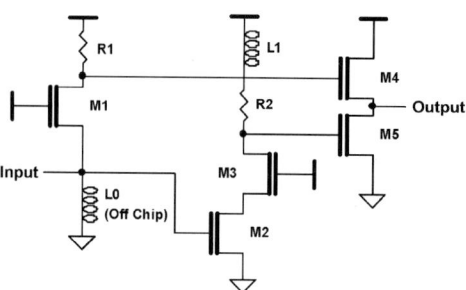

Figure 5. Simplified LNA schematic.

that noise temperature decreased with increasing gate width. However, gate-source capacitance (C_{gs}) and gate-drain capacitance (C_{gd}) also scale with gate width. As this capacitance is across the source, it degrades the power match and introduces significant gain slope. A compromise was reached at 700 μm gate width, this figure being distributed across 200 gate fingers to minimise gate resistance.

The cascode transistor (M3) improves gain somewhat by reducing the Miller capacitance. Some inductance (L1) in series with the load resistor (R2) on the amplifying side peaks the gain slightly, improving both the gain slope and the noise cancelling across the RF bandwidth.

A second stage (M4 and M5) gives additional gain and provides the differencing action between the amplifying and matching subcircuits. This stage also provides a high input impedance, maximising gain in the first stage and matching the amplifier to a 50 Ω load.

The amplifier is laid out in a six-layer process, as shown in Figure 6. It occupies 0.08 mm² of die. A 33 pF off-chip coupling capacitor and 22 nH shunt inductor provide a DC bias feed for the common-gate stage.

Simulated results for the amplifier are shown in Figure 7. Wiring parasitics account for approximately 2 dB gain and 8 K additional noise at 1.7 GHz, mainly through increased capacitance to substrate. Wiring parasitics also degrade the input return loss by approximately 1 dB. Note that the simulated performance includes the effect of minimum sized input and output bond pads, as well as bondwires and off-chip input shunt inductor losses. The amplifier has more than 18 dB gain across the 500–1700 MHz RF bandwidth, with better than 50 K noise temperature between 600 and 1400 MHz (as a comparison the mid-band figure without noise cancellation is approximately 180 K). Input return loss (relative to 50 Ω) is approximately 8 dB across the band, whilst output match is better than 15 dB. Input match may be traded for noise temperature by varying the size of M2, whilst adjusting R2 to ensure the cancelling criteria are met. The amplifier has a 1 dB compression point of −4 dBm, whilst drawing 23 mA with $V_{dd} = 1.8$ V. It is possible to extend the lower cutoff frequency to 300 MHz by increasing the off-chip shunt inductance. Note that with a future implementation in 0.13 μm CMOS, one may expect a noise temperature of approximately 20–30 K at band centre with similar input match.

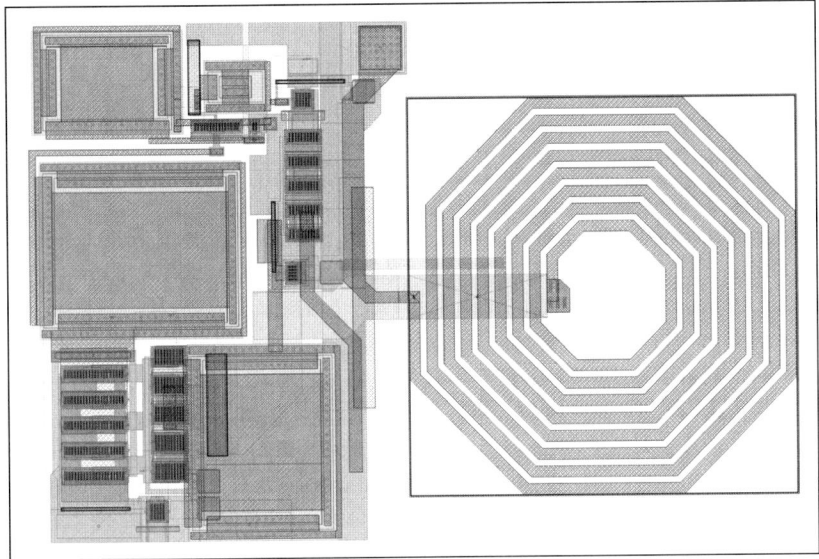

Figure 6. LNA layout.

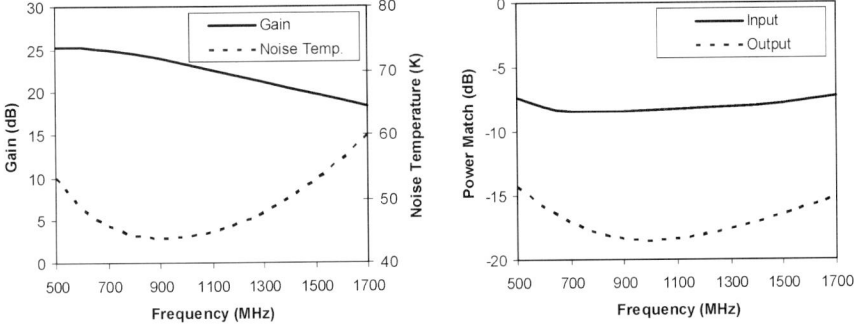

Figure 7. LNA simulated performance (including layout parasitics).

6. RF amplifier

In order to ensure that the full dynamic range of the LNA is preserved, a moderate gain, moderate noise temperature, relatively high power RF amplifier (Figure 8) was designed as a general purpose gain block. This amplifier utilises a simple inverter topology, with feedback to set the input power match at 50 Ω. M1 is biased with a conventional current mirror, whilst an additional DC feedback loop adjusts the bias on M2 to maintain the output at $V_{dd}/2$, ensuring maximum signal swing. M2 is sized to match a 50 Ω load.

The amplifier is laid out in the same six-layer process as the LNA, and occupies 0.03 mm^2 of die. It is fully self-biasing. On simulation, the amplifier achieves

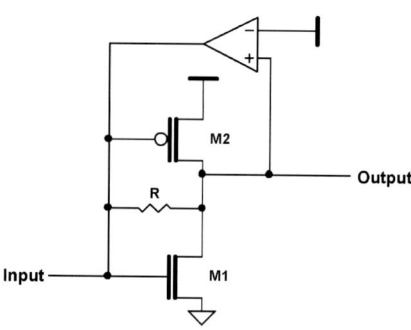

Figure 8. Simplified RF amplifier schematic.

10 dB gain across a full decade bandwidth (200 MHz to 2 GHz), with <200 K noise temperature and >8 dB input and output return losses. The 1 dB compression point is +6 dBm at 1 GHz. This corresponds to a peak to peak voltage swing of 1.3 V with V_{dd} set at 1.8 V; the current drain is then 12 mA.

7. RF high-pass filter

The likely presence of strong interfering signals, especially those in the FM broadcast band (88–108 MHz), requires the use of a high-pass filter early in the signal chain. This filter (Figure 9) comprises a fifth-order 300 MHz Chebychev filter, followed by a third-order 500 MHz Chebychev filter, the latter able to be bypassed by virtue of a CMOS switch. This arrangement is necessary because a potential application for the prototype receiver requires operation down to 300 MHz, with relaxed noise and input power match requirements.

The simulated performance for the filter is shown in Figure 10. The left hand plot shows the 300 MHz response, with the control input grounded, whilst the right hand plot shows the filter response with the control input held at 1.8 V, for a corner frequency of 500 MHz. Although the performance remains acceptable, the

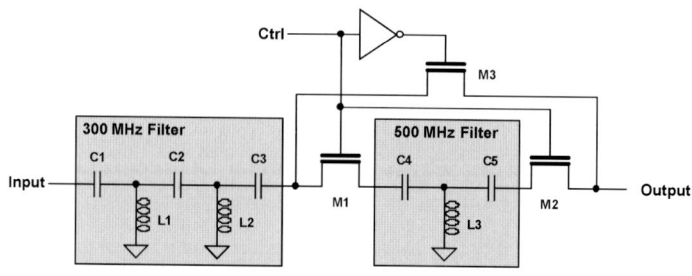

Figure 9. Simplified RF high-pass filter schematic.

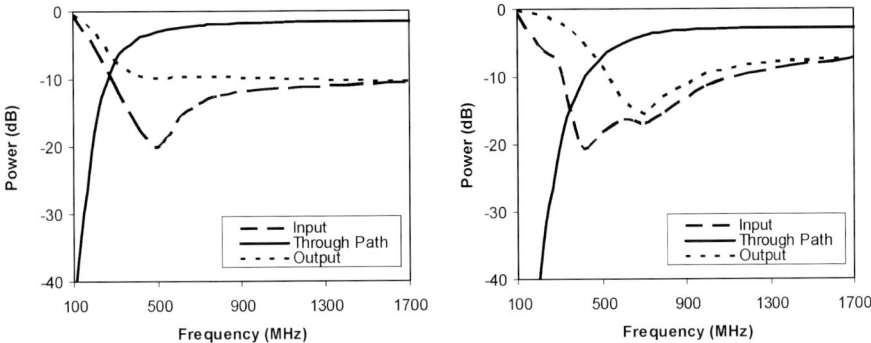

Figure 10. High-pass filter simulated performance.

additional shunt capacitance of the switch (M1 through M3) is evident in the output return loss, especially with the 500 MHz filter switched in.

8. Conclusion and further work

Results to-date of a development program for a CMOS integrated receiver, including an LNA with better than 50 K mid-band noise temperature and operation over a 500–1700 MHz bandwidth have been described. Work performed so far on the prototype receiver indicates that some of the limitations of CMOS, such as substrate loss and poor quality passives, may be overcome by utilising circuit topologies developed for low frequency analogue and video applications. Testing of fabricated devices will begin in late 2004 and the full integrated receiver is scheduled for tests in early 2006.

Acknowledgements

This work is supported by The Australia Telescope National Facility (ATNF) and Macquarie University. The Australia Telescope is funded by the Commonwealth of Australia for operation as a National Facility managed by CSIRO. This work is funded by a Major National Research Facilities grant. I thank my supervisors, Dr Jeffrey Harrison (Macquarie University) and Dr Peter Hall (ISPO) for their patient guidance and wisdom, and for encouraging me to explore unusual circuit topologies.

References

Bruccoleri, F.: 2004, *IEEE J. Solid-State Circuits*. **39**, 275–282.
Gray, P. R., Hurst, P. J., Lewis, S. H. and Meyer, R. G.: 2001, *Analysis and Design of Analog Integrated Circuits*, 4th Ed., Wiley, New York.

Hayashi, G., Kimura, H., Simomura, H. and Matsuzawa, A.: 1998, A 9 mW 900 MHz CMOS LNA With Mesh Arrayed MOSFETs, *Symposium on VLSI Circuits*, pp. 84–85.

IBM: 2004, RF-CMOS Process Specifications [online], IBM Semiconductor, available from http://www-306.ibm.com/chips/techlib/techlib.nsf/literature/Foundry (accessed 12 March 2004).

Lee, T. H.: 1998, *The Design of CMOS Radio-Frequency Integrated Circuits*, Cambridge University Press, Cambridge.

Leroux, P., Janssens, J. and Steyaert, M.: 2002, *IEEE J. Solid-State Circuits* **37**, 760–765.

Shaeffer, D. K. and Lee, T. H.: 1997, *IEEE J. Solid-State Circuits* **32**, 745–759.

Subbana, S. et al.: 2000, Review of Silicon-Germanium BICMOS Technology After 4 Years of Production and Future Directions, *Gallium Arsenide Integrated Circuit Symposium*, pp. 7–10.

DATA TRANSPORT

FIBER OPTIC NETWORK TECHNOLOGY FOR DISTRIBUTED LONG BASELINE RADIO TELESCOPES

D.H.P. MAAT and G.W. KANT
ASTRON, P.O. Box 2, 7990 AA Dwingeloo, The Netherlands
(e-mail: maat@astron.nl)

(Received 2 August 2004; accepted 28 February 2005)

Abstract. The construction costs of distributed radio telescopes are to a great extent determined by the deployment costs of the fiber optic data transport network that is needed to transport the received information to the data processor(s). As such, the baseline and data rates that are feasible for a specified amount of money are determined by the status of the technology and deployment costs of the communication network. In this paper the present day data transport status is described and, using a costing model, the most attractive data transport technologies are determined, taking the LOFAR telescope (ASTRON, 2005) as an example. In the outlook, the near-term data transport technology developments are described.

Keywords: CWDM, data transport network, distributed radio telescope, Ethernet, GbE, network design, network topology, radio telescope, WDM, wide area network, 10 GbE

1. Introduction

To meet the resolution and sensitivity demands that are required in present day radio astronomy, distributed radio telescopes are needed with long baselines and large numbers of stations with many antennas, each producing a large amount of data. For obtaining the required high quality imaging, various antenna distributions can be used, ranging from a random antenna distribution to non-uniform highly structured antenna arrangements. Which of these antenna distributions is chosen depends on construction related issues like, e.g., costs and geographical constraints. In addition, many options for the way the data processing is performed are available: both a completely centralized data processing and a fully distributed processing of the data (i.e., all data processing is performed at the stations) are among the possibilities. The actual situation will be in between both options.

Both the antenna arrangement and the way the data processing is handled have a strong influence on the data transport system in the radio telescope. For example, in the case where centralized data processing is applied in combination with a random antenna distribution, the amount of data that needs to be transported is much larger than when an antenna distribution having an increasing antenna concentration towards the central processor is used. If in this latter case part of the data that is produced at the antenna stations is also processed at these stations, even less data needs to be transported within the telescope. The configuration of the telescope also affects the way the data is handled. When centralized data processing is applied, the data traffic has a unidirectional, point-to-point characteristic, while in the case of a distributed correlator telescope, data routing is required. Which type of data handling is preferred for a distributed radio telescope strongly depends on the state of the available communication technologies during the time the telescope is designed and constructed.

2. Present day network technology for distributed radio telescopes

For the transportation of radio telescope data several types of networking technology are available, of which SONET/SDH (Bellcore, 1999) and Ethernet (IEEE SA, 2002) are the most important. The SONET/SDH and Ethernet network technology types differ strongly. The technology for SONET/SDH is specially designed for telecommunication networks, in which the various nodes should be capable of exchanging information synchronously. A large part of the technology in SONET/SDH networks is used for obtaining a very high availability level. This percentage of time a SONET/SDH network is available for data transport is in the order of 99.9999%. The SONET/SDH network topology consists of interconnected rings with bidirectional traffic that form a meshed network. This ring topology in combination with fast re-routing technology strongly reduces the risk that, e.g., a fiber-cut blocks the information exchange between two or more nodes. To establish connections between the different rings, and for connecting the subscribers to these rings, routing equipments like cross connects and add-drop nodes are used. The synchronization, high availability and routing features in SONET/SDH networks require relatively complex technology and fiber topologies, resulting in relatively high deployment costs. The SONET/SDH technology is well suited for transportation of large amounts of information (up to Tb/s) over long distances (>1000 km). The communication equipment in SONET/SDH is currently assembled with the use of "building blocks" having data rates up to $10\,\mathrm{Gb\,s^{-1}}$, which will be extended towards $40\,\mathrm{Gb\,s^{-1}}$ in the coming years.

In the past years the use of Ethernet in wide area networks (WAN's) has increased strongly. Unlike SONET/SDH, optical Ethernet is specially designed for data communication in which a synchronous data transport is not required. Several network complexity levels are available for the current Ethernet communication networks. The simplest (standard) version consists solely of point-to-point connections for which relatively simple technology is sufficient. In more complex optical Ethernet systems with a large number of subscribers the information exchange between the various nodes is performed with the use of routers.

In standard Ethernet only very basic availability functionality is present. An increased availability level is obtained by employing Resilient Packet Ring (RPR) technology (RPR Working Group, 2005) in combination with an optical Ethernet ring network. In almost all cases WAN Ethernet technology is less complex and less costly than SONET/SDH equipment.

Which type of technology is to be applied in a distributed radio telescope depends on the requirements for the telescope and the way it is constructed. In the LOFAR telescope, which is used as an example in this paper, optical Ethernet is employed. Ethernet is chosen because of its low cost, easy to handle, properties. The time synchronization of the data is obtained by time-tagging the data packets that are sent from the antenna stations to the processor. The routing equipment that is available for optical Ethernet is not needed in the LOFAR-WAN since only point-to-point connections from the antenna locations towards the data processor are employed.

The availability of point-to-point connections in a distributed telescope WAN can be improved by creating fiber rings in the network topology in combination with RPR technology. Since the construction of these rings requires the deployment of additional fiber tracks, the costs of improving the availability level of the telescope WAN are high, especially for long haul point-to-point connections. Since the very high availability levels of telecommunication networks are not required for distributed radio telescopes, it is likely that the use of relatively simple availability-improving technologies is sufficient for a telescope WAN.

3. Optical Ethernet technology

The current mainstream optical Ethernet core technology is $1\,\text{Gb}\,\text{s}^{-1}$ (OC-24) Ethernet (1 GbE), which employs a transmitter/receiver combination at both sides of a fiber pair. The transmission distance of the 1 GbE link depends on the type of laser/transmitter and the type of fiber that are used. For COTS Ethernet technology several transmitter/receiver component multi-source agreements (MSA's) exist for which a number of transmission distances are commercially available. Of all Ethernet tranceiver MSA's the SFP(1 GbE) (van Doorn, 2005) and XFP(10 GbE) (XFP MSA Group, 2004) MSA's are the latest and most advanced. In this paper SFP/XFP based technology is addressed. For short-range distances (<300 m) low cost source lasers (e.g., VCSEL's) and multimode fibers are used. The longest distance 1 GbE links (~100 km) employ high power DFB source lasers and sensitive receivers in combination with single mode fiber.

Where data rates larger than $1\,\text{Gb}\,\text{s}^{-1}$ are required along a link, a number of options are available. The most straightforward way is to combine a number of parallel fibers along the track, which is denoted as space division multiplexing (SDM). A more sophisticated way of increasing the data rate is to raise the number of wavelength channels in the fiber. By assigning a 1 GbE data stream to each wavelength (wavelength division multiplexing or WDM), a single fiber can be used for the transportation of a multiple amount of 1 GbE channels. Another single fiber technique is time division multiplexing (TDM), which uses the aggregation of a number of low data rate streams into a single high data rate stream. For present day

Ethernet this means that a number of (electrical) 1 GbE streams are combined into a single (optical) $10\,\text{Gb\,s}^{-1}$ (OC-192) Ethernet stream (10 GbE) with the use of an aggregation switch.

3.1. PRESENT DAY WDM ETHERNET

For 1 GbE small WDM systems are currently available that use a relatively large wavelength channel separation: coarse WDM. In these CWDM systems a maximum of eight wavelength channels can be employed with a maximum transmission distance of about eighty kilometers. The price of an eight-channel 80 km CWDM module is about \$28000 (US dollars are used throughout this paper).

3.2. PRESENT DAY 10 GbE TECHNOLOGY

$10\,\text{Gb\,s}^{-1}$ Ethernet equipment is currently available with a (relatively short) maximum transmission distance of 10 km. This equipment is relatively costly: \$5000 for an optical module in combination with \$9000 for the accompanying aggregation switch.

For extending the 1 GbE and 10 GbE transmission distances, existing optical amplification equipment like EDFA's (erbium doped fiber amplifier) or SOA's (semiconductor optical amplifier) can be used. However, since standardization is not available for optical amplification in Ethernet links, system research is necessary when this type of technology is to be applied. The only way to increase the transmission distance within the current Ethernet standards is to combine a commercially available Ethernet receiver and transmitter and to use it as a regenerator. This regeneration solution is, especially in WDM and 10 GbE systems, a relatively costly way of extending the transmission distance.

4. Multiplexing technologies in Ethernet links: A cost comparison

By applying multiplexing technologies like TDM and (C)WDM big bundles of fibers can be replaced by just a few fiber pairs, in which way a reduction of the deployment costs can be obtained. Since TDM and WDM equipment is costly, the use of multiplexing technologies will not always lead to a cost reduction. To determine which of the three technology options (SDM, WDM or TDM) will provide the lowest costs for a specific communication link, a model has been developed which is used to predict the deployment costs of a communication link or network. Apart from the costs of the optical communication equipment, the fiber cable and the trenching (existing runs are not used), this model also takes into account the costs of all other in-door equipment like patch cords and distribution frames. Also, the cost consequences of the required engineering, fiber testing, licensing and splicing are inserted in the model. The equipment pricing that is used in this

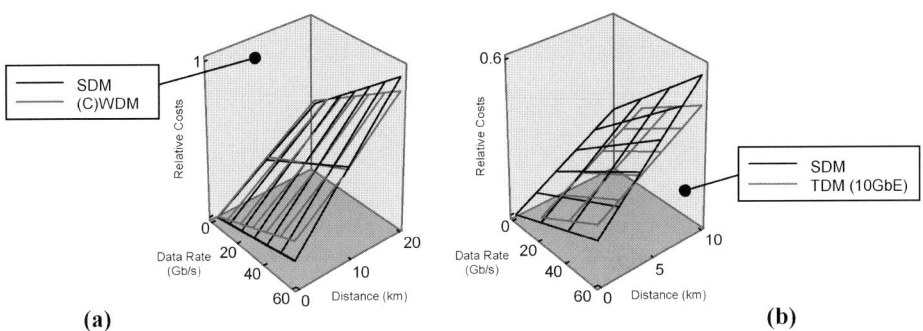

Figure 1. Deployment costs comparison; (a) SDM vs. (C)WDM; (b): SDM vs. TDM (10 GbE), both as functions of link data rate and distance.

tool is determined with the use of generally available information (internet) or via direct requests for list prices. All labor-involved costs only apply to the ASTRON location in the Netherlands and may be different at other places around the world. The pricing that is used in the model was updated up to the summer of 2004. Since the costs of Ethernet equipment changes rapidly, the situation after 2004 may differ from what is presented in this paper.

In Figure 1 the calculated relative deployment costs of a point-to-point link are depicted for the SDM, (C)WDM or TDM (10 GbE) technologies as functions of the data rate and the transmission distance. Figure 1a shows that for transmission distances up to about 10 km the application of SDM is less costly, while for larger distances and data rates exceeding $10\,\mathrm{Gb\,s^{-1}}$ the most attractive solution is provided by (C)WDM technology. A comparable situation can be seen on the right side in Figure 1b which shows that TDM (10 GbE) becomes economically attractive for distances larger than 3 km.

5. Costs of the data transport network in distributed radio telescopes

The costs related to an increase of the antenna station data rate or an increase of the telescope extent can also be determined with the use of the earlier mentioned cost model. In these calculations a distributed radio telescope topology is required. In the calculations presented in this paper a generalized topology of the LOFAR telescope is used (Bregman et al., 2002), in which the following WAN features can be distinguished:

- Antenna stations are distributed along a number of curved arms along which the stations are positioned, using an exponentially increasing distance from the center (0.4 km, 1.0 km, 1.8 km, 3 km, . . .).
- All data is processed at the center of the telescope, resulting in a unidirectional data transport along the arms towards the center.

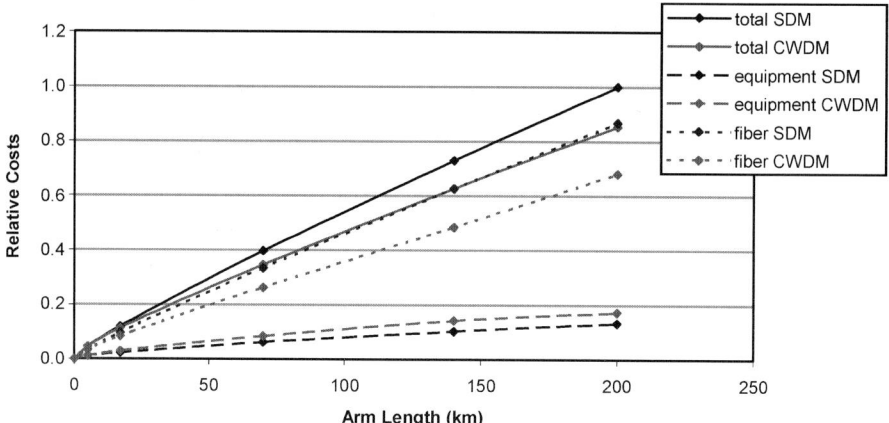

Figure 2. The calculated relative deployment costs of the data transport network along a LOFAR arm as a function of the length of the arm for both 1 GbE SDM and 1 GbE CWDM. In this calculation a station data rate of 10 Gb s^{-1} is used. For both SDM and CWDM the communication equipment costs and the costs for the deployment of the fiber are given.

- The data will be transported with point-to-point connections: ring and/or mesh topologies are not used such that costly routing is avoided.

In the calculations, the antenna stations are used as link nodes at which signal regeneration or multiplexing can take place. CWDM or TDM techniques are only applied in between the various nodes in case a cost reduction with respect to SDM is obtained.

In Figure 2 the calculation results for SDM and CWDM are given. Results for TDM (10 GbE) are omitted since these results differ only slightly from the SDM results. Figure 2 shows that for a station data rate of 10 GbE most of the WAN costs are related to the deployment of the fibers. For arm lengths exceeding 50 km the fiber related costs depend approximately linearly on the length of the arm. The increased costs per kilometer at arm lengths smaller than ∼10 km is caused by the strongly increasing amount of stations per kilometer towards the data processor, which results in increased costs, e.g., fiber management equipment and splicing.

The equipment costs per kilometer arm length get smaller for increasing arm lengths. This cost reduction per kilometer is caused by the more efficient use of communication equipment in the longer tracks, which holds for both the transmitter/receiver and for the WDM equipment. This latter observation is supported by the calculation results depicted in Figure 2: at an arm length of 17 km the CWDM/SDM (total) cost ratio is 0.95 while at 200 km this ratio is reduced to 0.85. For longer arm lengths this deployment cost ratio remains approximately the same.

In the evaluated WAN topology the use of CWDM instead of SDM is only favorable when long arms lengths are used (>10 km). Assuming a linear cost vs. arm length relation, a length doubling from, e.g., 200 km to 400 km results for SDM

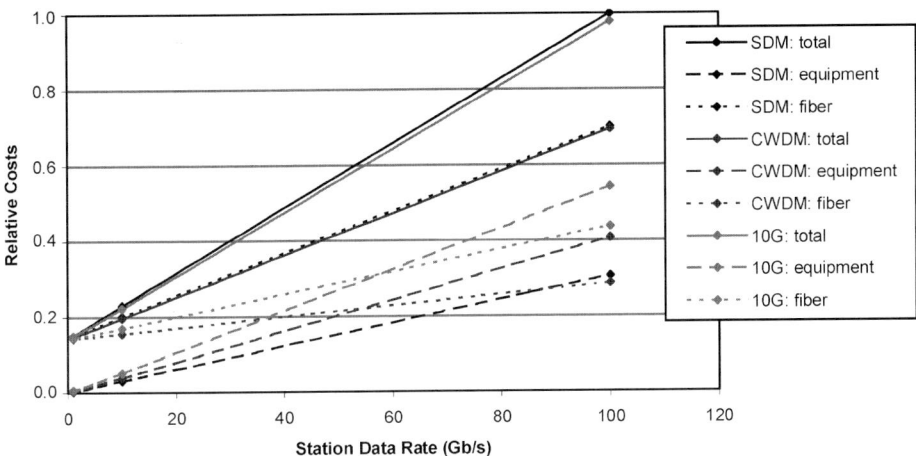

Figure 3. The calculated relative WAN deployment costs as a function of the station data rate using an arm length of 200 km. The total costs, the communication equipment costs and the costs for the deployment of the fiber are given for SDM, CWDM and TDM (10 GbE).

in (almost) a cost doubling (1.9), while for CWDM this length increase results in a cost increase by a factor of 1.6.

The amount of data that needs to be transported from each station to the data processor strongly influences the WAN deployment costs. Since relative costly trench digging is not involved when data rates are considered, the cost impact of a station data rate increase is not as big as the impact of an arm lengthening. In Figure 3 the calculation results of the WAN deployment costs vs. station data rate are depicted. It shows that the deployment costs grow linear with a station data rate increase. The relative short transmission distance (<10 km) of the current 10 GbE equipment makes 10 GbE unsuited for long range links and necessitates closely spaced regenerators. As a result the use of 10 GbE does not provide a cost reduction compared to the use of SDM. In case CWDM equipment is deployed substantial cost savings are obtained. The larger the data rate the more profitable the use of CWDM becomes: for a 10 Gb s^{-1} station data rate the CWDM/SDM (total) cost ratio is 0.85 while for a station data rate of 100 Gb s^{-1} this ratio has reduced to 0.69.

6. Outlook

In the coming years further optical Ethernet technology development will take place. For 10 GbE communication it is likely that long distance (∼80 km) optical communication equipment will become available. In addition, new Ethernet related WDM technology will become available. The goal in this development will be to introduce Optical Ethernet WDM systems with more wavelength channels (∼100) than in the current CWDM (~8) systems. The combination of the new WDM and 10 GbE technology leads to a strong improvement of the current Ethernet multiplexing techniques, resulting in further deployment cost reductions for high bit rate communication networks.

7. Conclusions

In this paper the deployment costs of a distributed radio telescope WAN have been investigated. In this investigation, a generalized LOFAR-WAN topology was used as an example data transport network. Optical Ethernet was determined to be the most suitable data transport technology for this network. Point-to-point Ethernet link deployment cost calculations show that the application of multiplexing techniques in high bit rate ($>10\,\text{Gb s}^{-1}$), long distance ($>3\,\text{km}$) communication links will lead to a reduction of the deployment costs for this link. Most of the network related costs for expanding the baseline of a distributed radio telescope concern the deployment of fibers. Deployment cost calculations for the generalized LOFAR-WAN topology show that the WAN expansion costs can be reduced by about 15% by applying CWDM equipment. The costs related to an increase of the station data rate can also be reduced by using CWDM technology: the larger the data rate the bigger the cost reduction, ranging from 15% for a station data rate of $10\,\text{Gb s}^{-1}$ to 31% for $100\,\text{Gb s}^{-1}$ per station. The application of currently available 10 GbE technology does not provide big cost savings due to its relative short transmission distance. In the coming years improvements are expected in this area.

References

ASTRON: 2005, http://www.lofar.org
Bellcore: 1999, *GR-253-CORE SONET Transport Systems: Common Generic Criteria.*
Bregman, J. D., Kant, G. W. and Ou, H.: 2002, Multi-terabit routing in the LOFAR signal and data transport networks, *Proceedings XXIV URSI GA.*
IEEE SA: 2002, http://standards.ieee.org/getieee802/802.3.html
RPR Working Group: 2005, http://www.ieee802.org/17/
van Doorn, S.: 2005, http://www.schelto.com/SFP/
XFP MSA Group: 2004, http://www.xfpmsa.org/

ALMA AND E-MERLIN DATA TRANSMISSION SYSTEMS: LESSONS FOR SKA

RALPH SPENCER*, ROSHENE MCCOOL, BRYAN ANDERSON,
DAVE BROWN and MIKE BENTLEY

Jodrell Bank Observatory, University of Manchester, Macclesfield SK11 9DL, U.K.
(*author for correspondence, e-mail: res@jb.man.ac.uk)*

(Received 16 July 2004; accepted 4 October 2004)

Abstract. ALMA, EVLA and e-MERLIN use high data rate optical fibre links based on commercially available 10 Gbps opto-electronics. This paper describes the systems designed by NRAO and JBO staff to be used in ALMA and e-MERLIN. ALMA has a requirement for a 120 Gbps data rate per telescope, requiring the use of 12 lasers in the 1550 nm telecommunication band, with maximum link lengths around 20 km. e-MERLIN has a lower bandwidth and requires 30 Gbps links per telescope and therefore three lasers per telescope; however, the link lengths reach up to around 400 km, and amplification, de-dispersion and regeneration are required. Dense wavelength division multiplexing is used to avoid fibre management problems and save fibre costs. The design criteria and experience gained in these projects is very relevant to SKA as the proposed configuration of the antenna elements maps well to the link lengths used here. The chosen data rate will be a major cost driver.

Keywords: data transmission, fibre-optics, radio astronomy, radio telescopes

1. Introduction

ALMA, the EVLA (Perley and Napier, 2004; Durand and McCool, 2004) and e-MERLIN (Garrington, 2004) have very similar IF data-transfer requirements with the exception that e-MERLIN has only a quarter of the bandwidth of the other two. A common design to satisfy all three instruments has been developed as a joint effort between the National Radio Astronomy Observatory in the USA and Jodrell Bank Observatory in the UK. NRAO is responsible for the data protocol (Freund, 2003) and the electronic formatting and deformatting elements. Jodrell Bank Observatory has been responsible for the optical-fibre elements.

The optimum technology for fibre-optic links depends very much on the lengths of link paths and the bandwidths to be carried. In the case of ALMA, EVLA and e-MERLIN, the path lengths are generally too long and the bit rates are too high to permit the use of inexpensive multimode fibre and directly-modulated lasers. Single-mode fibres and externally-modulated lasers have to be used. Dense wavelength division multiplexing (DWDM) techniques are also used to permit a single optical fibre to carry all the astronomical data. Analogue transmission methods were considered but were discarded largely on the grounds of stability, bandwidth and dynamic range as well as the availability of COTS digital devices. The range of path lengths in SKA however may require a combination of various techniques to give an optimal solution.

2. The ALMA links

ALMA, the Atacama Large Millimetre Array, is being built on the Chajnantor plateau in Chile. The Atacama high desert at 5000 m altitude is one of the driest regions on Earth, and so is ideal for millimetre-wave astronomy. The array is planned to consist of 64 high performance 12-m diameter parabolic antennas designed to work at frequencies up to 850 GHz, with maximum baselines of around 10 km. Receivers operating in cryogenic dewars cooled to around 4 K and covering bands from 30 GHz up to 850 GHz will produce four IF signals each 2 GHz wide in two polarisations, giving a total bandwidth of 16 GHz. Roads, power and fibre connections are to be made to each antenna with the fibres brought back to a central location housing the correlator.

The IF data transmission system (DTS) transfers sampled and digitised astronomical data from the telescopes of ALMA to the correlator. The eight 2-GHz-wide IF bands are sampled and digitised at 4 GHz rates with 3 bits of precision. Four formatter boards each demultiplex, format and multiplex two 12 Gbps input streams into three output data streams at 10 Gbps each. The large framing (30/24) overhead stemmed from an earlier design which incorporated 8–10 bit encoding to eliminate long runs of 1's or 0's and kept due to the availability of COTS components for digital links. The 12 output streams at 10 Gbps modulate the optical outputs of twelve laser diodes operating at distinct wavelengths near 1540 nm and the resulting, modulated optical signals are multiplexed onto a single optical fibre. At the correlator, the process is reversed in 12 optical receivers and four deformatter boards which each deliver a pair of 96-bit words of 32 packed samples to the correlator at a rate of 125 Mwps. The total, astronomical data rate from each telescope is 96 Gbps and the net rate on an optical fibre carrying digitised astronomical data to the correlator is 120 Gbps.

Figure 1 is a schematic diagram of the optical system for a single telescope-to-correlator link. The locations of the optical components are shown. There are optical fibres from more than 216 antenna pads, pedestals on which to mount antennas, to the patch panel. An operator will manually connect occupied antenna pads through the patch panel to one of the 64 correlator inputs. More pads could be added to cater for an added compact array. Figure 1 shows an erbium doped fibre amplifier (EDFA) and an integrated optical demultiplexer module though there is some question of whether any optical amplification is really necessary. The maximum fibre length is assumed to be 20 km. The array can be reconfigured slowly by moving an antenna and altering the patch arrangement. It is not envisaged that antennas will be moved at a rate significantly greater than one per day. Optical attenuators will be added

Fibre Optic Links for the ALMA IF DTS

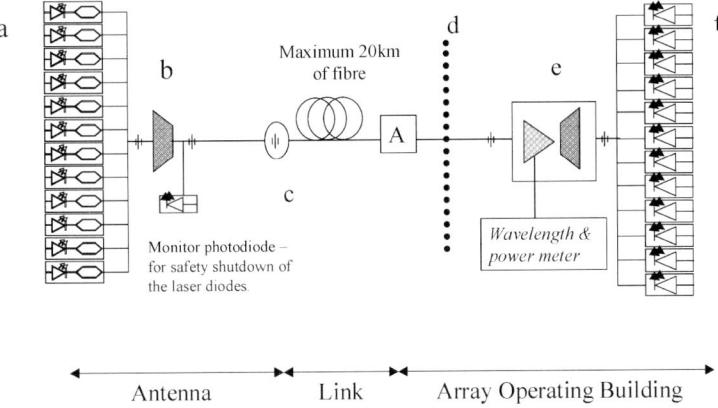

Figure 1. ALMA DTS optical sub system, a: electro-absorption modulated laser diodes, b: optical multiplexer and monitor, c: all weather connector to fibre network, d: patch panel, e: EDFA and de-multiplexer, f: photodiode receivers, A: optical attenuator.

to the shorter fibre segments to equalise roughly the power levels into the optical receivers.

The basic requirements for a link are that the overall bit-error rate should not exceed 10^{-6} at the end of life having started with a margin of 6 dB more optical power than the minimum necessary and the components used should be commercial off-the-shelf devices from multiple, independent sources of supply. The link design takes into account signal-to-noise ratios, dispersion and non-linear effects in the fibres which tend to increase the error rates. Consideration of power dissipation and cooling were major factors in the design given the reduced air density and pressure at the altitude of operation. The permitted error rates are high by normal telecommunication standards but are acceptable in radio astronomical applications where results are obtained by averaging billions of samples.

The laser transmitters have integrated electro-absorption modulators. In the current design, a controller card allows laser current, laser-diode temperature, modulator bias and optical power levels to be monitored and controlled remotely. An alternative solution of using the transmitter half of a more highly integrated, commercial transponder unit is being considered, as this would reduce the number of assemblies and avoid carrying 10 Gbps signals on coaxial cables.

Each DTX transmitter module has a pair of sampler-digitisers, a demultiplexer, a formatter, multiplexers, three laser transmitter modules, associated power supplies and controllers. They are enclosed in metal boxes with honeycomb-filter ports which allow cooling air to be blown through the module. There are four such modules per antenna. Since they are mounted in the vertex cabin of the telescope, close to the antenna feeds, special attention has been paid by NRAO in controlling unwanted electromagnetic leakage from the internal components.

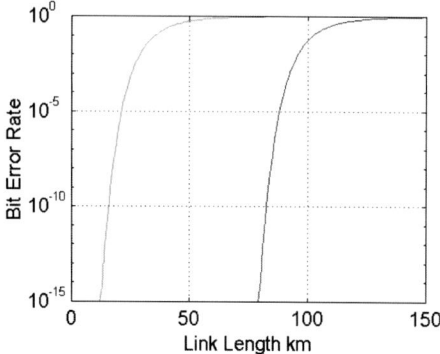

Figure 2. Theoretical bit-error rate allowing for a 6 dB start-of-life margin, link losses and dispersion. The right-hand curve is for a link including an EDFA with gain of 20 dB.

The link power budget includes allowances for link loss, a 6 dB start-of-life margin and the effects of dispersion, polarisation-dependent loss, transmitter modulator extinction ratio, cross talk between optical channels and other non-linear effects. Factoring in noise at the receiver (thermal, shot noise and spontaneous beat noise from the EDFA) allows the bit-error rate to be predicted (Agrawal, 1997). Figure 2 shows the expected error rate as function of link length with and without an optical amplifier. Without optical amplifiers and with links lengths up to 20 km, the predicted error rates only just meet the specification showing that the use of amplifiers may be desirable. Typical power levels are 0 dBm at the output of the laser/modulators, −12 dBm at the input to the EDFA, with a minimum of −16 dBm start of life input to the photodiode receivers. If the link losses are kept under control and the component losses are better than worse case specification then it will be possible to use avalanche photodiodes in the receivers and remove the EDFAs. However this situation will not be known until after the fibre cables have been installed on site.

The cost for the opto-electronics (i.e. excluding the digital formatters and deformatters and fibre connections) at current prices (January 2004) is around £41000 per 120 Gbps link. The bulk of the expenditure is in the lasers.

3. e-MERLIN

A significant upgrade to the MERLIN Array operated at Jodrell Bank Observatory is underway. The telescope feeds, polarisers and receivers are being replaced by broadband alternatives and the signals from the telescopes will be digitised near the front ends and carried on optical fibres to a new correlator (McCool et al., 2002).

The existing narrow-band microwave links will be discarded. In conjunction with the upgraded surface on the 76-m Lovell telescope, an improvement in sensitivity of a factor of ∼30 is expected.

The distances from the most distant telescopes in MERLIN to the correlator at Jodrell Bank are over 100 km and the terrain between the telescopes is densely populated. Laying a private fibre network is not a viable option. Fortunately, it has been possible to negotiate a deal giving access to existing dark fibre through a leasing agreement with Fujitsu, Global Crossing and Hutchinson, and to lay new fibre for the connections between the telescopes and Jodrell Bank to the network. Estimated costs of fibre dig in the UK are around £25 per metre in rural areas rising to several times that in cities. About 100 km of new dig is required, mostly in the soft verges of minor roads. Most of the existing commercial fibre to be used in e-MERLIN is laid alongside railway tracks. The total length of fibre in the system is around 600 km. The initial plan was to use only a single fibre (with no redundancy) on cost grounds, however most telescopes (with the exception of Cambridge) will be equipped with two fibres.

e-MERLIN will have (at least in its initial phase) receivers for L band (1.2–1.8 GHz), C band (4.5–6 GHz) and K band (21–23 GHz). The IF bandwidth will be 2 GHz per polarisation, with two polarisations and 3 bit digitisation, the same as for a single IF in ALMA and the EVLA. In addition, as in the EVLA, a lower bandwidth (500 MHz in this case) polarisation pair can be digitised to 8 bits allowing better interference immunity in the severely RFI-plagued L-band. We are using the same, common design of digitiser and fibre optic transmitter and optical receiver as ALMA and the EVLA to save design effort. The resulting digital signals at 3×10 Gbps appear on three optical carriers which can be wavelength-division-multiplexed onto one fibre for each telescope. The fibre optic system is outlined in Figure 3.

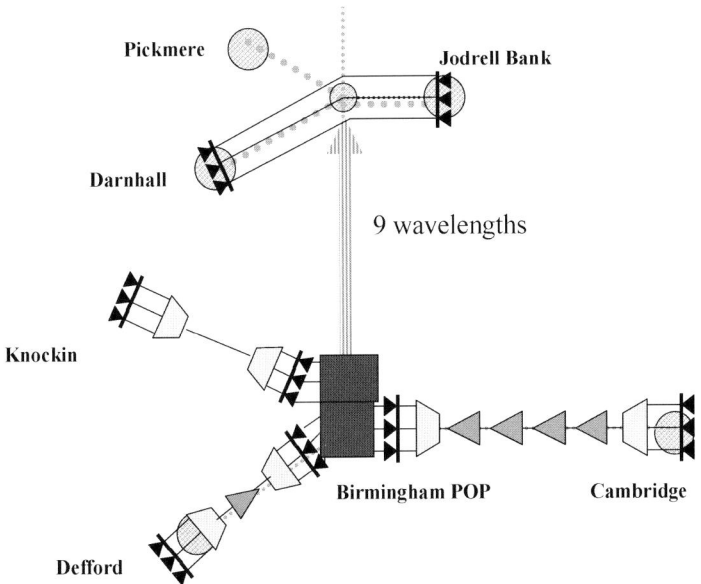

Figure 3. Fibre optic links in e-MERLIN.

Wavelength division multiplexing is used, except over the short links where it is not cost effective. The distances involved are such that optical amplification is needed as previously shown in Figure 2.

The typical distance limit between amplifiers is around 80 km, with the amplifiers sited at intermediate co-location sites. The installed fibre is standard single mode, with a dispersion coefficient of 17 ps nm^{-1} km^{-1} and so de-dispersion, using dispersion compensating fibre situated with the amplifiers, is used. Digital regeneration is also required at a point of presence (POP) site in Birmingham.

The fibre cable is being laid at the time of writing (June 2004), with installation to be completed by early 2005. The instrument is expected to be operational in 2007.

4. SKA and fibre links

So what can the design of these links tell us about SKA? The problem here is in predicting what the optimal technology will be in 2015. All we can do is to estimate what could be done based on current developments.

Data transmission rates for SKA have been estimated by Veidt and Dougherty (2003). Using their values for the array parameters and assuming a one square degree field-of-view, the data rate for two polarisations with 3 bit digitisation and 4 GHz bandwidth (i.e. the parameters for operation at frequencies of more than 11 GHz) the data rate per station is 54 Tbps for the small-N array and approximately 30% less for the large-N array. The large-N array assumes 13 antenna elements per station. The data rates become roughly 60 Tbps for the small-N array and 40 Tbps for the large N array if we allow for formatting, headers, check-sums etc.

The configuration specification for SKA deduced from science requirements (Jones 2004) suggests that the antennas should be distributed such that 20% of the collecting area is within 1 km, 50% within 5 km and 75% within 150 km, with a maximum baseline of at least 3000 km.

Interestingly enough, these distances roughly correspond to points where data transmission techniques change. Local area networks currently make use of multimode fibre interconnections for distances up to about 1 km. Inexpensive systems exist using Gigabit Ethernet, with 10 Gb/s capable devices beginning to appear.

An interesting development is the use of vertical cavity laser devices (VCSELs) that can be directly modulated and manufactured in arrays. This may offer a solution to the high data rate problem in the Canadian Large Adaptive Reflector design where the focal plane structure could have 2500 receivers each requiring Gbps data transmission from the aerostat down to the ground. VCSEL arrays containing 64 devices are available, and modulation rates of 10 Gb/s have been used for short (300-m) links (Michalzik et al., 2001). Quantum dot devices may also offer a cost-saving solution, and 10 Gbps operation has been shown to be possible (Markus and Fiore, 2004). There is every indication that high data rates and inexpensive systems should be available for the SKA, which will make heavy use of computer network technology for these short intra-station distances. SKA data rates are high compared with current practice, but use of the techniques available in future local area networks is attractive.

For path lengths greater than about 1 km and high data rates, it is necessary to use single-mode fibre in which the diameter of the fibre core is such that only one waveguide mode can propagate. It is still possible to directly modulate the laser but the result is frequency chirp which limits the bandwidth and maximum range of devices to sub-10 Gbps and sub-30 km. Improvements in the range and bandwidth of these devices are continually being made and so it may be possible to do better in future. However, multiple fibres or DWDM techniques with wavelength-stable, externally modulated devices will probably be required to reach Tbps rates.

Links beyond a few tens km currently require externally-modulated lasers and single mode fibre. As we can see from the cases above, spans of more than about 20 km using WDM require optical amplification. The amplification can be by EDFAs or by Raman amplifiers or a combination of the two. Currently, the costs of amplification are only a fraction of the total costs in DWDM systems and that is not likely to change in the future.

Currently, the sweet spot for communications links is at 40 Gbps per wavelength, though 80 and 160 Gbps systems have been tested. It has been demonstrated that 10 Tbps links are possible on a single fibre (e.g. Binh, 2002) if novel modulation techniques and DWDM are used. At the moment the device costs are such that these links are much more expensive than multiple 10 Gbps links using DWDM. We estimate the current, opto-electronics costs of a 1-Tbps link over 150 km using 10 Gbps devices to be around $ 0.5M. Thus the links in the inner 150 km for SKA would cost $0.75 \times 32 \times 0.5 \times 60 = \$720M$, assuming 32 stations in the small-N design, and $0.75 \times (8900/13) \times 0.5 \times 40 = \$10280M$ for the large-N design, using the Veidt-and-Dougherty parameters where there are 13 elements per station and 8900 dishes. No doubt, the costs of high-data-rate devices will have decreased significantly by 2015. The cost of 10 Gbps devices has decreased by a factor of around five in the last four years but is now levelling off. However, it is clear that, at the moment, the costs of links for the large-N design are astronomically high. Note that the costs given here exclude the costs of the associated digital electronics.

The costs of laying fibre over large distances are such that it is much cheaper to lease or buy existing, under-utilised resources if possible. This could be via dark fibre as in e-MERLIN or as managed bandwidth. Managed bandwidth costs are subject to commercial pressure and could be high. The necessary use of routers and switches in a telecommunication system also adds unnecessary expense. On the other hand, the need for amplification, dispersion compensation and the use of and payment for co-location facilities make the e-MERLIN dark-fibre solution unattractive even before the issues of maintenance are considered. Recent developments in wavelength switching may however be helpful in that it is easier to manage other forms of traffic. For example a research network (UKLight) has been set up in the UK to test the protocols for and the performance of such a system, though it emulates wavelength switching using conventional routers. The test network has a data rate of 10 Gbps. The expectation is that all-optical networks will be installed in the next few years and these should have the capacity that SKA requires.

5. Conclusion

The many advantages of fibre-optic communication systems have made their use ubiquitous in new radio astronomy instruments. The three major instruments currently in construction, ALMA e-MERLIN and EVLA would not be possible without the wide bandwidth capability of optical fibres, and it is clear that SKA will need to make extensive use of fibre technology. It is also clear that the opto-electronic costs could be prohibitive unless new technology reduces the prices for multi-parallel devices considerably. A more detailed study of data rates within SKA is needed to optimise costs. However one should note that the electronics industry expects major advances to come out of the transition from the continuum to the quantum physical level by further miniaturisation. This argues for leaving the opto-electronics design as late as possible. We acknowledge the huge contributions that staff at NRAO have made to these studies.

References

Agrawal, G. P.: 2002, *Fibre Optic Communication Systems*, 2nd Edn., Wiley, New York.
Binh, L. N.: 2002, *Proc. 27th URSI Gen. Assem.*, p. 0585.
Garrington, S. T.: 2004, 'Astronomical Telescopes and Instrumentation', *SPIE Paper 5489-22*, Glasgow.
Freund, R., VLA Memo Number 420, http://www.alma.nrao.edu/memos/html-memos/alma420/memo420.pdf.
Durand, S. and McCool, R.: 2004, 'Astronomical Telescopes and Instrumentation', *SPIE Paper 5496-64*, Glasgow.
Jones, D. L., SKA Memo 45, available at http://www.skatelescope.org/pages/p_docsand pres.htm.
McCool, R., Anderson, B., Spencer, R., and Garrington, S.: 2002, Proc. 27th URSI Gen. Assem., p. 1138.
Markus, A. and Fiore, A.: 2004, *Phys. Status Solidi A* **201**, 338.
Michalzik, R., et al.: 2001, *IEICE Trans. Electron.* **E84-C**, 629.
Perley, R. A. and Napier, P. N.: 2004, 'Astronomical Telescopes and Instrumentation', *SPIE Paper 5489-64*, Glasgow.
Veidt, B. and Dougherty, S., SKA Memo 42, available at http://www.skatelescope.org/pages/p_docsand pres.htm.

SIGNAL PROCESSING

ARRAY SIGNAL PROCESSING FOR RADIO ASTRONOMY

ALLE-JAN VAN DER VEEN,[1] AMIR LESHEM[2] and ALBERT-JAN BOONSTRA[3,*]
[1]*Department Electrical Engineering/DIMES, Delft University of Technology, Mekelweg 4, 2628 CD Delft, The Netherlands;* [2]*Department Electrical Engineering, Bar-Ilan University, Tel-Aviv, Israel;* [3]*ASTRON, R&D Department, P.O. Box 2, 7990 AA Dwingeloo, The Netherlands*
(*author for correspondence, e-mail: boonstra@astron.nl)

(Received 14 July 2004; accepted 17 January 2005)

Abstract. Radio astronomy forms an interesting application area for array signal processing techniques. Current synthesis imaging telescopes consist of a small number of identical dishes, which track a fixed patch in the sky and produce estimates of the time-varying spatial covariance matrix. The observations sometimes are distorted by interference, e.g., from radio, TV, radar or satellite transmissions. We describe some of the tools that array signal processing offers to filter out the interference, based on eigenvalue decompositions and factor analysis, which is a more general technique applicable to partially calibrated arrays. We consider detection of interference, spatial filtering techniques using projections, and discuss how a reference antenna pointed at the interferer can improve the performance. We also consider image formation and its relation to beamforming.

Keywords: array signal processing, interference mitigation, radio astronomy

1. Introduction

The future of radio astronomical discoveries depends on achieving better spatial resolution and sensitivity while maintaining immunity to terrestrial interference which is rapidly growing. The last two demands are obviously contradicting as improved sensitivity implies receiving more interfering signals. RFI detection and removal is now an important topic in radio astronomy. A promising track here is to switch to massive phased array technology, where we will gain both in terms of resolution and sensitivity while increasing the flexibility to filter out interference. The international efforts in this direction are coordinated under the framework of the Square Kilometre Array (SKA) programme. The first example of a flexible massive phased array radio telescope is LOFAR (13,000 elements) which is currently under construction in The Netherlands.

The principle of interferometry has been used in radio astronomy since 1946 when Ryle and Vonberg constructed a radio interferometer using dipole antenna arrays (Ryle, 1952). In 1962 the principle of aperture synthesis using earth rotation was proposed (Ryle, 1962), and applied for example in the 5 km Cambridge radio telescope, the 3 km Westerbork Synthesis Radio Telescope (WSRT) in The Netherlands and the 36 km Very Large Array (VLA) in the USA.

In this paper, we present a signal processing data model (Section 2) and subsequently give an overview of several problems in radio astronomy where array signal processing can make a contribution, namely calibration using factor analysis (Section 3), detection of interference (Section 4), interference removal using spatial filtering (Section 5), and image formation (Section 6).

1.1. NOTATION

Superscript T denotes matrix transpose, H denotes complex conjugate transpose, vec(·) denotes the stacking of the columns of a matrix in a vector, and unvec(·) is the inverse operation of vec(·). The Kronecker product is denoted by $\otimes$, **I** is the identity matrix, and **1** is a vector with all ones.

2. Data model

2.1. RECEIVED DATA MODEL

Assume we have a telescope array with p elements. We consider the signals $x_i(t)$ received at the antennas $i = 1, \ldots, p$ in a sufficiently narrow subband. For the interference free case the array output vector $\mathbf{x}(t)$ is modeled in complex baseband form as

$$\mathbf{x}(t) = \mathbf{v}(t) + \mathbf{n}(t) \tag{1}$$

where $\mathbf{x}(t) = [x_1(t), \ldots, x_p(t)]^T$ is the $p \times 1$ vector of telescope signals at time t, $\mathbf{v}(t)$ is the received sky signal possibly due to many astronomical sources, assumed on the time scale of (order) 10 s to be a stationary Gaussian vector with covariance matrix $\mathbf{R}_v = \mathcal{E}\{\mathbf{v}(t)\mathbf{v}(t)^H\}$ (the astronomical 'visibilities'), and $\mathbf{n}(t)$ is the $p \times 1$ Gaussian noise vector with covariance matrix $\mathbf{D}$. We assume that the noise is Gaussian, and uncorrelated among the sensors, which means that $\mathbf{D}$ is diagonal. Usually identically distributed noise is assumed, for which $\mathbf{D} = \sigma^2 \mathbf{I}$, but this implies accurate calibration as discussed in Section 3.

Suppose there are q interfering sources, stationary only over short time intervals, with signals $s_i(t)$ for $i = 1 \ldots q$, and spatial signatures $\mathbf{a}_i$. Without loss of generality, we can absorb the unknown amplitude of $s_i(t)$ into $\mathbf{a}_i$ and thus set the power of $s_i(t)$ to 1. Let $\mathbf{A}$ be a $p \times q$ matrix where the q columns represent the q interferer spatial signature vectors $\mathbf{a}_i$, and let $\mathbf{s}(t)$ be a vector with the q signals $s_i(t)$. The output vector, extended with interference, is modeled as

$$\mathbf{x}(t) = \mathbf{v}(t) + \mathbf{A}(t)\mathbf{s}(t) + \mathbf{n}(t) \tag{2}$$

We assume that the processing bandwidth is sufficiently narrow, meaning that the maximal propagation delay of a signal across the telescope array is small compared to the inverse bandwidth, so that this delay can be represented by a phase shift of the signal. If the assumption is not satisfied, as for many existing telescopes, a form of subband processing has to be implemented.

2.2. COVARIANCE MODEL

Suppose that we have obtained observations $\mathbf{x}[m] := \mathbf{x}(mT_s)$, where T_s is the sampling period. We assume that $\mathbf{A}(t)$ is stationary at least over intervals of MT_s, and construct short-term covariance estimates $\hat{\mathbf{R}}_k$,

$$\hat{\mathbf{R}}_k = \frac{1}{M} \sum_{m=kM+1}^{(k+1)M} \mathbf{x}[m]\mathbf{x}[m]^H \tag{3}$$

where M is the number of samples per short-term average. Several filtering algorithms in this paper are based on applying operations to each $\hat{\mathbf{R}}_k$ to remove the interference, followed by further averaging over the resulting matrices to obtain a long-term average.

Consider the $\mathbf{A}_k := \mathbf{A}(kMT_s)$ as deterministic, and denote $\mathcal{E}\{\hat{\mathbf{R}}_k\}$ by $\mathbf{R}_k$. According to the assumptions, $\mathbf{R}_k$ has model

$$\mathbf{R}_k = \mathbf{\Psi} + \mathbf{A}_k \mathbf{A}_k^H = \mathbf{R}_v + \mathbf{D} + \mathbf{A}_k \mathbf{A}_k^H \tag{4}$$

where $\mathbf{\Psi}$ is the interference-free covariance matrix, $\mathbf{\Psi} = \mathbf{R}_v + \mathbf{D}$.

So far, the formalism considered only single polarization arrays. The models are easily extended to the polarization case. Let $\tilde{\mathbf{x}}(t) \equiv [x_{1x}(t), x_{1y}(t), \ldots, x_{px}(t), x_{py}(t)]^T$, where the subscript ix and iy for the ith telescope denote the two orthogonal polarizations. Then the $2p \times 2p$ polarization covariance matrix $\tilde{\mathbf{R}}$ is defined by $\tilde{\mathbf{R}} \equiv \mathcal{E}\{\tilde{\mathbf{x}}\tilde{\mathbf{x}}^T\}$. The resulting polarization data model is described in Hamaker (2000) and Boonstra and van der Veen (2003). Although the data model is straightforward, extending the non-polarization signal processing to polarization processing is complicated. In this overview paper we focus on single polarization signal processing.

3. Covariance matrix factorization

3.1. EIGENVALUE DECOMPOSITION

The internal structure of the covariance matrix $\mathbf{R}_k$ can be exploited for calibration purposes, for interference mitigation and imaging. Suppose that the noise covariance is equal for each sensor, $\mathbf{R}_n = \sigma_n^2 \mathbf{I}$, assume that the visibilities are much weaker than the noise powers, and assume that $q < p$. Then $\mathbf{R}_k$, dropping the index k, can be decomposed using an eigenvalue analysis as

$$\mathbf{R} = \mathbf{U}\mathbf{\Lambda}\mathbf{U}^H = [\mathbf{U}_s \ \mathbf{U}_n] \begin{bmatrix} \mathbf{\Lambda}_s + \sigma_n^2 \mathbf{I}_q & 0 \\ 0 & \sigma_n^2 \mathbf{I}_{p-q} \end{bmatrix} \begin{bmatrix} \mathbf{U}_s^H \\ \mathbf{U}_n^H \end{bmatrix} \quad (5)$$

where $\mathbf{U}_s$ is the interferer subspace. It is a $p \times q$ matrix containing the eigenvectors corresponding to the q eigenvalues in the $q \times q$ diagonal matrix $\mathbf{\Lambda}_s$. $\mathbf{U}_n$ is a $p \times (p-q)$ matrix containing the eigenvectors corresponding to the noise subspace. Note that the signal subspace and the noise subspace span the entire space, $\mathbf{U} = [\mathbf{U}_s \ \mathbf{U}_n]$. Note also that this technique only works for noise matrices $\mathbf{D}$ with identical diagonal entries. An more general technique is factor analysis which is described next.

3.2. FACTOR ANALYSIS DECOMPOSITION

Factor analysis is a statistical technique with origins in psychometrics and biometrics (Lawley and Maxwell, 1971; Mardia et al., 1979). It assumes a collection of data $\mathbf{X} = [\mathbf{x}(1), \ldots, \mathbf{x}(N)]$ with covariance

$$\mathbf{R} = \mathcal{E}\{\mathbf{x}(k)\mathbf{x}(k)^H\} = \mathbf{A}\mathbf{A}^H + \mathbf{D} \quad (6)$$

where $\mathbf{R}$: $p \times p$ Hermitian, $\mathbf{A}$: $p \times q$ and $\mathbf{D}$: $p \times p$ diagonal. The objective of factor analysis is, for given $\mathbf{R}$, to identify $\mathbf{A}$ and $\mathbf{D}$, as well as the factor dimension q. We can furthermore model $\mathbf{R}$ in terms of noise subspace $\mathbf{U}_n$ and signal subspace $\mathbf{U}_s$ (Leshem et al., 2000)

$$\mathbf{R} = \mathbf{U}\mathbf{\Lambda}_0\mathbf{U}^H + \mathbf{D} = [\mathbf{U}_s \ \mathbf{U}_n] \begin{bmatrix} \mathbf{\Lambda}_s & \\ & 0 \end{bmatrix} \begin{bmatrix} \mathbf{U}_s^H \\ \mathbf{U}_n^H \end{bmatrix} + \mathbf{D} \quad (7)$$

where $\mathbf{U} = [\mathbf{U}_s \ \mathbf{U}_n]$, and where $\mathbf{\Lambda}_s$ is the diagonal eigenvalue matrix containing the interferer powers. Here we assume $q \leq p$. Thus, the "factor analysis decomposition" (FAD) can be viewed as a generalization of the eigenvalue decomposition.

This decomposition is relevant in case the noise covariance is unknown but diagonal, $\mathbf{R}_n = \mathbf{D}$, which corresponds to the noise being uncorrelated among the sensors. In contrast, the usual eigenvalue decomposition for estimating $\mathbf{U}_s$ is only valid if the noise powers are equal among sensors ($\mathbf{R}_n = \sigma^2 \mathbf{I}$), which is generally true only after accurate calibration and noise whitening.

In general we can not estimate $\mathbf{A}$ uniquely, since $\mathbf{A}$ can be replaced by $\mathbf{A}\mathbf{V}$ for an arbitrary unitary matrix $\mathbf{V}$. If the eigenvalues are not repeated and we sort them in descending order, then $\mathbf{U}_s$ and $\mathbf{\Lambda}_s$ can be uniquely determined. There are other ways to constrain $\mathbf{A}$ to be a unique factor, e.g. by taking it to be a lower-triangular rectangular Cholesky factor with positive real diagonal entries.

3.3. ESTIMATION OF THE FAD

Assume that the factor rank q is known. Given $\hat{\mathbf{R}} = \frac{1}{N}\mathbf{X}\mathbf{X}^H$, and a sufficiently small q, we wish to estimate $\hat{\mathbf{A}}$ and $\hat{\mathbf{D}}$ such that $\hat{\mathbf{R}} \approx \hat{\mathbf{A}}\hat{\mathbf{A}}^H + \hat{\mathbf{D}}$. There are several approaches for this.

A maximum likelihood (ML) estimate of the factors $\mathbf{A}$: $p \times q$ and $\mathbf{D}$ is dependent on the choice of q. The largest permissible value of q is that for which the number of degrees of freedom $v = (p-q)^2 - p \geq 0$, or $q \leq p - \sqrt{p}$. For larger q, there is no identifiability of $\mathbf{A}$ and $\mathbf{D}$: any sample covariance matrix $\hat{\mathbf{R}}$ can be fitted. Even for smaller q, $\mathbf{A}$ can be identified only up to a $q \times q$ unitary transformation at the right, i.e., we can identify span($\mathbf{A}$). Luckily, this is sufficient for many applications.

For $q > 0$, there is no closed form solution to the estimation of the factors $\mathbf{A}$ and $\mathbf{D}$ in the ML estimation of $\hat{\mathbf{R}}_q = \hat{\mathbf{A}}\hat{\mathbf{A}}^H + \hat{\mathbf{D}}$. There are several approaches for obtaining an estimate.

A technique known as alternating least squares, is to alternatingly minimize the least squares cost function[1] $\|\hat{\mathbf{R}} - (\mathbf{A}\mathbf{A}^H + \mathbf{D})\|_F^2$ over $\mathbf{A}$ keeping $\mathbf{D}$ fixed, and over $\mathbf{D}$ keeping $\mathbf{A}$ fixed. This technique tend to converge very slowly but may be used for fine-tuning.

A fast converging technique is Gauss–Newton iterations on the original ML cost function, or on the (weighted) least squares cost. This requires an accurate starting point.

Ad-hoc techniques exist for solving the least squares problem, possibly followed by a Gauss-Newton iteration. These techniques try to modify the diagonal of $\hat{\mathbf{R}}$ such that the modified matrix is low-rank q, hence can be factored as $\mathbf{A}\mathbf{A}^H$. The case $q = 1$ was studied in more detail in Boonstra and van der Veen (2003).

3.4. RADIO ASTRONOMY APPLICATION EXAMPLES

We mention two applications of the decompositions described above.

The first example is signal subspace estimation in the presence of uncorrelated noise (Leshem and van der Veen, 2001), e.g., for the purpose of spatial filtering of interference. An example is shown in Figure 1. Here, the data model is $\mathbf{R} = \mathbf{A}\mathbf{A}^H + \mathbf{D} + \mathbf{R}_v$, where $\mathbf{A}$ corresponds to the interfering signals, $\mathbf{D}$ is the diagonal noise covariance matrix, and $\mathbf{R}_v \ll \mathbf{D}$ is the sky covariance. Using factor analysis, the number of interferers q is detected, and a basis $\mathbf{U}_A \sim \text{ran}(\mathbf{A})$ is estimated, as well as its orthogonal complement $\mathbf{U}_A^\perp$. Applying a projection matrix $\mathbf{P}_A^\perp = \mathbf{U}_A^\perp \mathbf{U}_A^{\perp H}$ to $\mathbf{R}$ will cancel the interference, as $\mathbf{P}_A^\perp \mathbf{A} = 0$:

$$\tilde{\mathbf{R}} = \mathbf{P}_A^\perp \hat{\mathbf{R}} \mathbf{P}_A^\perp \tag{8}$$

---

[1] A cost function describes the error associated with the difference between a data model and its observation sample estimate.

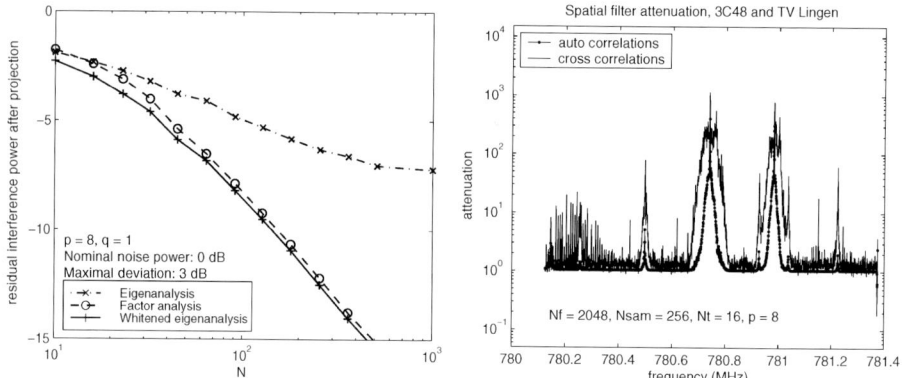

Figure 1. The left figure shows the residual interference power after projections in a simulation: $\mathbf{U}_A$ is estimated from (*i*) eigenvalue decomposition, (*ii*) factor analysis, and (*iii*) eigendecomposition after whitening by $\mathbf{D}^{-1/2}$, assuming true $\mathbf{D}$ is known. The right figure shows spatial projection filter attenuation results of television sound carrier waves observed at the WSRT. The projection filter was applied after whitening by $\mathbf{D}^{-\frac{1}{2}}$; the diagonal noise term $\mathbf{D}$ was estimated by using factor analysis.

The left figure shows $\|\tilde{\mathbf{R}} - \mathbf{P}_A^\perp (\mathbf{D}+\mathbf{R}_v) \mathbf{P}_A^\perp \|_F$. Clearly, the solution using eigenvalue decompositions is not suitable if the noise covariance is not a multiple of the identity matrix. The right figure shows an application of factor analysis on observed data at the WSRT. It shows projection filter attenuation curves of television sound carrier waves after whitening by $\mathbf{D}^{-\frac{1}{2}}$. The diagonal noise term $\mathbf{D}$ was estimated by using factor analysis.

A second example is gain calibration (Boonstra and van der Veen, 2003). Initially the antenna gains and noise powers of the telescopes are unknown. To estimate them, a common procedure is to point the telescopes at a strong sky source and make an observation. This produces a rank-1 factor model $\mathbf{R} = \mathbf{g}\sigma_s^2 \mathbf{g}^H + \mathbf{D}$, where σ_s^2 is the source power (assumed to be known from tables), $\mathbf{g}$ is the antenna gain vector, and $\mathbf{D}$ is a diagonal matrix containing the noise powers of each antenna. These can be estimated from $\mathbf{R}$ using rank-1 factor analysis. Figure 2, left, shows the principle of

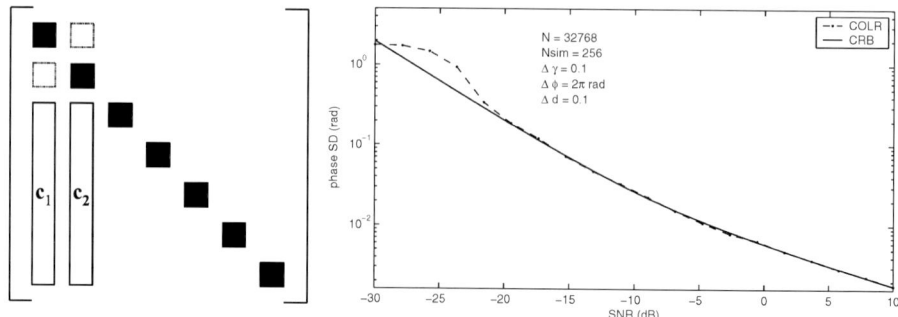

Figure 2. Column ratio factor estimation, principle (left) and estimation accuracy of the method compared to the Cramer–Rao bound (right).

a rank-1 column ratio factor analysis (COLR). Two columns $\mathbf{c}_i$ and $\mathbf{c}_j$, excluding the diagonal, are related by: $\mathbf{c}_i = \alpha_{ij}\mathbf{c}_j$. The complex "ratio" $\alpha_{ij} = \mathbf{c}_j^\dagger \mathbf{c}_i$ can then be used to estimate the diagonal terms of $\mathbf{g}\sigma_s^2\mathbf{g}^H$, which subsequently yields an estimate of $\mathbf{D}$. Applying an eigenvalue decomposition on $\mathbf{R} - \mathbf{D}$ gives a consistent estimate of the vector $\mathbf{g}$. The right figure shows a simulation of a rank-1 gain estimation problem, where the phase estimation accuracy of the COLR method was compared with the theoretical Cramer–Rao lower bound.

4. Detection

The detection problem is given by a collection of hypotheses

$$\begin{aligned}\mathcal{H}_q &: \mathbf{x}(k) \sim \mathcal{CN}(0, \mathbf{R}_q) \\ \mathcal{H}' &: \mathbf{x}(k) \sim \mathcal{CN}(0, \mathbf{R}'), \quad q = 1, 2, \dots \end{aligned} \quad (9)$$

where $\mathcal{CN}(0, \mathbf{R})$ denotes the zero-mean complex normal distribution with covariance $\mathbf{R}$, $\mathbf{R}_q$ is the covariance matrix of the model with q interferers,

$$\mathbf{R}_q = \mathbf{A}\mathbf{A}^H + \mathbf{D}, \quad \text{where} \quad \mathbf{A} : p \times q, \quad \mathbf{D} \text{ diagonal} \quad (10)$$

and $\mathcal{H}'$ corresponds to a default hypothesis of an arbitrary (unstructured) positive definite matrix $\mathbf{R}'$. The generalized likelihood ratio test (GLRT) detector for this problem tests $\mathcal{H}_q$ versus $\mathcal{H}'$, where the unknown parameters are replaced by maximum likelihood estimates under each of the hypotheses.

4.1. IDENTICAL NOISE POWER

In case the noise matrix $\mathbf{D}$ can be written as $\mathbf{D} = \sigma_n^2 \mathbf{I}$, where the noise power σ_n^2 is known, then the test statistic (Box, 1949) can be written as

$$T(\mathbf{X}) = -Np \log \prod_{i=1}^{p} \frac{\hat{\lambda}_i}{\sigma_n^2} \quad (11)$$

where N is the number of samples, $\hat{\lambda}_i$ is the ith eigenvalue estimate, and $\mathbf{X} = [\mathbf{x}(1), \dots, \mathbf{x}(N)]$ is the data. The statistic $T(\mathbf{X})$ is χ^2 distributed, which allows us to select the threshold for a desired false alarm rate (Leshem et al., 2000).

A relatively simple to implement test is an eigenvalue threshold test based on an asymptotic formula for the largest singular value of a $p \times N$ white Gaussian noise matrix (Edelman, 1988)

$$\gamma = \sigma_n^2 \left(1 + \frac{\sqrt{p}}{\sqrt{N}}\right)^2 \quad (12)$$

4.2. ARBITRARY NOISE POWER

In case the noise matrix **D** is diagonal with unknown entries, we can use a more general factor analysis approach (Mardia et al., 1979; Lawley and Maxwell, 1971), resulting in a maximum likelihood test statistic (van der Veen et al., 2004) given by

$$T(\mathbf{X}) = N \log |\hat{\mathbf{R}}_q^{-1} \hat{\mathbf{R}}| \tag{13}$$

If $\mathcal{H}_q$ is true and N is moderately large (say $N - q \geq 50$), then $2T_q(\mathbf{X})$ has approximately a χ_v^2 distribution with $v = (p - q)^2 - p$ degrees of freedom. This provides a threshold for a test of $\mathcal{H}_q$ versus $\mathcal{H}'$ corresponding to a desired probability of "false alarm" (here the probability of rejecting $\mathcal{H}_q$ when it is true).

Figure 3 shows an illustration of an eigenvalue distribution of pre-whitened data obtained at the (60 antenna-element) LOFAR phased array test station (ITS). A physical interpretation of the eigenvalue decomposition is that the eigenvectors give an orthogonal set of "directions" or spatial signatures, and the eigenvalues give the power of the signals coming from those directions. If there are q interferers, then q eigenvalues will be above the noise power levels, and $p - q$ eigenvalues will be distributed around the average noise power level. Not that if $q > 1$, the eigenvectors in general do not correspond to physical directions, cf. Leshem et al. (2000). Clearly visible is that at three frequencies only a single transmitter can be detected; at 26.36 MHz multiple transmitters are present.

The detection theory can be applied to mitigate intermittent interference. Results concerning detection probabilities and residual interference after detection and excision can be found for example in Leshem et al. (2000). Interference detection

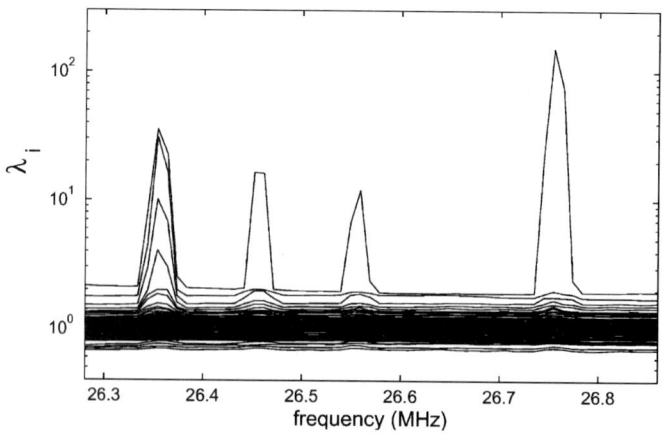

Figure 3. Eigenvalue distribution of the covariance matrix after whitening for an observation at the LOFAR radio telescope test station (ITS). The figure shows multiple transmitters at 26.36 MHz, and three frequencies with single transmitters. The frequency resolution is 10 kHz.

can also improve spatial filtering approaches, by avoiding application of spatial filters (and resulting distortions) in cases when there is no interference detected.

5. Spatial filtering

Interference cancellation is becoming increasingly important in radio astronomy. Depending on the interference and the type of instrument, several kinds of RFI mitigation techniques are applicable (Leshem et al., 2000; Fridman and Baan, 2001). For intermittent interference, the most effective techniques are based on detection and "blanking": omitting the contaminated samples from the covariance estimate, using a single sensor (Fridman, 1996; Weber et al., 1997) or multiple sensors (Leshem et al., 2000). For continually present interference at an array of p telescope dishes, spatial filtering is possible. The desired instrument outputs in this case are $p \times p$ correlation matrices, integrated to several seconds (e.g., 10 s). Based on short-term correlation matrices (integration to e.g., 10 ms) and narrow subband processing, the array signature vector of an interferer can be estimated and subsequently projected out (Raza et al., 2002) – we describe this technique below.

An interesting option is to utilize a reference antenna which picks up only the interference, so that adaptive cancellation techniques can be implemented (Barnbaum and Bradley, 1998; Ellingson et al., 2001). Spatial filtering on extended arrays was first considered by Briggs et al. (2000) for a single dual-polarized telescope (two channels) and two reference antennas. Jeffs et al. (2003, 2004) propose spatial filtering algorithms along the lines of Raza et al. (2002).

5.1. SPATIAL FILTERING USING PROJECTIONS

Suppose that an orthogonal basis $\mathbf{U}_k$ of the subspace spanned by interferer spatial signatures span($\mathbf{A}_k$) is known. We can then form a spatial projection matrix $\mathbf{P}_k^\perp := \mathbf{I} - \mathbf{U}_k \mathbf{U}_k^H$ which is such that $\mathbf{P}_k^\perp \mathbf{A}_k = 0$. When this spatial filter is applied to the data covariance matrix all the energy due to the interferer will be nulled. Let

$$\hat{\mathbf{Q}}_k := \mathbf{P}_k^\perp \hat{\mathbf{R}}_k \mathbf{P}_k^\perp \tag{14}$$

then

$$\mathcal{E}\{\hat{\mathbf{Q}}_k\} = \mathbf{P}_k^\perp \mathbf{\Psi} \mathbf{P}_k^\perp \tag{15}$$

where $\mathbf{\Psi} := \mathbf{R}_v + \sigma^2 \mathbf{I}$ is the interference-free covariance matrix. When we subsequently average the modified covariance matrices $\hat{\mathbf{Q}}_k$, we obtain a long-term estimate

$$\hat{\mathbf{Q}} := \frac{1}{N}\sum_{k=1}^{N} \hat{\mathbf{Q}}_k = \frac{1}{N}\sum_{k=1}^{N} \mathbf{P}_k^{\perp} \hat{\mathbf{R}}_k \mathbf{P}_k^{\perp}. \quad (16)$$

$\hat{\mathbf{Q}}$ is an estimate of $\mathbf{\Psi}$, but it is biased due to the projection. A bias correction matrix $\mathbf{C}$ can be derived using the relation $\text{vec}(\mathbf{EFG}) = (\mathbf{G}^t \otimes \mathbf{E})\text{vec}(\mathbf{F})$ (Raza et al., 2002)

$$\mathbf{C} = \frac{1}{N}\sum_{k=1}^{N} \mathbf{P}_k^{\perp t} \otimes \mathbf{P}_k^{\perp} \quad (17)$$

leading to the following (bias-corrected) estimate of $\hat{\mathbf{\Psi}}$:

$$\hat{\mathbf{\Psi}} := \text{unvec}(\mathbf{C}^{-1}\text{vec}(\hat{\mathbf{Q}})). \quad (18)$$

If the interference was completely projected out then $\hat{\mathbf{\Psi}}$ is an unbiased estimate of the covariance matrix without interference. A detailed analysis of this algorithm will appear in van der Tol and van der Veen (2004). The main conclusion is that the variance of the estimate of $\mathbf{\Psi}$ is equal to $(1/N)\mathbf{C}^{-1}\sigma^4$, whereas for "clean" data it would be $(1/N)\sigma^4$. For interferers which are sufficiently moving, $\mathbf{C}^{-1}$ is well conditioned[2] and the penalty is comparable to a loss in number of samples. Even for stationary interferers, $\mathbf{C}^{-1}$ might be well conditioned due to the motion of the telescopes, but it depends on the integration length and the location of the sky source which is being tracked. Cases where an interferer enters only on a single telescope always lead to a singular $\mathbf{C}$ and cannot be resolved by this algorithm.

Figure 4 shows observed condition numbers of $\mathbf{C}^{-1}$ for different transmitters as a function of long-term integration time. For fixed location transmitters such as television (TVL) and amateur broadcasts (amat), the condition number decreases to low values (<5) after about 100 s, as is expected from an analysis of the telescope instrumental fringe rotation. The condition number for the satellite GPS signal decreases more rapidly, because of its motion. Airplane radar (DME) transmits in bursts, the integrated covariance matrices therefore contain many short-time full-rank noise matrices. As a result, the long-term correction matrix $\mathbf{C}$ will have a low condition number.

An alternative to projection filtering is filtering by subtraction. This type of filtering will lead to comparable results. The subtraction filter however will also be biased (Leshem et al., 2000), and needs correction. The attenuation for both projection and subtraction filtering is limited by the spatial signature estimation accuracy, which is described in Leshem and van der Veen (2000).

---

[2] The condition number of matrix is the ratio of the largest and the smallest eigenvalues, and is a measure for the amount of noise enhancement caused by matrix inversion. In the best case, the condition number is equal to 1 (unitary matrix).

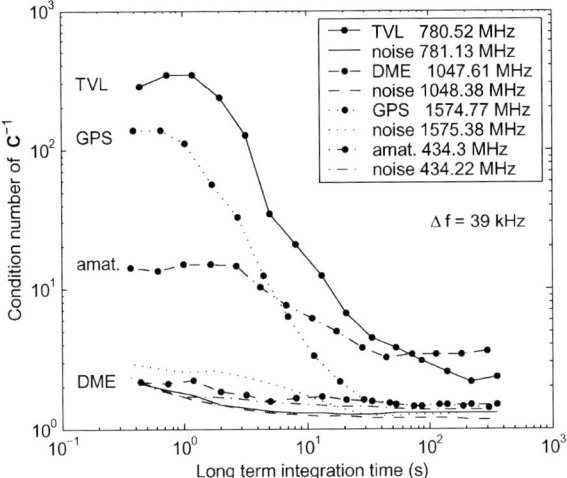

Figure 4. Observed spatial filter correction matrix condition numbers for different observed transmitters at the WSRT telescope.

5.2. SPATIAL FILTERING WITH AN EXTENDED ARRAY

If the telescope array is extended with one or more reference antennas, we can follow the same procedure as described before. Let p_0 be the number of primary antennas, and p be the total number of antennas. The data covariance matrix can be partitioned accordingly as

$$\mathbf{R}_k = \begin{bmatrix} \mathbf{R}_{00,k} & \mathbf{R}_{01,k} \\ \mathbf{R}_{10,k} & \mathbf{R}_{11,k} \end{bmatrix}. \tag{19}$$

where $\mathbf{R}_k$ has model

$$\begin{aligned}\mathbf{R}_k &= \mathbf{\Psi} + \mathbf{A}_k\mathbf{A}_k^H = \mathbf{R}_v + \mathbf{\Sigma} + \mathbf{A}_k\mathbf{A}_k^H \\ &= \begin{bmatrix} \mathbf{R}_{v,0} + \mathbf{A}_{0,k}\mathbf{A}_{0,k}^H + \sigma_0^2\mathbf{I} & \mathbf{A}_{0,k}\mathbf{A}_{1,k}^H \\ \mathbf{A}_{1,k}\mathbf{A}_{0,k}^H & \mathbf{A}_{1,k}\mathbf{A}_{1,k}^H + \sigma_1^2\mathbf{I} \end{bmatrix}\end{aligned} \tag{20}$$

and where $\mathbf{\Psi}$ is the interference-free covariance matrix, and $\mathbf{\Sigma} := \text{diag}[\sigma_0^2\mathbf{I}, \sigma_1^2\mathbf{I}]$ is the diagonal noise covariance matrix (assumed known). The objective is to estimate the interference-free covariance submatrix $\mathbf{\Psi}_{00} := \mathbf{R}_{v,0} + \sigma_0^2\mathbf{I}$.

Following the preceding algorithm applied to $\mathbf{R}_k$, the reconstructed covariance matrix is size $p \times p$, and we can simply select the $p_0 \times p_0$ submatrix in the top-left corner, $\mathbf{\Psi}_{00}$ (Jeffs et al., 2003, 2004). An improved algorithm would not reconstruct the other blocks of $\mathbf{\Psi}$ (van der Veen and Boonstra, 2004). Indeed, let the projected estimates $\hat{\mathbf{Q}}$ be as before in (16). Then it can be shown that:

$$\mathcal{E}\{\text{vec}(\hat{\mathbf{Q}})\} = \mathbf{C}\text{vec}(\boldsymbol{\Psi}). \tag{21}$$

where $\mathbf{C}$ is given by (17). Partition $\boldsymbol{\Psi}$ as in (20) into four submatrices. Since we are only interested in recovering $\boldsymbol{\Psi}_{00}$, the other submatrices in $\boldsymbol{\Psi}$ are replaced by their expected values, respectively $\boldsymbol{\Psi}_{01} = \mathbf{0}$, $\boldsymbol{\Psi}_{10} = \mathbf{0}$, $\boldsymbol{\Psi}_{11} = \sigma_1^2 \mathbf{I}$. This corresponds to solving the reduced-size covariance model error minimization problem,

$$\boldsymbol{\Psi}_{00} = \arg\min_{\boldsymbol{\Psi}_{00}} \left\| \text{vec}(\hat{\mathbf{Q}}) - \mathbf{C}\text{vec}\left(\begin{bmatrix} \boldsymbol{\Psi}_{00} & \mathbf{0} \\ \mathbf{0} & \sigma_1^2 \mathbf{I} \end{bmatrix} \right) \right\|^2. \tag{22}$$

The solution of this problem reduces to a standard LS problem after separating the knowns from the unknowns. Partition $\mathbf{C}$ in $\mathbf{C}_1$ (corresponding to $\text{vec}(\boldsymbol{\Psi}_{00})$) and $\mathbf{C}_2$ (corresponding to $\text{vec}(\sigma_1^2 \mathbf{I})$), then the solution is (van der Veen and Boonstra, 2004):

$$\text{vec}(\boldsymbol{\Psi}_{00}) = \mathbf{C}_1^\dagger \left(\text{vec}(\hat{\mathbf{Q}}) - \sigma_1^2 \mathbf{C}_2 \mathbf{1} \right) \tag{23}$$

If σ_1^2 is unknown, then it can be estimated using a straightforward modification. The advantage compared to the preceding algorithm is that $\mathbf{C}_1$ is a tall matrix, and better-conditioned than $\mathbf{C}$. This improves the performance of the algorithm in cases where $\mathbf{C}$ is ill-conditioned, e.g., for stationary interferers, or an interferer entering on only a single telescope. Asymptotically for large INR of the reference array, the algorithm is seen to behave similar to the traditional subtraction technique.

5.3. EXPERIMENT

A reference signal is useful only if it has a better SNR than the primary antennas. Therefore, an omnidirectional antenna combined with a custom-off-the-shelf amplifier is not good enough. To be versatile, we have tested the preceding technique on a reference signal obtained from the beamformed output of a wideband phased array of 64 elements. This system has a bandwidth of 600–1700 MHz, a baseband of 20 MHz, two digital beamforming outputs, and it is part of the "thousand elements array" (THEA), developed by ASTRON. The reference signal is correlated along with the telescope signals as if it was an additional telescope, and spatial filtering algorithms can be applied to the resulting short-term integrated covariance matrices.

The test data is an observation of the strong astronomical source 3C48 contaminated by Afristar satellite signals. The primary array consists of $p_0 = 6$ of the 14 telescope dishes of the WSRT. The reference signals are $p_1 = 2$ beamforming outputs of the THEA system. One beam was pointed approximately to the

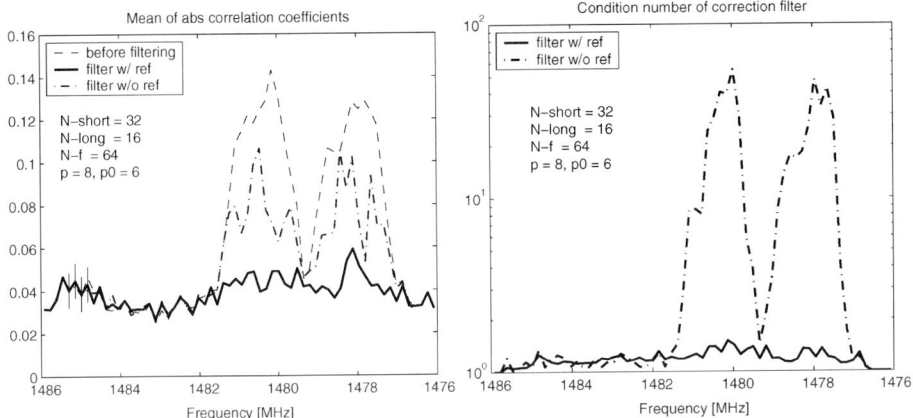

Figure 5. Averaged cross-correlation spectrum before and after filtering (left), and condition number of correction filter with and without reference antenna (right).

satellite, the other was used for scanning. We recorded 65 kSamples at 20 MS/s, and processed these offline. After short-term windowed Fourier transforms, the data was split into 64 frequency bins, correlated, and averaged over 32 samples to obtain 16 short-term covariance matrices. The resulting cross-correlation spectra after filtering, shown in Figure 5 (left), demonstrate that the spatial filtering with reference antenna has done much better to remove the interference than the case without reference antenna. The residual correlation of about 4% is known to be the SNR of the astronomical source. The lines are noisy due to the finite sample effect. The figure (right) also shows the condition number of the filter with and without the reference antenna. The condition number, as expected, improves dramatically when a reference antenna is used.

6. Imaging

6.1. MATRIX FORMULATION

As described in more detail in Leshem et al. (2000), image formation is also a fruitful area for array signal processing techniques. Astronomers try to estimate the intensity (brightness) $I_f(\mathbf{s})$ of the sky as a function of the location $\mathbf{s}$ and frequency f. They do this by measuring the correlation (called the "visibility" V_f) between identical sensors i and j with locations $\mathbf{r}_i$ and $\mathbf{r}_j$, corresponding to a baseline $\mathbf{r}_i - \mathbf{r}_j$. Let (ℓ, m) denote normalized coordinates of the sky source ($-1 \leq \ell, m \leq 1$), and (u, v, w) the baseline vector of the antenna pair measured in wavelengths. Assuming a planar array, w can be removed from the equations via geometrical delay compensation. Under certain approximations, the "measurement equation" is given by (Perley et al., 1994)

$$V_f(u,v) = \iint I_f(\ell,m) e^{-j2\pi(u\ell+vm)} d\ell dm. \quad (24)$$

It has the form of a Fourier transformation. The function $V_f(u,v)$ is sampled at various coordinates $(u_{ij}(t), v_{ij}(t))$ by first of all taking all possible sensor pairs i, j or baselines $\mathbf{r}_i - \mathbf{r}_j$, and second by realizing that the sensor locations $\mathbf{r}_i, \mathbf{r}_j$ are actually time-varying since the earth rotates. Given a sufficient number of samples in the (u,v) domain, the relation can be inverted to obtain an image (the 'map').

Assume that the sky consists of a large number (d) point sources. Equation (24) can then be written slightly differently as

$$V(u_{ij}(t), v_{ij}(t)) = \sum_{l=1}^{d} e^{-j2\pi(u_{i0}(t)\ell_l+v_{i0}(t)m_l)} \cdot I(\ell_l, m_l) \cdot e^{j2\pi(u_{j0}(t)\ell_l+v_{j0}(t)m_l)}. \quad (25)$$

where (u_{i0}, v_{i0}) are coordinates of the ith antenna with respect to a common reference point. The connection to our previous framework is obtained by collecting the visibilities into correlation matrices $\mathbf{R}$, where $\mathbf{R}_{ij}(t) = V(u_{ij}(t), v_{ij}(t))$. The above equation can then be written as

$$\mathbf{R}_k = \mathbf{A}_k \mathbf{B} \mathbf{A}_k^H \quad (26)$$

where $\mathbf{R}_k \equiv \mathbf{R}(t_k)$, $\mathbf{A}_k = [\mathbf{a}_k(\ell_1, m_1), \ldots, \mathbf{a}_k(\ell_d, m_d)]$, and

$$\mathbf{a}_k(\ell, m) = \begin{bmatrix} e^{-j2\pi(u_{10}(t_k)\ell+v_{10}(t_k)m)} \\ \vdots \\ e^{-j2\pi(u_{p0}(t_k)\ell+v_{p0}(t_k)m)} \end{bmatrix} \quad (27)$$

$$\mathbf{B} = \begin{bmatrix} I(\ell_1, m_1) & & 0 \\ & \ddots & \\ 0 & & I(\ell_d, m_d) \end{bmatrix}$$

where $\mathbf{a}_k(\ell, m)$ is recognized as the array response vector. As usual, the array response is frequency dependent. The response is also slowly time-varying due to the earth rotation.

6.2. INVERSE FOURIER IMAGING

6.2.1. *Classical inverse Fourier imaging*

The relation between sky brightness $I(\ell, m)$ and visibilities $V(u,v)$ (where u, v are taken at frequency f) is given by the measurement Equation (24). We have measured V on a discrete set of baselines $\{(u_i, v_i)\}$. The "dirty image" (a lumpy image obtained via direct Fourier inversion possibly modified with some weights c_i) is defined by

$$I_D(\ell, m) := \sum_i c_i\, V(u_i, v_i)\, e^{j2\pi(u_i \ell + v_i m)} \tag{28}$$

It is equal to the 2D convolution of the true image $I(l, m)$ with a point spread function $B_0(l, m)$ known as the "dirty beam":

$$I_D = I * B_0, \qquad B_0(\ell, m) := \sum_i c_i\, e^{j2\pi(u_i \ell + v_i m)} \tag{29}$$

B_0 is the dirty beam, centered at the origin. The weights $\{c_i\}$ are arbitrary coefficients designed to obtain an acceptable beam-shape, with low side lobes, in spite of the irregular sampling.

Specializing to a point source model, $I(\ell, m) = \sum_l I_l \delta(\ell - \ell_l, m - m_l)$ where I_l is the intensity of the source at location (ℓ_l, m_l), gives

$$V(u, v) = \sum_l I_l\, e^{-j2\pi(u\ell_l + vm_l)} \tag{30}$$

$$I_D(\ell, m) = \sum_l I_l\, B_0(\ell - \ell_l, m - m_l) \tag{31}$$

Thus, every point source excites the dirty beam centered at its location (ℓ_l, m_l).

From the dirty image I_D and the known dirty beam B_0, the desired image I is obtained via a deconvolution process. A popular method for doing this is the CLEAN algorithm (Hogbom, 1974). The algorithm assumes that B_0 has its peak at the origin, and consists of a loop in which a candidate location (ℓ_l, m_l) is selected as the largest peak in I_D, and subsequently a small multiple of $B_0(\ell - \ell_l, m - m_l)$ is subtracted from I_D. The objective is to minimize the residual, until it converges to the noise level.

6.2.2. Inverse Fourier imaging after projections

If we take projections or any other linear combination $[c_{ij}]$ of the visibilities $\{V(u_i, v_i)\}$ during measurements we have instead available

$$Z(u_i, v_i) = \sum_j c_{ij}\, V(u_j, v_j) \tag{32}$$

Suppose we compute the dirty image in the same way as before, but now from Z, then it can be shown (van der Veen et al., 2004) that the dirty image is obtained via a convolution, but the dirty beam is now space-varying. Nonetheless, they are completely known if we know the linear combinations that we took during observations. Thus, the CLEAN algorithm can readily be modified to take the varying beam shapes into account: simply replace $B_0(\ell, m)$ by $B_l(\ell, m)$ everywhere in the algorithm.

6.3. IMAGING VIA BEAMFORMING TECHNIQUES

6.3.1. *CLEAN and sequential beamforming*
Using a parametric point-source model, the image deconvolution problem can be interpreted as a direction-of-arrival (DOA) estimation problem, e.g., as

$$[\{\hat{s}_l\}, \hat{\mathbf{B}}] = \arg \min_{\{s_l\}, \mathbf{B}} \sum_{k=1}^{K} \left\| \hat{\mathbf{R}}_k - \mathbf{A}_k(\{s_l\}) \mathbf{B} \mathbf{A}_k^H(\{s_l\}) - \sigma^2 \mathbf{I} \right\|_F \tag{33}$$

(**B** is constrained to be diagonal with positive entries.) This is recognized as the same model as used for DOA estimation in array processing. Note however that the array is moving ($\mathbf{A}_k$ is time-dependent), and that there are many more sources than the dimension of each covariance matrix.

In this notation, the image formation in Section 6.2.1 can be formulated as follows. If we write $I_D(\mathbf{s}) \equiv I_D(\ell, m)$ and $\mathbf{a}_k(\mathbf{s}) \equiv \mathbf{a}_k(\ell, m)$, we can rewrite the dirty image (28) as (van der Veen et al., 2004)

$$I_D(\mathbf{s}) = \sum_k \mathbf{a}_k^H(\mathbf{s}) \mathbf{R}_k \mathbf{a}_k(\mathbf{s}) \tag{34}$$

We omitted the optional weighting. Also note that, with noise, we have to replace $\mathbf{R}_k$ by $\mathbf{R}_k - \sigma^2 \mathbf{I}$. The iterative beam removing in CLEAN can now be posed as an iterative LS fitting between the sky model and the observed visibility (Schwarz, 1978). Finding the brightest point $\mathbf{s}_0$ in the image is equivalent to trying to find a point source using classical Fourier beamforming, i.e.,

$$\hat{\mathbf{s}}_0 = \arg \max_{\mathbf{s}} \sum_{k=1}^{K} \mathbf{a}_k^H(\mathbf{s})(\mathbf{R}_k - \sigma^2 \mathbf{I}) \mathbf{a}_k(\mathbf{s}). \tag{35}$$

Thus, the CLEAN algorithm can be regarded as a generalized classical sequential beamformer, where the brightest points are found one by one, and subsequently removed from $\mathbf{R}_k$ until the LS cost function (33) is minimized. An immediate consequence is that the estimated source locations will be biased, a well known fact in array processing. When the sources are well separated the bias is negligible compared to the standard deviation, otherwise it might be significant. This gives an explanation for the poor performance of the CLEAN in imaging extended structures (see e.g., Perley et al., 1994).

6.3.2. *Minimum variance beamforming approaches*

Once we view image formation/deconvolution as equivalent to DOA estimation with a moving array, we can try to adapt various other DOA estimators for handling the image formation. In particular the deflation approach used in the CLEAN algorithm can be replaced by other source parameters estimators. One approach that seems particularly relevant in this context is the minimum-variance distortionless response (MVDR) method of beamforming (Capon, 1969). The major new aspect here is the fact that the array is moving and that there are more sources than sensors.

Instead of working with the dirty image $I_D(\mathbf{s}) = \sum_k \mathbf{a}_k^H(\mathbf{s}) \mathbf{R}_k \mathbf{a}_k(\mathbf{s})$, the basis for high-resolution beamforming techniques is to look at more general "pseudo-spectra"

$$I'_D(\mathbf{s}) := \sum_k \mathbf{w}_k^H(\mathbf{s}) \mathbf{R}_k \mathbf{w}_k(\mathbf{s}) \tag{36}$$

Here, $\mathbf{w}_k(\mathbf{s})$ is the beamformer pointing towards direction $\mathbf{s}$, and $I'_D(\mathbf{s})$ is the output energy of that beamformer. Previously we had $\mathbf{w}_k(\mathbf{s}) = \mathbf{a}_k(\mathbf{s})$; the objective is to construct beamformers that provide better separation of close sources.

A generalized MVDR follows by defining a weight vector $\mathbf{w}_k$ which minimizes the output power at time k subject to the constraint that we have a fixed response towards the look direction $\mathbf{s}$ of the array. Solving this (well-known) problem leads to the overall spectral estimator

$$I'_D(\mathbf{s}) = \sum_{k=1}^{K} \frac{1}{\mathbf{a}_k^H(\mathbf{s}) \hat{\mathbf{R}}_k^{-1} \mathbf{a}_k(\mathbf{s})} \, . \tag{37}$$

The locations of the strongest sources are given by the maxima of $I'_D(\mathbf{s})$. It is known that the MVDR has improved resolution compared to the classical beamformer which is the basis for the CLEAN algorithm. The algorithm is readily extended to handle the "space-varying" beamshapes that occur after spatial filtering. It is also possible to use more advanced forms of beamforming, e.g., "robust Capon beamforming" (RCB) (Lorenz and Boyd, 2003; Vorobyov et al., 2003).

Figure 6 illustrates this by comparing a dirty image produced in the classical way to the dirty image corresponding to (37) and to robust Capon beamforming (Stoica et al., 2003; van der Veen et al., 2004; van der Tol and van der Veen, 2005). The measurement data is a "snapshot" collected from a 60-element test station for the LOFAR telescope. Since this is a two-dimensional array, it does not depend on earth rotation to enable imaging. Due to the limited integration time, the sky sources are not yet observed and only interference shows up, which is visible at the horizon. All other features are due to the sidelobes of the dirty beam. A disadvantage of MVDR beamformers is that due to array calibration errors, the

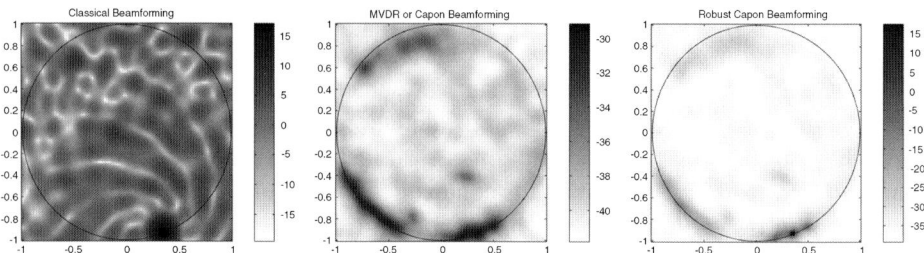

Figure 6. Examples of "dirty images" from the LOFAR test station: classical beamforming (left), MVDR image (middle), and robust Capon beamforming (right).

scaling of the spectral estimator is affected. This problem is remedied in the robust Capon beamformer, also shown in Figure 6.

7. Conclusion

Technological advances in the last decade have created interesting possibilities for large distributed interferometric radio telescopes with very large receiving areas and a sensitivity which is one to two orders of magnitude better than the current generation. Two proposed instruments in this context are LOFAR (LOFAR, 2005; Bregman, 1999; Butcher, 2004) and SKA (SKA, 2005; van Ardenne, 2002); the previously mentioned THEA system is a small scale step-up in the design of SKA. Prominent among the challenges of designing and building these telescopes (apart from the costs) are the mitigation of radio interference, the calibration of the system, and the sheer signal processing complexity. Signal processing techniques such as discussed in this paper will enable observations when it would normally be impossible due to interference. These techniques are vital for the next generation of radio telescopes.

Acknowledgements

We would like to thank S. van der Tol at TU Delft and our project partners at ASTRON for the very useful collaboration. This work was supported in part by the STW under DEL77-4476, DTC.5893. This overview paper covers material from several prior publications from the authors.

References

Ardenne van, A. and Butcher, H.: 2002, *Square Kilometer Arrray Project: Concepts and Technologies*, URSI General Assembly, Maastricht.

Barnbaum, C. and Bradley, R. F.: 1998, *Astron. J.* **115**, 2598.

Boonstra, A. J. and Veen van der, A. J.: 2003, *IEEE Trans. Signal Process.* **51**(1), 25.
Boonstra, A. J. and Veen van der, A. J.: 2003, *IEEE Int. Conf. Acoustics, Speech, Signal Process. (ICASSP)*.
Bregman, J. D.: 1999, in A. B. Smolders and M. P. Haarlem van (eds), *Perspectives on Radio Astronomy – Technologies for Large Antenna Arrays*, ASTRON, Dwingeloo, p. 23.
Briggs, F. H., Bell, J. F. and Kesteven, M. J.: 2000, *Astron. J.* **120**, 3351.
Butcher, H. R.: 2004, *Proc. SPIE* **5489**, 537.
Capon, J.: 1969, *Proc. IEEE*, 1408.
Edelman, A.: 1988, *SIAM J. Matrix Anal. Appl.* **9**(4).
Ellingson, S. W., Bunton, J. and Bell, J. F.: 2001, *Astrophys. J. Suppl.* **135**(1), 87.
Fridman, P. A.: 1996, 'A Change Point Detection Method for Elimination of Industrial Interference in Radio Astronomy Receivers', in *Proceedings of the 8th IEEE Signal Processing Workshop on Statistical Signal and Array Processing*, Corfu, p. 264.
Fridman, P. A. and Baan, W. A.: 2001, *Astron. Astrophys.* **378**, 327.
Hamaker, J. P.: 2000, *A&A Suppl. Ser.* **143**, 515.
Hogbom, J. A.: 1974, *Astron. Astrophys. Suppl.* **15**, 417.
Jeffs, B. D., Warnick, K. and Li, L.: 2003, *IEEE Int. Conf. Acoustics, Speech, Signal Process. (ICASSP)* **5**, 77.
Jeffs, B. D., Li, L. and Warnick, K.: 2005, *IEEE Trans. Signal Process.* **53**(2), 439.
Lawley, D. N. and Maxwell, A. E.: 1971, *Factor Analysis as a Statistical Method*, Butterworth & Co, London.
Leshem, A., Veen van der, A. and Boonstra, A. J.: 2000, *Astrophys. J. Suppl. Ser.* **131**(1), 355.
Leshem, A. and Veen van der, A. J.: 2000, *IEEE Tr. Iformation Th.* **L16**(5).
Leshem, A. and Veen van der, A. J.: 2001, 'Multichannel Detection and Spatial Signature Estimation with Uncalibrated Receivers', in *Proceedings of the 11th IEEE Workshop on Statistical Signal Processing*, Singapore.
Lorenz, R. G. and Boyd, S. P.: 2003, *Thirty-Seventh Asilomar Conf. Signals, Syst. & Comput.* **2**, 1345.
Mardia, K. V., Kent, J. T. and Bibby, J. M.: 1979, *Multivariate Analysis, Probability and Mathematical Statistics Series*, Academic Press, London.
Perley, R. A., Schwab, F. R. and Bridle, A. H.: 1994, *Synthesis Imaging in Radio Astronomy*, Astronomical Society of the Pacific Conference Series, Vol. 6.
Raza, J., Boonstra, A. J. and Veen van der, A. J.: 2002, *IEEE Signal Process. Lett.* **9**(2), 64.
Ryle, M.: 1952, *Proc. Royal Soc. A* **211**, 351.
Ryle, M.: 1962, *Nature* **194**, 517.
Schwarz, U. J.: 1978, *Astron. Astrophys.* **65**, 345.
Stoica, P., Wang, Z. and Li, L.: 2003, *IEEE Signal Process. Lett.* **10**, 172.
Tol van der, S. and Veen van der, A. J.: 2004, *IEEE Trans. Signal Process.* **53**(3), 896.
Tol van der, S. and Veen van der, A. J.: 2005, 'Application of Robust Capon Beamforming to Radio Astronomical Imaging', in *IEEE Int. Conf. Acoustics, Speech, Signal Process. (ICASSP)*, Philadelphia (PA) March 2005.
Veen van der, A. J. and Boonstra, A. J.: 2004, *IEEE Int. Conf. Acoustics, Speech, Signal Process. (ICASSP)*, Montreal.
Veen van der, A. J., Leshem, A. and Boonstra, A. J.: 2004, 'Signal Processing for Radio Astronomical Arrays', *IEEE Sensor Array and Multichannel Signal Procesing Workshop (SAM)*, Sitges, Barcelona.
Vorobyov, S. A., Gershman, A. B. and Luo, Z.-Q.: 2003, *IEEE Trans. Signal Process.* **51**(2), 313.
Weber, R., Faye, C., Biraud, F. and Dansou, J.: 1997, *Astron. Astrophys. Suppl.* **126**(1), 161.
LOFAR web site, http://www.lofar.org.
SKA web site, http://www.skatelescope.org.

SKA CORRELATOR ADVANCES

JOHN D. BUNTON
CSIRO ICT Centre Epping, NSW, 1710, Australia
(e-mail: john.bunton@csiro.au)

(Received 6 August 2004; accepted 1 November 2004)

Abstract. When the SKA was proposed, a major technical obstacle to its feasibility was the cost of the correlator. Significant advances made in correlator design since then are described. These advances have made SKA correlators possible within reasonable cost constraints. At the same time performance issues with the proposed FX architecture have been addressed.

Keywords: correlator, FX, XF, filterbank, radio astronomy, SKA

1. Introduction

When the idea of the SKA was first mooted, the correlator was recognised as one of the major economic challenges to its viability. The decreasing cost of electronics, coupled with technical advances since that time, have answered this economic challenge. In this paper the design of the SKA correlator is addressed and the technical advances that make it possible are described.

2. XF or FX?

Major correlators being built now are the ALMA and EVLA correlators, each costing ~$10M. The SKA correlator will have a similar bandwidth but about 100 times as many inputs, assuming a 12-m antenna based design. Thus, the SKA correlator has ~10 000 times the performance of correlators currently being built. Increasing the size of the antenna reduces correlator requirements but the added specification for a wide field-of-view negates this.

The traditional method of building a correlator (for example, VLA (Napier et al., 1983), AT (Wilson and Brown, 1987) and BIMA (Urry and Hudson, 1995) and the first version at ALMA (Escoffier, 1997)) is to use an XF correlator (D'Addario, 1989) which directly measures the cross correlation function. An example of an XF correlator for one baseline is shown in Figure 1. The delay nD is needed to allow the measurement of the two-sided cross correlation function. With complex input data one multiplier and accumulator is needed for each frequency channel.

The alternative to measuring the cross correlation function is to directly measure the cross power spectrum which greatly reduces the computational load. The basic

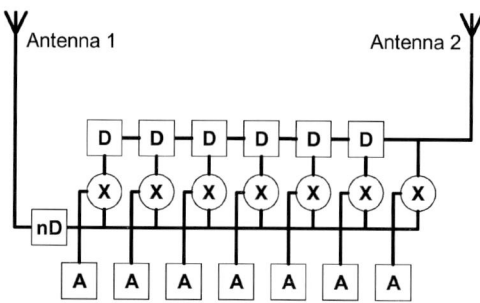

Figure 1. Basic XF correlator, D is a delay, X a multiplier, and A an accumulator.

design for this correlator is shown in Figure 2. The data from each antenna is first transformed into the frequency domain and then the cross power in each spectral channel is measured to give the cross power spectrum. This type of correlator is called an FX correlator (Chikada et al., 1984; D'Addario, 1989) because the frequency transformation (F) precedes the cross multiply operation (X). Traditionally, the FFT has been used to perform the frequency transformation (originally the F in the FX nomenclature stood for FFT (Chikada et al., 1984)). As the FFT is an isomorphism the number of input samples is equal to the number of output samples. Thus only one complex multiplier and adder is needed per baseline when the input data is processed as non-overlapping blocks. This is independent of the number of frequency channels.

But each frequency channel still needs a register to store the correlations and there is no reduction in the memory requirements compared to an XF correlator.

The disadvantage of the FX correlator is the cost of the frequency transformation, which adds ~100 operations per data sample. For the SKA this is a small

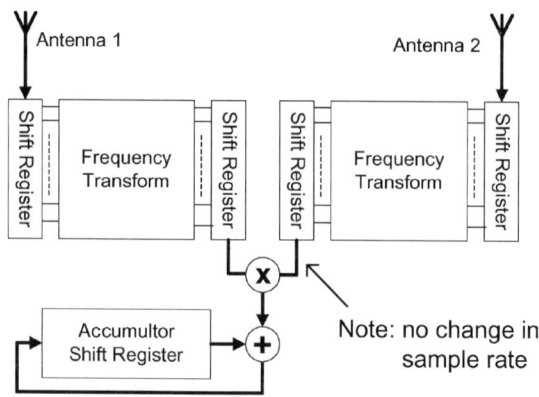

Figure 2. Basic FX correlator.

overhead as the data must be correlated against hundreds or even thousands of antennas.

3. FX correlator problems

Except for systems with a small number of antennas, the FX correlator has a computational advantage leading to its implementations in a number of correlators, e.g. Nobeyama (Chikada et al., 1984), VLBA (Benson, 1993) and GMRT (Tatke, 1998). However, they have not been universally popular because they have suffered from:

1. higher interconnection cost due to bit and data rate increase (Escoffier, 1997),
2. degraded signal-to-noise performance (Chikada et al., 1987) and
3. higher implementation costs.

The last of these is largely because of the need to implement a real-time digital filterbank. As has been mentioned above, this is not a significant computational load for the SKA. However, there is still the extra design cost. This is now comparatively low as the filterbanks can now be implemented in FPGAs, effectively eliminating major NRE costs associated with the ASICs that have been used previously. Solutions to the other two problems have also been found.

4. Bit-rate increase

In XF correlators, a 2-bit sampling of the antenna signal is standard for most applications. This minimises the data transmission cost. The crude sampling reduces the signal-to-noise ratio (D'Addario, 1989) but this is more than offset by the increased bandwidth that can be processed. To overcome the non-linear correlation gain with input level changes, and to achieve the best signal-to-noise ratio, the decision levels of the A/D are controlled by the statistics of the sample data (Automatic Level Control or ALC) (D'Addario, 1989).

In FX correlators the crude analogue-to-digital sampling has normally been retained, but after the frequency transformation the data in each frequency bin is commonly represented by 6 bits (Escoffier, 1997; Bhatnagar, 2001). This increases the cost of cabling to the cross-multiply accumulate units by a factor of 3. The solution to this is to treat each frequency channel of the FX correlator as if it was a narrowband XF correlator. In the XF correlator, the quantiser is normally preceded by a down-conversion stage and IF filter as shown in Figure 3. This is identical to the processing needed to generate one channel of the filter bank. In the FX correlator this is a digital process and the filter is preceded by an A/D converter. If the A/D converter is chosen so that it introduces little noise and distortion then the performance of the two signal chains can be made identical by including an

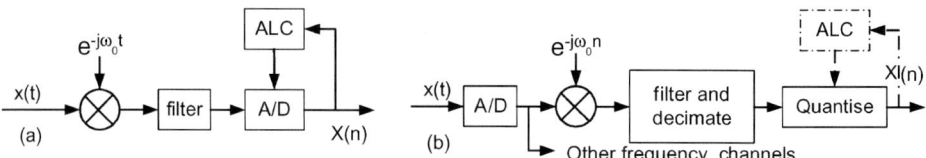

Figure 3. (a) Quantiser in an XF correlator, (b) one channel of an FX correlator filterbank with preceding A/D converter and output quantisation.

ALC loop around the final quantiser in the FX correlator signal chain (Bunton, 2001b)

5. Data rate increase and signal-to-noise degradation

FX correlators have traditionally used the FFT for the frequency transformation. This causes the FX correlator to suffer from a signal-to-noise degradation or a data rate increase. If the FFT processes contiguous blocks of data the input and output sample rates are identical. But taking the product of two spectra to generate the correlation causes aliasing in the lag domain. This decreases the signal-to-noise ratio by an estimated 1.22 (Okamura et al., 2001) for spectral line observations. Use of contiguous FFT processing also leads to errors in the measurement of narrow-band sources. A source at the band edge has an amplitude of 0.41 compared to one at the centre of the band. When the power in the adjacent band is included, the total power is 0.82 and in error by 18%. Overlapping data blocks (e.g. D'Addario, 1989) or averaging over adjacent frequency channels (Okamura et al., 2001) can reduce this degradation. The first method increases the data rate and the second reduces frequency resolution. If frequency channel averaging is used the cost of the filterbanks and amount of memory in the correlator increases for a given frequency resolution. Overlapping the data blocks increases signal transport cost and the compute load of the correlator. For previous implementations overlaps of 50% or more have been used, doubling the correlator cost for a given total bandwidth.

What is needed is a channel response that is flat across the passband with a small transition band, while at the same time not increasing the data rate. This is achieved by use of polyphase filterbanks instead of the FFT to generate the frequency data for the correlator (Bunton, 2000). The structure of a polyphase filterbank (Crochiere and Rabiner, 1983) is shown in Figure 4.

The FFT is preceded by a polyphase filter allowing the channel impulse response to be longer than the FFT length. Thus the channel impulse response can be chosen arbitrarily. An example of the channel response possible is shown in Figure 5, which shows a response where the channel impulse response is 12 times the length of the FFT. In the right-hand plot the frequency response of the channel is shown together with the response of an adjacent channel, dotted.

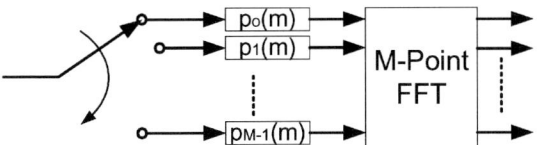

Figure 4. Polyphase filterbank.

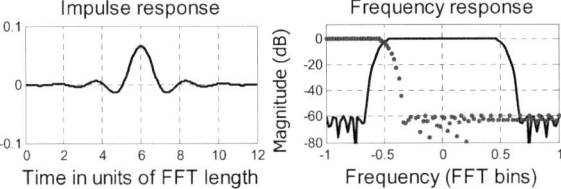

Figure 5. Example of prototype lowpass filter and resulting channel response. Adjacent channel shown dotted.

The response has 0.1 dB passband ripple and 60 dB stopband attenuation. The response of each channel can be made as good as or better than any analogue filters with the added benefit that the filter response is absolutely stable. It is seen that there is very little overlap between the two channels. The small amount of aliasing that results from the overlap can be removed by operating the polyphase filterbank in an oversampling mode. However, the filter quality is such that just a 15% increase in sampling frequency is needed to prevent aliasing to any frequency component in the passband (-0.5 to 0.5 FFT bins). With this increase the filter response ensures the measurement of the power of narrowband sources to an accuracy of 1%.

Thus, the high-quality channel response possible with the polyphase filterbank together with a minimal data rate increase ensures that the FX correlator can work to the same level of performance as an XF correlator. In practice the small errors introduced by not oversampling the data will be acceptable for most astronomy applications.

6. High resolution

An FX correlator has a resolution equal to the bandwidth divided by the number of frequency channels. To achieve a high resolution in frequency, millions of frequency channels can be implemented (Okamura et al., 2001). In this design the enormous output data rates are reduced by summing over adjacent frequency channels. Another strategy to increase frequency resolution is to reduce the input bandwidth, as is done in XF correlators. However, the resulting resolution is proportional to bandwidth, not quadratic as it is in XF correlators. The WIDAR correlator (Carlson and Dewdney, 2000) implements an FXF structure where the data is first broken down into sub-bands and then each sub-band is analysed by an XF correlator. This retains the high-resolution advantages of the XF correlator but only partially realises the computational advantages of the pure FX approach. If an FX correlator is reprogrammable a large fraction of the frequency channels could be discarded and the resources freed up used to implement XF correlation on the remaining channels (Bunton, 2001c).

A more flexible approach takes individual channels and passes them through a second or even third polyphase filterbank (Bunton, 2003). With two stages of filterbanks, resolutions of better than 1 kHz are possible in a 1 GHz band, while still providing full bandwidth low-resolution correlations. The total correlator output data rate can be constrained by processing a subset of the first stage frequency channels or by using adjacent channel averaging.

7. Implementation

FX correlators are more complex than XF correlators due to the need for a real-time filterbank. As has been discussed already FPGAs provide a low-cost path for the design of these filterbanks. Even though this approach leads to higher component costs than an ASIC design this will not be significant for the SKA because filterbanks form only a small part of the compute load of the correlator. An FPGA design also provides greater flexibility as many filterbank resolutions can be implemented in the same hardware. An example of a filterbank design that is already being built is the 2 GHz implementation for the AT (Ferris et al., 2001). By the time of SKA construction the cost of component for such a filterbank will be less than $1000.

A second implementation problem that is common to all correlators with a large number of antennas is the cabling to the cross-multiply accumulate units (XMACs). If each XMAC can process only a small subset of all correlations this would result in the same antenna data being sent to many XMACs, greatly increasing the interconnection cost. For an FX correlator the solution to this problem is to route the data so that the data streams into the correlator modules contain data for all antennas over a small frequency range (Urry, 2000). An example of the resulting correlator structure is shown in Figure 6.

Another common problem for FX correlators is the memory associated with each XMAC, which may require significantly more silicon area than the cross multiplier. If this memory can be located externally to the XMAC chip or FPGA then the on-chip processing power can be maximised. However, the off-chip memory bandwidth can limit the processing capacity. What is needed is a reduction in the rate at which the XMAC units generate data that needs to be sent off chip. A re-ordering of the data before it is processed by the XMAC solves this problem (Bunton, 2001a). Instead of processing data as it arrives, buffer it and supply a set of inputs corresponding to a common baseline and frequency channel to the XMAC. The resulting correlations

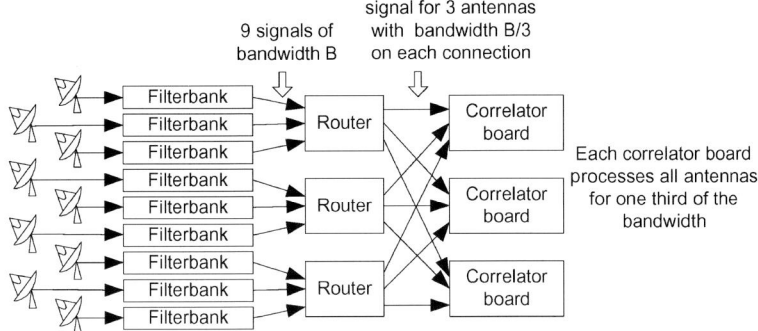

Figure 6. Data routing in a 9-antenna, 3-correlator board FX correlator.

can then be accumulated on-chip in a single accumulator. The output data rate is reduced in direct proportion to the length of the data sets. Moving the memory chip interface off-chip (Bunton and Adams, 2004) maximises the I/O bandwidth as there is now no on-chip overheads for memory reads or addressing.

The input data rate to the XMAC can also be a problem. Each chip or FPGA will process correlation between two subsets of the antennas at any one time. Storing multiple subsets on the chip allows them to be reused (Bunton and Adams, 2004). With N subsets, $N(N-1)/2$ correlation between subsets can be formed. Compared to sending two subsets for each correlation this reduces the input data rate by $N-1$.

8. Conclusion

The pace of developments towards a viable SKA correlator has been great over the last couple of years. It is now accepted that an FX correlator architecture will be used and that it will lose little in terms of performance when compared to an XF correlator. Problems with the implementation continue to be addressed and it is now generally accepted that the correlator will be only a small part of the total cost (D'Addario and Timoc, 2002; Bunton, 2002).

Acknowledgements

The techniques and improvements described in this paper has mainly been the work of the correlator groups at Berkeley, Nobeyama and Sydney. The author is aware that some concepts have been simultaneously and independently discovered by the different groups. As the author's knowledge of these cases is incomplete only the first published result is given.

References

Bhatnagar, S.: 2001, *Radio Study of Galactic Supernova Remnants and the Interstellar Medium,* PhD dissertation, University of Pune, India. Available at http://www.ncra.tifr.res.in/~sanjay/thesis/node31.html.

Benson, J.: 1995, in J. A. Zensus, P. J. Diamond and J. Napier (eds.), *ASP Conference Series, Vol. 82,* Astronomical Society of the Pacific, San Francisco.

Bunton, J. D.: 2000, An Improved FX correlator, *Alma Memo 342,* Nov. 2000. Available at http://www.alma.nrao.edu/memos/html-memos/abstracts/abs342.html.

Bunton, J. D.: 2001a, A Cross Multiply Accumulate Unit for FX Correlators, *Alma Memo 392,* Sep. 2001. Available at http://www.alma.nrao.edu/memos/html-memos/abstracts/abs342.html.

Bunton, J. D.: 2001b, 'FX Correlators for the SKA', *The SKA: Defining the Future, Berkeley Workshop, July 2001.* Available at http://www.skatelescope.org/skaberkeley/.

Bunton, J. D.: 2001c, 'Prototype SKA Technologies at Molonglo: 3.Beamformer and Correlator', *The SKA: Defining the Future, Berkeley Workshop, July 2001.* Available at http://www.skatelescope.org/skaberkeley/.

Bunton, J. D.: 2002, 'Correlator Costs for Arrays and Filled Aperture Antenna', *International SKA Conference 2002, Groningen, The Netherlands, 13–15 August 2002.* Available at www.lofar.org/ska2002/pdfs/Bunton_correlator_costs.pdf.

Bunton, J. D.: 2003, Multi-resolution FX Correlator, Alma Memo 447, Feb 2003. Available at http://www.alma.nrao.edu/memos/.

Bunton, J. D. and Adams, T.: 2004, 'SKAMP Spectral Line Correlator', *SKA2004, Penticton, B.C., Canada, 18–22 July.*

Crochiere, R. E. and Rabiner, L. R.: 1983, Multirate Digital Signal Processing, Prentice-Hall, New Jersey.

Carlson, B. R. and Dewdney, P. E.: 2000, 'Efficient Wideband Digital Correlation', Electronic Letters, Vol. 36, No. 11, 25 May 2000, pp. 987–988.

D'Addario, L. and Timoc, C.: 2002, 'Digital Signal Processing for the SKA', *International SKA Conference 2002, Groningen, The Netherlands, 13–15 August 2002.* Available at http://www.lofar.org/ska2002/pdfs/D'Addario-correlator.pdf.

D'Addario, L. R.: 1989, in Perley, Schwab and Bridle (eds.), *Synthesis Imaging in Radioastronomy,* Astronomical Society of the Pacific, San Francisco.

Chikada, Y. et al.: 1984, in J. A. Roberts (ed.), *Indirect Imaging,* Cambridge University Press, Cambridge, U.K., pp. 387–404.

Chikada, Y et al.: 1987, 'A 6 × 320-Mhz 1024-Channel FFT Cross-spectrum Analyzer for Radio Astronomy', *Proc IEEE 75, 9,* pp. 1023–1209, September 1987.

Escoffier, R.: 1997, The MMA Correlator, *Alma Memo 166,* July 1997. Available at http://www.alma.nrao.edu/memos/html-memos/abstracts/abs166.html.

Ferris, R. H., Bunton, J. D. and Stuber, B.: 2001, 'A 2GHz Digital Filterbank Correlator', *The SKA: Defining the Future, Berkeley California, July 9–12, 2001.* Available at http://www.skatelescope.org/skaberkeley/.

Napier, P. J., Thompson, A. R. and Ekers, R. D.: 1983, 'The Very Large Array: Design and Performance of a Modern Synthesis Radio Telescope', *Proc. IEEE 71, 11,* pp. 1295–1320, Nov. 1983.

Okamura, S. K., Chikada, Y., Momose, M. and Iguchi, S.: 2001, Feasibility Study of the Enhanced Correlator for 3-way ALMA I., *Alma Memo 350,* February 2001, http://www.nrao.alma.edu/memos/.

Tatke, V. M.: 1998, *A Digital Spetral Correlator for GMRT,* Master's Thesis, Indian Institute of Science, Bangalore, India.

Urry, W. L. and Hudson, J. A.: 1995, *Correlator Reference Manual.* Available at http://astron.berkeley.edu/~bosco/corman.html.

Urry, W. L.: 2000, A Corner Turner Architecture, *ATA Memo 14*, November 2000. Available at http://astron.berkeley.edu/ral/ATAMemo14.pdf.

Wilson, W. E. and Brown, D. R.: 1987, 'The Australia Telescope Correlator Chip', *21st International Electronics Convention and Exhibition of the IREE Aust., Sydney '87*, pp. 399–402.

RFI MITIGATION AND THE SKA

STEVEN W. ELLINGSON
Virginia Polytechnic Institute and State University, U.S.A.
(e-mail: ellingson@vt.edu)

(Received 18 August 2004; accepted 1 November 2004)

Abstract. Radio frequency interference (RFI) has plagued radio astronomy from its inception. The *Workshop on the Mitigation of Radio Frequency Interference in Radio Astronomy* (RFI2004) was held in Penticton, BC, Canada in July 2004 in order to consider the prognosis for the RFI problem, in particular as it impacts the planned Square Kilometre Array (SKA). This paper concludes that RFI is unlikely to be a "showstopper" in achieving SKA science goals, but that improved RFI mitigation technology may nevertheless be essential in order to take advantage of the vastly improved sensitivity, bandwidth, and field-of-view. Reported results provide some optimism that the desired improvements in RFI mitigation technology are possible, but indicate that much more work is required.

Keywords: radio astronomy, radio frequency interference, Square Kilometre Array

1. Introduction

The Square Kilometre Array (SKA) is an international project, currently in the early planning stages, to build a next-generation radio telescope that is approximately two orders of magnitude more sensitive than any existing instrument at centimeter wavelengths.[1] Radio astronomy at these wavelengths has, since its inception, been plagued by man-made radio frequency interference (RFI) which potentially might be as bad or worse when the SKA is commissioned. Thus, there is great interest in understanding potential limitations that RFI might impose on SKA-enabled science, and how RFI might be mitigated so as to remove those limitations. To this end, the *Workshop on the Mitigation of Radio Frequency Interference in Radio Astronomy* was held in Penticton, BC, Canada in July 2004. In this paper, I attempt to summarize the findings of the workshop as they relate to the SKA. As the proceedings of this conference are publicly available,[2] this paper does not attempt to the review the detailed proceedings of the workshop, but instead presents some conclusions reached by the author in the context of both this workshop and the existing literature on the topic of RFI characterization and mitigation.

---

[1] SKA Project website, http://www.skatelescope.org.
[2] The interim proceedings website is http://www.ece.vt.edu/~swe/rfi2004, to be transferred to a permanent URL in the Fall 2004. Also, papers from this workshop will appear as a special section in the journal *Radio Science*, expected publication date Fall 2005.

2. RFI and radio astronomy

It is convenient to classify man-made RFI as being either *internal*, i.e., generated by the instrument; or *external*, i.e., originating from intentional and unintentional radio emissions generated by our civilization. Internal RFI can be managed through careful design and (undesirable though it may be) modifications after commissioning; whereas external RFI typically cannot, in particular when it appears in the observed bandwidth. Although small bands of frequencies are reserved for radio astronomy by international agreement (e.g., the band around the 1420 MHz rest frequency of HI), science requirements routinely motivate observations outside these bands. For this reason, the focus of this paper is on external RFI, which is clearly a threat that cannot be mitigated solely through traditional radio telescope design principles.

A sample survey of the external RFI environment as perceived by a centimeter-wave radio telescope is that by Beaudet et al. (2003); other site surveys have generated similar results, differing primarily in levels as opposed to utilization of spectrum. In general, external RFI below 3 GHz consists primarily of modulated carriers with bandwidths between 3 kHz and 20 MHz. Ironically, external RFI occupies only a tiny percentage of the radio frequency spectrum, and many of the carriers are associated with mobile radio systems which transmit only intermittently. This has much to do with why astronomers frequently are able to observe outside bands reserved for radio astronomy. In fact, reported cases in which external RFI completely precludes a scientific observation are remarkably few; an example being (Weintroub, 1998). Some observing modes inherently suppress RFI by virtue of the signal processing involved; for example the RFI decorrelating effect of fringe stopping in aperture synthesis imaging (Thompson, 1982), and the tendency of RFI to appear as undispersed (dispersion measure ~ 0) clutter in pulsar surveys. However, only limited mitigation is provided by these effects, and spectroscopy remains relatively vulnerable because RFI is often sufficiently strong that on-source integration alone is unable to satisfactorily suppress it to the desired level of sensitivity.

The more common impact of RFI is that it limits the overall productivity of radio astronomy, making desirable observations prohibitively difficult or expensive. An example is the manual post-observation editing of visibilities to remove RFI, as is sometimes practiced in aperture synthesis imaging. While quite effective, it is difficult to automate and therefore becomes extraordinarily tedious as the observation length and observed bandwidth increase. Another example is the efforts of radio astronomers to observe in the presence of interference from the Russian *GLONASS* navigation satellite constellation (Galt, 1991; Combrinck et al., 1994), which requires a complex scheduling regimen to avoid main beam transits, and which is disruptive to telescope operations (and eventually contaminates spectroscopy in long integrations through sidelobe reception anyway). Such indirect consequences of external RFI may translate into dramatically increased requirements for labor and telescope time, and thus can be as limiting to science as RFI which directly "jams" an observation.

3. Techniques for mitigating external RFI

The study of techniques for mitigating external RFI contaminating the analog output of radio telescopes receivers has been a topic of heightened interest in recent years, spurred on by technological advances that enable signal processing approaches to RFI mitigation. The implicit and overarching principle in such techniques is that RFI can often be effectively mitigated in whatever domain (time, frequency, angular position, polarization, etc.) it is most easily localized. A concise taxonomy of mitigation techniques might be organized as follows[3]: (1) *Excision* (in the sense of "cutting out"); for example by placing nulls in the antenna pattern (Ellingson and Hampson, 2002; Raza et al., 2002), or blanking contaminated time-frequency pixels (Leshem et al., 2000; Ellingson and Hampson, 2003); (2) *Canceling* (in the sense of subtracting), nominally providing a "look through" capability as in reference-signal-driven adaptive canceling (Barnbaum and Bradley, 1998), parametric estimation and subtraction of RFI waveforms (Ellingson et al., 2001), and post-correlation estimation and subtraction of contaminated spectra using reference antennas (Briggs et al., 2000); and (3) *Anticoincidence*, broadly meaning discrimination of RFI by exploiting the fact that widely separated antennas should perceive astronomy identically, but RFI differently. Any classification of techniques however is somewhat misleading, as the specific techniques can be applied similarly in different domains and in various combinations; and some canceling techniques degrade into excision techniques under certain conditions. What is certain is that the effectiveness of any given technique depends on (1) the instrumental configuration (or SKA design concept), (2) the observing mode (e.g., spectroscopy, continuum aperture synthesis imaging, pulsar/dispersion searching), and (3) the nature of the RFI itself (e.g., persistent or intermittent, spatially coherent or scattered by multipath, etc.) such that there is no single technique which can address all possible scenarios.

4. Why RFI mitigation is important for the SKA

The effect of RFI on the SKA can be expected to be comparable to the effect of RFI on existing instruments. This is because the planned 100-fold increase in collecting area, by itself, does not by imply increased sensitivity to external RFI, because the peaks of the far sidelobes of a highly directional antenna pattern tend to be independent of directivity (specifically, the far sidelobes tend to exhibit about the same gain as an isotropic antenna). Furthermore, mitigating effects benefiting existing instruments – such as decorrelation due to fringe stopping in aperture synthesis,

---

[3] References are provided to indicate published examples of techniques mentioned as well as to provide some context for understanding the more detailed presentations and papers of the RFI2004 workshop.

and dispersion measure discrimination in pulsar searches – can be expected to be equally effective for SKA. On this basis, it might be concluded that SKA can make do with RFI mitigation which is no more developed than what is already employed in present-day radio astronomy, and that the result will be scientific output which is at least as good as what is obtained today, scaled to the sensitivity of SKA. However, there are at least three considerations which make this an unwise assumption:

- *SKA key science drivers are likely to motivate observations in which RFI – even at existing levels – will be a bigger problem.* For example, an important use of the SKA will be to observe the 21-cm line of hydrogen at cosmological distances; i.e., high redshifts. The sensitivity of present day observations limits observations to relatively low redshifts, which can be observed in the protected band or detected in searches of modest (10's of MHz) bandwidths. With the SKA, the 21-cm line will be detectable at relatively high redshifts with associated frequencies covering 100's of MHz extending through frequencies used for aviation distance measuring equipment (DME) and ground based radar, both of which are especially problematic for spectroscopy. As described above, RFI mitigation may be required not because the observations are *impossible* without it, but rather because the observations are impractically tedious and expensive (in terms of telescope utilization) without it.
- *RFI mitigation may be required to effectively exploit the increased productivity and flexibility envisioned for SKA.* The SKA is envisioned to be an instrument featuring unprecedented levels of automation, with the ability to conduct multiple observing programs simultaneously and continuously. Combined with the anticipated enormous increases in bandwidth, field-of-view, etc. this leads to an orders-of-magnitude increase of raw data output. Since it is not reasonable to expect proportional increases in the pool of available labor to post-process and analyze this data using present-day methods, these steps must be eliminated or simplified; the penalty for not doing this will be to reduce the science "throughput" for the SKA to a fraction of what is obtained from a single existing radio telescope today. With respect to post-processing and analysis to mitigate RFI, this means that either existing manual excision techniques must be automated, or that the RFI must be mitigated by other means "upstream" in the signal path.
- *External RFI might get worse.* It is difficult to anticipate how the radio spectrum will be used over the years that the SKA is operational, but history suggests that it is likely that the general trend of *increasing* utilization of the radio frequency spectrum by communications and navigation systems will continue. If so, the consequence will be that the time-frequency distribution of external RFI perceived by SKA will become more dense, perhaps overwhelming the mitigating effects which make radio astronomy feasible in the present. In this sense, RFI mitigation is also *risk* mitigation.

There are a variety of other, more specific reasons why RFI mitigation techniques of increased sophistication might be important or essential for SKA. For example, one of the daunting challenges in SKA design is data transport. Because large-N designs entail transport of vast numbers of sample data streams, it is important to minimize sample bit-widths, so as to constrain data rates. Although radio astronomical data can nominally be encoded with as few as 2–4 bits per sample, this can increase to 6–14 bits when strong RFI is present. Thus, RFI mitigation at this point allows sample bit-widths to be minimized, simplifying an already-difficult data transport problem. A second example is the impact of RFI on the emerging area of "transient" science – detecting and characterizing astrophysical signals which may have durations on time scales from nanoseconds to hours. Because RFI is also variable on these timescales, transient science may be particularly vulnerable to RFI and may require active RFI mitigation in order to be yield useful data.

5. RFI2004: Outlook for RFI mitigation technology

Among the objectives of the RFI2004 workshop were to present and discuss the "latest and greatest" in the field of RFI mitigation technology, and to consider the implications for SKA. Readers are encouraged to review the workshop proceedings and make their own judgments; however the impressions of the author are as follows:

- Good news is that RFI mitigation technology continues to show great promise. In the proceedings are demonstrations of effective mitigation against problematic RFI such as L-band radar, DME, Digital TV, Iridium, Instrument Landing Systems (ILS), High-Frequency broadcast signals, and so on.
- On the other hand, optimism about the ultimate effectiveness of sophisticated new techniques is probably premature. The successes described above are anecdotal and for the most part were achieved under carefully controlled conditions known to be germane to the technique applied, so in some sense successes are not surprising. A challenge in future work will be to demonstrate the effectiveness of RFI mitigation in routine observing conditions and with little or no intervention from observers.
- Another concern is that demonstrations of RFI mitigation to date appear to be targeting the strongest forms of RFI; whereas (ironically) the most problematic RFI may be that which has levels detectible only at the extreme limits of sensitivity. Spectroscopy in particular is likely to be vulnerable in this respect. The ability to obtain useable noise-limited spectra at the limits of sensitivity (say, for example, in an 8 hour integration from a large dish) in RFI-afflicted frequency bands has not yet been demonstrated.
- The "tried and true" methods – blanking, exploiting RFI phase closure, editing of visibilities, and so on – still apply and can be greatly improved. Simply

automating tasks which currently must be done manually is likely to address a significant fraction of SKA RFI mitigation needs.

So, while much has been accomplished, there is still much to do in order develop RFI mitigation technology suitable for SKA, and some uncertainty remains as to whether RFI mitigation requirements (vaguely defined as they currently are) can be met.

6. Taking steps to ensure that SKA is not RFI-limited

So what can be done to help assure that SKA reaches first light without being unduly limited by RFI? Some suggested measures include

- *Avoidance.* Locate SKA in a maximally "RFI quiet" location and take steps to minimize internal RFI.
- *Regulation.* Take steps to minimize or control the levels and time-frequency density of external RFI through regulatory activity.
- *Exploit, improve, and extend "tried and true" RFI mitigation techniques.* For example, as described above, automating the task of editing visibilities to remove RFI is likely to be quite useful in meeting SKA RFI mitigation requirements for aperture synthesis imaging.
- *Identify and exploit unique features of the SKA that might facilitate more effective RFI mitigation.* For example, in "large-N" SKA design concepts, the collecting area is divided among thousands to millions of individual "receptors", which can be individually manipulated to null large numbers of interferers (for example, satellites) with only marginal and deterministic degradation of main beam sensitivity. Another example is exploiting fast correlator integration "dump times" to improve the sensitivity and selectivity of time-frequency blanking algorithms, or exploiting configurable correlator architectures to flexibly accommodate post-correlation RFI mitigation techniques.
- *Develop multiple lines of defense.* Today it seems clear that there is no single "cocktail" of RFI mitigation techniques that is effective for all instrumental configurations, observing modes, and forms of RFI. This argues for a "toolbox" strategy, in which a suite of RFI mitigation techniques is available to the observer and which can be implemented, customized, or disabled as observing requirements dictate.

Acknowledgements

The helpful insights and suggestions of the attendees of RFI2004 are acknowledged and appreciated; F. Briggs, R. Ekers, and R. Fisher in particular. Also, A.-J.

Boonstra, B. Jeffs, and W. van Driel (among many others) contributed significantly to success of the RFI2004 workshop and influenced some of the content of this paper. This work was supported in part by the National Science Foundation (NSF) under Award AST-0138263 *via* subcontract to Cornell University.

References

Barnbaum, C. and Bradley, R. F.: 1998, *AJ* **116**, 2598.
Beaudet, C. M. et al.: 2003, RFI Survey at the ALMA Site at Chajnantor. ALMA Memo 470, 15 July 2003. Available at http://www.alma.nrao.edu/memos/.
Briggs, F. H., Bell, J. F. and Kesteven, M. J.: 2000, *AJ* **120**, 3351.
Combrinck, W. L., West, M. E. and Gaylard, M. J.: 1994, *PASP* **106**, 807.
Ellingson, S. W., Bunton, J. D. and Bell, J. F.: 2001, *ApJS* **135**, 87.
Ellingson, S. W. and Hampson, G. A.: 2002, *IEEE Trans. Ant. Prop.* **50**(1), 25–30.
Ellingson, S. W. and Hampson, G. A.: 2003, *ApJS* **147**, 167.
Galt, J.: 1991, in D. L. Crawford (ed.), *ASP Conf. Ser. 17, Light Pollution, Radio Interference, and Space Debris*, ASP, San Francisco, p. 13.
Leshem, A., van der Veen, A.-J. and Boonstra, A.-J.: 2000, *ApJS* **131**, 355.
Raza, J., Boonstra, A.-J. and van der Veen, A. J.: 2002, *IEEE Sig. Proc. Let.* **9**(2), 64–67.
Thompson, A.: 1982, *IEEE Trans. Ant. Prop.* **30**(3), 450–456.
Weintroub, J.: 1998, *Radio Spectroscopy Applied to a Search for Highly Redshifted Protogalactic Structure*, Ph.D. Thesis, Harvard University, 1998. Available at http://seti.harvard.edu/grad/jpdf/thesis.pdf.

A 256 MHz BANDWIDTH BASEBAND RECEIVER/SPECTROMETER

R.H. FERRIS* and S.J. SAUNDERS
CSIRO ATNF, PO Box 76, Epping, NSW, 1710, Australia
(*author for correspondence, e-mail: dick.ferris@csiro.au*)

(Received 19 July 2004; accepted 22 April 2005)

Abstract. Wideband untuned baseband receivers incorporating high-resolution digitisers and polyphase digital filter banks (PDFBs) have been proposed for SKA and LOFAR, and PDFBs are being actively developed as core elements of beamformers, spectrometers and cross correlators for a number of other radio telescopes. Early on-air test results of a 1024-channel, 0–256 MHz baseband receiver, demonstrate excellent spectral purity and robustness against strong in-band RFI.

Keywords: baseband receiver, polyphase digital filter bank, RFI, SKA, spectrometer

1. Introduction

The current development of wideband, even multi-octave coverage, radio telescopes brings with it the requirement of robust receiver technology, which can tolerate the presence of strong in-band signals from both terrestrial and space-born transmitters. The combination of very wide frequency range and wide fractional bandwidth is difficult to achieve with traditional heterodyne receiver designs, and the prospect of building and maintaining thousands of receivers in extensive arrays such as the SKA and LOFAR is a strong driver for finding a simpler architecture.

One solution is the baseband receiver, in which the antenna signal is merely low-pass (or bandpass) filtered, amplified and then sampled and digitised in a moderate to high resolution ADC. Subsequent digital processing provides frequency selection, bandlimiting, etc., and this may often be integrated with the channelisation and other signal processing inherent in the telescope's back-end instrumentation.

Since the astronomy signal is essentially smooth-spectrum Gaussian noise, wide bandwidth means that it is sufficient to provide only the relatively small gain required to lift its integrated power over the inherent noise of the ADC, i.e. to the order of a few quantising steps. The rest of the converter's range remains free to linearly accommodate strong man-made transmissions ("RFI"). Subsequent DSP can then isolate these to their own frequency channels without corrupting the clean spectrum in between, or allowing them to modulate the gain of the instrument.

The DSP of choice is the polyphase digital filter bank (PDFB) (Bellanger and Daguet, 1974; Bellanger et al., 1976), a uniform filter bank equivalent to an array of equally spaced quadrature down-converters, but with a total computational requirement less than that of two individual converters, even for very large number of channels. The logical computation rate is at the relatively slow Nyquist rate of the output channels, rather than the high input rate necessitated by the wide analog signal bandwidth, and the naturally parallel process is readily implemented in contemporary platform FPGAs.

In practically all wideband applications, the high quality channelisation, such as can be provided by a PDFB, is an essential part of the back-end signal processing. Placing the filter bank at the front of the signal path allows the one channelisation process to facilitate many other functions. These include frequency selection, RFI excision and subsequent data compression, hybrid wideband beamforming of sub-arrays including null-steering, and local calibration, as well as the frequency analysis ultimately required. Such utility has made the PDFB an attractive processing element for a number of proposed and current radio telescope developments, involving both baseband and heterodyne receiver front ends. For example, baseband receivers with PDFBs have been proposed for the SKA (Hall, 2002; Ferris, 2002) and LOFAR (Hinteregger, 2003), and PDFB back-end processors have been proposed for the SKA (Ferris et al., 2001) and are actively under development for the ATA (Durkin, 2004) and the ATCA (Ferris and Wilson, 2002).

In this paper, we report test results from a "proof of concept" PDFB spectrometer (PoCS) developed as part of the ATCA project. It achieves 1024-channel, 256 MHz bandwidth operation, with the eventual objective being a 2 GHz bandwidth correlator for the ATCA. For the purpose of demonstration the spectrometer, preceded by an 8-bit ADC, was incorporated as a DC-256 MHz baseband receiver and attached to an MIT prototype LOFAR "Hi-Band" active antenna. With comparable frequency coverage and resolution, this configuration also serves to demonstrate a suitable LOFAR receiver technology, in RFI signal level conditions considerably more demanding than at the preferred site at Mileura. The formal PoCS demonstration involves processing 2-bit digitised IF data from a mm-wavelength receiver on the Mopra radiotelescope. This operation is but briefly reported here as the more interesting performance characteristics achieved in the PDFB are obscured by the effects of such low-resolution digitisation.

2. Hardware description

A simplified model of the spectrometer configured as a baseband receiver is illustrated in Figure 1.

The antenna (Corey and Kratzenberg, 2003) is designed to cover the band 110–240 MHz. It comprises a broadband drooping dipole with integral filter and LNA, with the net gain of the structure falling at lower frequencies to counteract the rising cosmic background level. Its response continues smoothly beyond the high end but falls away sharply below a broad resonance at 80 MHz. It is sufficient for present

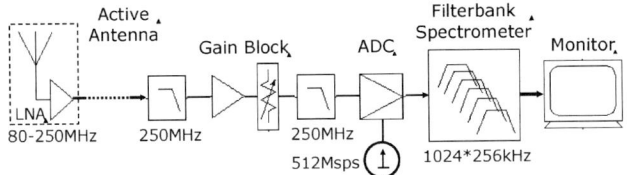

Figure 1. Filterbank spectrometer as baseband receiver.

purposes that there is some response evident throughout the DC-256 MHz band of interest.

Analog gain and lowpass filtering between the antenna and ADC was provided by stock laboratory items with suitable specifications. Approximate net gain was 10 dB for the Sydney data and 35 dB (including 16 dB cable loss) at the Mopra test site. The ADC was a MAX104EVKIT (Maxim, 1999) card providing 8-bit real sampled data at 512 Msps. At this rate the converter is rated at >7.5ENOB.

The equivalent circuit for the filterbank at the heart of the spectrometer is shown in Figure 2. After digitisation the input band is fed to a parallel array of quadrature down-converters (QDCs, a.k.a. 'baseband converters') in each of which it is mixed with a complex local oscillator. This shifts the entire spectrum down by an amount equal to the LO frequency in each case. Real lowpass filters with cutoff $b/2$ then provide a symmetrical bandpass function of width b centred on DC. Since the negative frequency component of the input spectrum is also attenuated by the filter, the signal that emerges from each QDC is complex, and represents a segment of the input spectrum of width b and centre frequency equal to that of the LO. In a spectrometer these signals are not required as such – only their mean power levels. The data streams are also considerably oversampled after their reduction of bandwidth, and can be decimated by a factor M. Usually, as here, $M = K$ so that

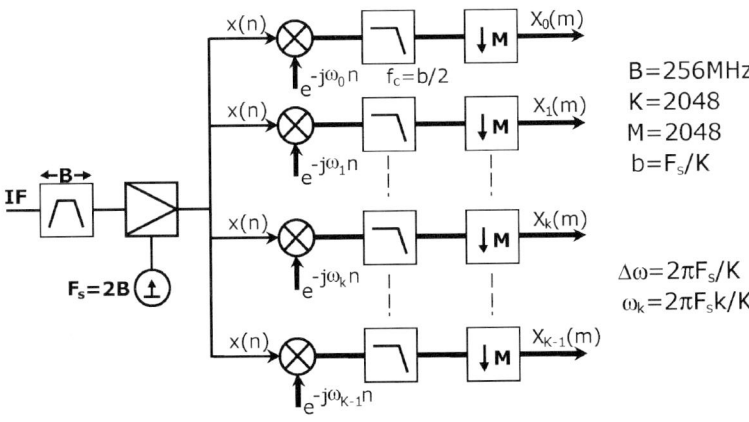

Figure 2. Filterbank modelled as array of quadrature down-converters.

the signals are critically sampled. The modulus squared of each retained sample is then calculated and accumulated to form the spectrum.

The direct implementation of Figure 2 would involve large amounts of expensive high-speed hardware but a number of major simplifications are possible. In each QDC the direct form of the channel filter, operating at the ADC clock rate, is replaced by its polyphase equivalent operating at the decimated rate, so that only the desired output samples are produced. This reduces computation by a factor of M ($=K$). Second, the order of mixing and filtering can be reversed, and since identical filters now process the one ADC data stream, only one such filter is required to serve all the QDCs, yielding another factor of K reduction in filter computations. Finally, the algebra of the mixer products for each QDC is recognisable as one term of a DFT, so by evaluating the entire set together the efficiency of an FFT can be exploited, reducing the computation of this part by a factor of $K/\log K$. The net result is that for filterbanks of a few thousand channels, the total computation load, or hardware requirement, is rather less than that required to implement two instances of the channel filters.

The resulting PDFB structure in Figure 3 shows the high speed ADC data divided into K relatively low speed data streams by a cyclic switch or commutator. Each data stream is processed by one of the subfilters of the decomposed channel filter. The subfilter outputs then comprise the inputs to an FFT, which implements the oscillator-mixer parts of the QDC array. Note that the *underlying* filterbank always spans the frequency range from DC to the sampling frequency in order to use the DFT/FFT. However in the case of real sampled data, the Nyquist bandwidth is only half of this, so the upper half of the channels are redundant. Fortunately it is possible to evaluate both the filter and FFT parts to produce only the useful channels in exactly half the number of significant operations, so computational efficiency is maintained.

The efficiency of the PDFB has allowed the PoCS to be implemented in one three-million gate and two four-million gate Virtex2 FPGAs (Xilinx, 2004), installed in Dime-II development boards (Nallatech, 2002). The hardware is clocked at 128 MHz.

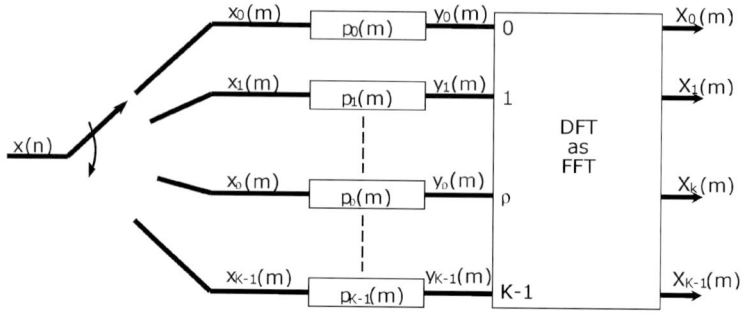

Figure 3. Factorised form of Figure 2. The p_ρ (*m*) together constitute a *single* LPF.

The 1024-channel PDFB comprises a 16384-tap FIR function and a 2048-point DFT. The FIR part is arranged as 2048 distinct 8-tap subfilters in the familiar polyphase structure. 18-bit coefficients ensure the 88 dB stopbands of the prototype filter, which was an equiripple design generated by the Parks–McClellan algorithm. Channel frequency responses overlap at the −6 dB points and full stopband attenuation is achieved about half way through the adjacent channel. This is not necessarily an optimum filter shape for normal filterbank operation, but it is easy to generate, and the deep flat stopbands proved valuable in exposing improper artefacts while debugging the model design in the simulator.

The transform part of the PDFB is implemented as two 2048-point complex FFTs operating in parallel, each performing two real transforms simultaneously to maintain the 512 Msps aggregate throughput. Each 250 kHz channel produces a complex output data stream at 250 ksps. Thus a full, short-time spectrum is available every 4 μs. These are integrated in 72-bit buffers for the nominated period, then rounded to 32 bits for output to the host computer.

3. Sydney test results

Engineering tests verified correct operation of the digital filterbank circuitry and demonstrated that the system dynamic range was limited by the ADC, consistent with its specifications. After adding the other elements of a baseband receiver (Figure 1), measurements of the local radio spectrum were made to provide a more realistic demonstration of the instrument's performance.

The first data presented was obtained outside the laboratory door in suburban Sydney, about 10 km from the site of several TV and FM broadcast transmitters. Receiver gain was increased until the composite antenna signal was just short of saturating the ADC, i.e. "0 dBFS" on the scale in Figure 4.

The spectrum is dominated by various strong FM, DAB, and analog plus digital TV transmissions, but more important are the immediately adjacent areas of clean spectrum at −76 dBFS. These are testament to the ability of the system to accommodate such strong signals while preserving otherwise clean spectrum for astronomy. We note here that adjacent channel rejection is −88 dB by design, and other tests indicate that accumulated roundoff noise is about −85 dB. The recorded background spectrum is a little above the ADC noise floor observed during bench tests.

4. Mopra test results

The second radio spectrum was obtained as the spectrometer was being installed on the Mopra radiotelescope for its formal demonstration. The rural environs of Mopra were expected to exhibit a much lower ambient noise level than the Sydney

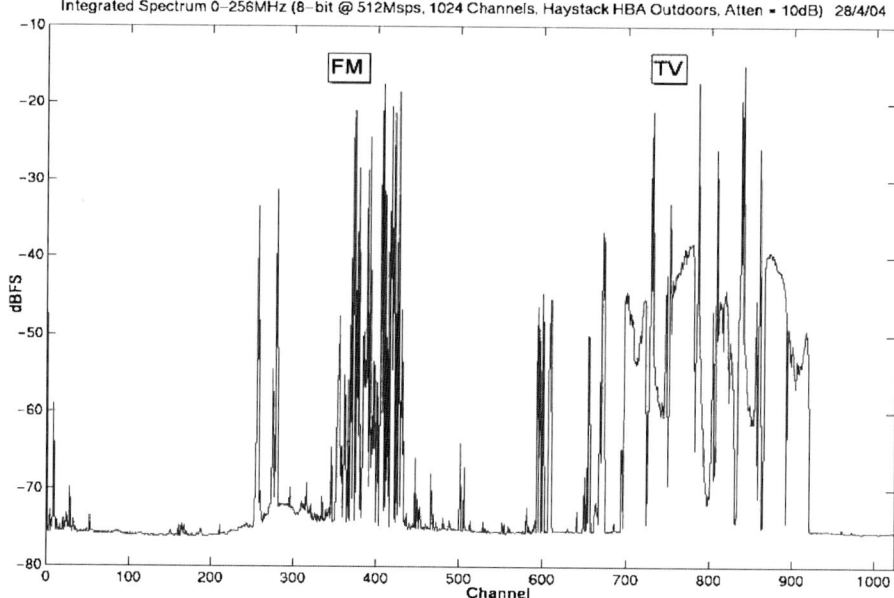

Figure 4. Suburban Sydney spectrum 0–256 MHz.

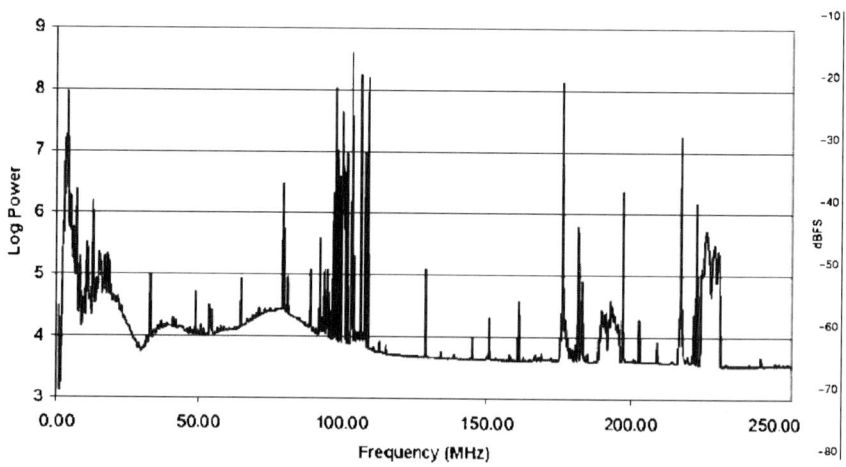

Figure 5. Rural spectrum at Mopra telescope using 8-bit ADC.

city region. Nonetheless the local community is still well served with FM and TV services as the accompanying spectrum shows in Figure 5. It was necessary to greatly reduce the amplitude of the strongest signals by moving the antenna into the lee of a hill so that receiver gain could be increased to the point where the cosmic background spectrum was revealed. Unfortunately the losses associated with the 90 m of thin coax used to effect this deployment tripled the system noise

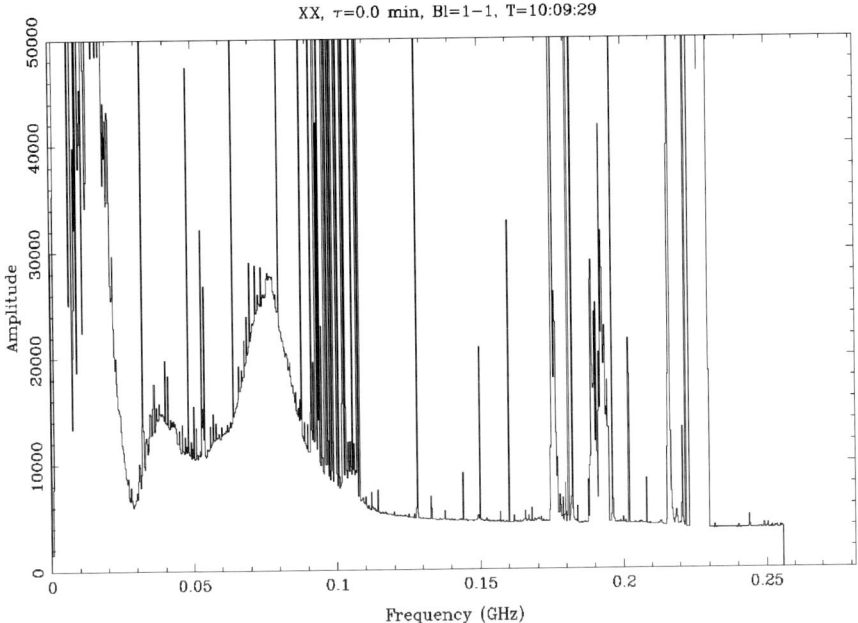

Figure 6. Linear zoom of Mopra baseline.

temperature, and prevented any useful measurement of the true antenna sensitivity. Remote activation of the LNA showed that ~30% of the background spectrum above 110 MHz came from the antenna, consistent with a mean sky temperature of about 300 K at that time.

The ability to see the background noise amongst strong transmissions is again well demonstrated in Figure 6, a linear scaled 'zoom' plot of the baseline region, where at a scale expansion factor of more than 10^5 there is still no evidence of leakage or intermodulation products from the transmissions into the adjacent channels. (The gain peak at 80 MHz corresponds to the antenna resonance, while most of the energy below 50 MHz was independently identified as direct leakage from the site machinery, computers, etc., into the long antenna cable.)

A second experiment at Mopra involved replacing the 8-bit ADC with a 2-bit, 3-level sampler (and a suitably modified version of the spectrometer). Such 2-bit encoders have been the backbone of astronomy instruments for the past three decades, and undoubtedly function well in an environment of pure Gaussian noise. There has been speculation that they will continue to be useful in the presence of many mutually incoherent non-Gaussian signals as these would dither out the obvious non-linearities and their by-products. As the spectrum in Figure 7 shows this is far from true. Whereas the strong signals around 100 and 175–225 MHz are present unchanged (the strong signal near DC is rejected by an internal HPF), the background spectrum is totally obscured by a carpet of harmonics and intermodulation products, and such an instrument is useless in this environment.

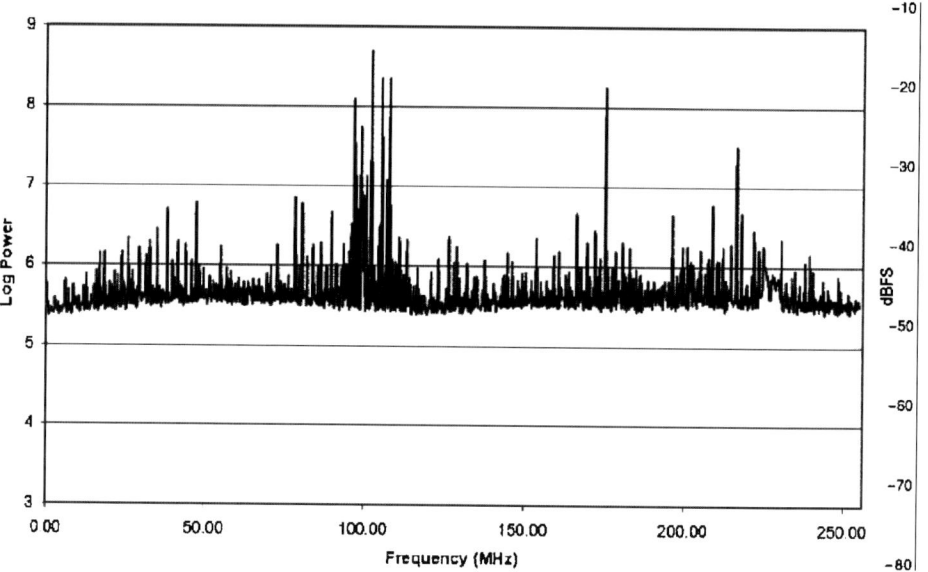

Figure 7. Mopra spectrum with 2-bit digitiser.

5. Conclusion

A PDFB spectrometer with state-of-the-art specifications has been built and shown to operate according to expectations. When driven by a digitiser with sufficient resolution it can provide an instrument that is extremely robust against strong in-band RFI. Whether as part of a low-frequency baseband receiver or on the back-end of a conventional receiver tuned to the increasingly challenging microwave bands, the PDFB can provide an excellent form of spectral analysis.

Acknowledgement

Funding for the upgrade of the ATCA comes from a Major National Research Facility grant from the Australian Federal Government.

References

Bellanger, M.G. and Daguet, J.L.: 1974, *IEEE Trans. Commun.* **Com-22**(9), 1199–1205.
Bellanger, M.G., Bonnerot, G. and Coudreuse, M.: 1976, *IEEE Trans. Acoust. Speech Signal Process.* **ASSP-24**(2), 109–114.
Corey, B. and Kratzenberg, E.: 2003, *High Band Antenna*, LOFAR Delta-PDR Meeting, Alexandria, Virginia, 28–29 October 2003, available: http://www.lofar.org/archive/meetings/oct2003/.
Durkin, T.: 2004, *Xcell J.* **48**, Q1-2004, available: http://www.xilinx.com/publications/xcellonline/xcell_48/xc_toc48.htm.

Ferris, R.H., Bunton, J. and Stuber, B.: 2001, *A 2 GHz Digital Filterbank Correlator*, SKA: Defining the Future, Berkeley, California, 9–12 July 2001, available: http://www.skatelescope.org/skaberkeley/.

Ferris, R.H.: 2002, *A Baseband Receiver Architecture for Medium-N SKA*, International SKA Meeting, 13–15 August 2002, Groningen, Netherlands, available: http://www.lofar.org/ska2002/.

Ferris, R.H. and Wilson, W.E.: 2002, *A Wideband Upgrade for the Australia Telescope Compact Array*, URSI XXVIIth General Assembly, Maastricht, 17–24 August, 2002, poster 1629, available: http://www.atnf.csiro.au/technology/electronics/.

Hall, P. (ed.): 2002, *Eyes on the Sky: A Refracting Concentrator Approach to the SKA*, SKA Memo No. 22, available: http://www.skatelescope.org/.

Hinteregger, H.: 2003, *Wideband Architecture Option(s)*, Haystack LOFAR Memo 12, available: http://web.haystack.mit.edu/lofar/.

Maxim Integrated Products: 1999, *Data Sheet 19-1503*, available: http://www.maxim-ic.com/.

Nallatech: 2002, available: http://www.nallatech.com/.

Xilinx: 2004, *Data Sheet DS031*, available: http://www.xilinx.com/.

THE SQUARE KILOMETRE ARRAY MOLONGLO PROTOTYPE (SKAMP) CORRELATOR

TIMOTHY J. ADAMS[1,*], JOHN D. BUNTON[2] and MICHAEL J. KESTEVEN[3]
[1]*School of Physics, University of Sydney, Australia;* [2]*CSIRO ICT Centre, Australia;*
[3]*CSIRO ATNF, Australia*
(**author for correspondence, e-mail: tadams@physics.usyd.edu.au*)

(Received 18 July 2004; accepted 28 February 2005)

Abstract. An FX correlator implementation for the SKAMP project is presented. The completed system will provide capabilities that match those proposed for the aperture plane array concept for the SKA. Through novel architecture, expansion is possible to accommodate larger arrays such as the 600-station cylindrical reflector proposals. In contrast to many current prototypes, it will use digital transmission from the antenna, requiring digital filterbanks and beamformers to be located at the antenna. This will demonstrate the technologies needed for all long baseline antennas in the SKA.

Keywords: correlator, FX, SKA, SKAMP

1. Introduction

The Square Kilometre Array Molonglo Prototype (SKAMP) project aims to demonstrate Square Kilometre Array (SKA) technology on the Molonglo Observatory Synthesis Telescope (MOST), located in rural New South Wales, Australia. The MOST is an 18 000 m² collector, consisting of two cylindrical paraboloids, 778 × 12 m, separated by 15 m and aligned East–West (Mills, 1981). Each paraboloid is divided into 44 bays, with each bay using a line feed consisting of 88 circular dipoles. Currently each bay is beamformed to a single polarisation signal channel, thus in total 88 signal channels are returned for processing.

The SKAMP project aims to achieve 300[1]–1400 MHz dual polarisation operation with 100 MHz bandwidth and 2048 spectral channel capability by 2007. In order to achieve this the project has been divided into three stages. Stage 1, continuum operation on a single polarisation, will demonstrate SKA processing technology using the current antenna, with a 3 MHz bandwidth centred at 843 MHz. To simplify down conversion, a correlator with 7.33 MHz bandwidth for one spectral channel and one polarisation will be demonstrated. Stage 2, spectral line operation, will see the system bandwidth increased to 30 MHz and the number of

---
[1] The lower frequency limit depends on linefeed developments.

spectral channels increased to 2048. Stage 3, wide band operation, requires a new linefeed at the focus of the cylinder, which allows dual polarisation observations within the 300 to 1400 MHz band. At this stage the observing bandwidth will be increased to 100–200 MHz.

At the core of the SKAMP is the correlator, responsible for forming baseline visibilities from the incoming antenna data. Two generations of correlator are proposed for SKAMP these being the continuum correlator and the spectral line correlator (an expanded spectral line correlator will be used for wide band operation). This article describes the correlator architecture, outlines current work, and demonstrates how this architecture provides for easy expansion to accommodate other SKA demonstrators and, eventually, the SKA.

2. Correlator architecture

The SKAMP requires a large correlator: 88 signal channels along with 8 Radio Frequency Interference (RFI) mitigation channels, amounting to a correlator capable of processing 96 dual polarisation signal channels. Each signal channel will be filtered to 2048 spectral channels. Figure 1 shows the complete Stage 3 correlator system. An FX architecture, where the frequency transformation (F) precedes the cross multiplication operation (X) (Chikada, 1984), is implemented using a polyphase filterbank and digital beamformer at the antenna, and cross multiplication and accumulation (XMAC) in the control room. The two components communicate through a digital optical link. Figure 1 also displays the facilities available for handling an increased number of antennas, using corner turning and channel re-ordering to distribute all antennas over a small frequency range, to multiple processing units (Urry, 2000). Each component of the system is now summarised.

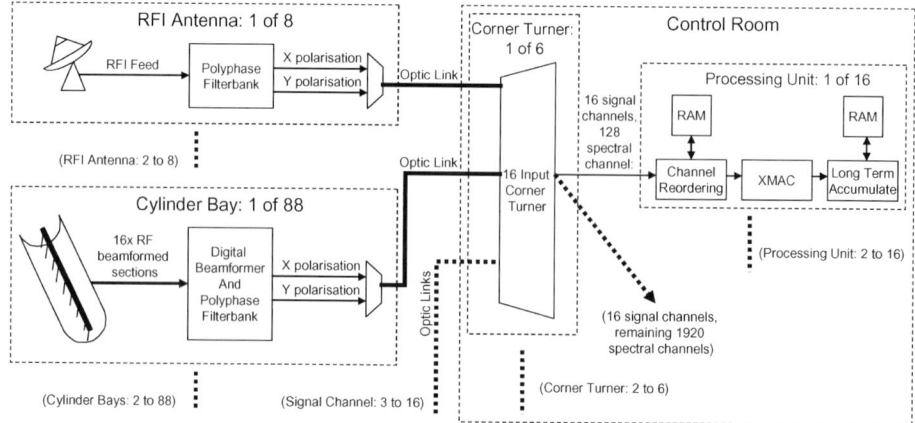

Figure 1. Stage 3 correlator block diagram.

2.1. DIGITAL BEAMFORMING

The cylindrical reflector concept offers the prospect of large fields-of-view for the SKA. Beamformers distributed along the cylinder are required to achieve this. The economics of data transport and processing will set the limit on the size of the processed field-of-view. SKAMP will demonstrate beamforming technology. For the current single polarisation feed, the linefeed in each bay will be divided into four sections. Each section will be RF beamformed and the four outputs will be digitally beamformed to form a single beam. In Stage 3 the linefeed for each bay will be subdivided into 16 sections, increasing the field-of-view and allowing multiple beams to be formed. Only one of these beams will be correlated.

2.2. DIGITAL POLYPHASE FILTERBANK

One dual polarisation, 2048 channel digital polyphase filterbank will be required per signal channel. The polyphase filterbank has been chosen to perform the frequency transformations as it provides a channel response that is flat across the passband, with a small transition band, while at the same time not increasing the data rate (Bunton, 2004). This ensures that the FX architecture will achieve similar performance to an XF architecture. It will be implemented using a field programmable gate array (FPGA), as already demonstrated by CSIRO Australia Telescope National Facility (Ferris, 2001) and the University of California, Berkeley, SETI project (Werthimer, 2004).

2.3. DIGITAL TRANSMISSION

For SKAMP there are 16 analogue signals from each bay. Digitising these and beamforming at the antenna greatly reduces the amount of data to be transmitted. Thus it is advantageous to beamform at the antenna and transmit the data digitally. The dual polarisation digital output of the polyphase filterbanks is serialised onto one single mode optic fibre per bay operating at 1.6 Gbits s^{-1}.[2] The data is transmitted back to the control building, where corner turning followed by cross multiplication and long-term accumulation take place.

2.4. CORNER TURNER

To allow for increased correlator capacity, the cross multiplication and long-term accumulation function is performed by multiple processing units. A method is required to distribute the computational load to the processing units without significantly increasing data interconnections. The SKAMP will achieve this by using

---

[2] For 8 bit sampling at 100 MHz.

multiple corner turners to distribute all antennas over a small frequency range, to multiple processing units (Urry, 2000). For wide band operation, 16 processing units will be required (see Section 3). Therefore for simplicity, the 96 signal channels arriving at the control room at 100 MS s^{-1}, will be organised into groups of 16. Each group will be connected to a corner turner board, resulting in a total of 6 corner turner boards. Each board will be equipped with 16 fibre optic receivers (including serial to parallel converters) and a Xilinx Spartan III XC3S1000 FPGA.[3] The FPGA will act as a switch matrix to distributes 128 spectral channels from each signal channel to the 16 processing units. This is achieved by skewing the incoming data in time from one antenna to the next, then multiplexing all antennas to each of the 16 output lines. This results in the incoming data format X_{ant}(Freq, Time) being re-organised to the output data format format X_{corr}(Ant, Freq, Time). The 16 output lines each operate at 100 MS s^{-1} and are each connected to a processing unit. Therefore, each processing unit will have 6 input data lines, each containing 16 antennas for the same 128 spectral channels.

2.5. Channel reordering

In order to gain most efficient use of the processing hardware, it is operated at a frequency much greater than the data rate, and multiplexed to process multiple spectral channels. The channel reordering facilitates the multiplexing by buffering 64 timestamps from the 16 signal channels passed from the corner turner, to double buffered RAM. Once 64 timestamps are buffered, a sample containing all antennas for all timestamps, for one spectral channel at a time, is sequentially passed to the cross multiply accumulator for processing. All of the 128 buffered spectral channels can be processed by one processing unit before a new sample of 64 timestamps is ready for processing.

2.6. Cross multiply accumulator (XMAC)

The cross multiply accumulate unit uses a double buffered input, and a multiplexed complex cross multiplier array to achieve optimal usage of the resources available (Adams and Bunton, 2003). Data is processed in packets of 64 containing all antenna data at a particular frequency. Antennas are allocated to groups, where the group size corresponds to the dimensions of the XMAC array. For the wideband correlator, each processing unit will have an XMAC with a 16 × 16 array of complex cross multipliers. Therefore, the antennas are placed in groups of 16, and the groups are cross multiplied until all combinations are processed. A total of 18 cycles are needed to generate all 4560 correlations (Bunton, 2001). The results of each complex cross

---

[3] The particular FPGA was chosen for its high I/O count. We require 16 channels × 8 bits × 2 (data input and output) = 256 I/O for data alone.

multiplication are accumulated on the XMAC, and passed to external long-term accumulation RAM at the end of each packet of 64 for longer (i.e., one second) accumulation.

3. Current status

Stage 1 (continuum operation) of the SKAMP is almost complete. Continuum operation required new signal channel samplers, digital delay lines, fringe tracking and a 96 station, 7.33 MHz bandwidth, continuum correlator. Each of these components has been designed and built, and testing is currently underway. Continuum operation is expected to commence in early 2005.

The continuum correlator processing unit is built using 21 Xilinx Spartan IIE 300 000 gate FPGA's, with 16 arranged in a 4×4 array for XMAC processing, and the remaining 5 implementing input buffering and computer interfacing. The correlator is shown in Figure 2. The array is 18 times multiplexed and processes at an array frequency of 132 MHz. All channel reordering and long-term accumulation is implemented on chip to simplify the design (Adams and Bunton, 2003). With advances in technology, it is now possible to implement the entire SKAMP continuum correlator in one Xilinx Virtex II Pro device (i.e. XC2VP70) using the built-in 18×18 bit multipliers and logic resources to create the complex cross multipliers, and onboard RAM for buffering. The array frequency could also be increased, allowing for increased input bandwidth.

Figure 2. SKAMP continuum correlator.

SKAMP continuum operation demonstrates the cross multiply and long-term accumulation features of the correlator architecture. For spectral line and wide band operation, the XMAC FPGA software design can remain unchanged but the hardware will be upgraded to facilitate increased bandwidth and dual polarisation operation. The channel reordering and long-term accumulation will be coordinated by separate FPGA's (i.e. Xilinx Spartan III) to the XMAC using external RAM. The complete processing unit will be duplicated, with the number of processing units required dependant on the observed bandwidth and number of polarisations. For spectral line operation, using a Xilinx Virtex II Pro XC2VP70 to implement the XMAC, we could simply use 3 processing units with 32 MB of RAM each,[4] and for wideband operation we would require 16 processing units with 16 MB of RAM each. Fewer processing units would be required if the XMAC operating frequency was increased, or the number of complex multiplier units within the XMAC FPGA was increased. This could be achieved by using a more complex FPGA (i.e. Xilinx Virtex 4 XCE 4VSX55).

4. Expansion for SKA

As has been shown in Section 3, the correlator architecture easily accommodates expansion of observed bandwidth and spectral coverage. It can also be shown that an increase in the number of antennas is also comfortably accommodated. This flexibility makes the SKAMP correlator architecture an excellent option for the SKA.

A comparison of correlator resource requirements for various SKA options is shown in Table I. This comparison uses station quantity data from the 2003 IEMT report (Hall, 2003), to calculate the magnitude of SKAMP architecture processing hardware required to process one beam from each station. It assumes the use of a single Xilinx Virtex II Pro XC2VP100 FPGA with 400 complex cross multipliers operating at 233 MHz for the XMAC unit, 2048 spectral channels, 64 sample buffering, two Stokes parameters and a 250 MHz observed bandwidth.

It can be seen that as the number of antennas increases, the processing resources required simply duplicate to accommodate the additional computational load. To process multiple beams from each station, the hardware would simply duplicate for each additional beam. However, as the number of processing units increases, the interconnect complexity and I/O requirements of the corner turners also increase. Thus for large antenna quantities, a more powerful processing unit is desirable, to reduce the number of processing units required.

Other SKA concepts demonstrated by the SKAMP project include the locating of digital filterbanks and beamforming at the antenna. All SKA concepts will require

---

[4]Assumptions: 256 complex cross multipliers per XMAC operating at 233 MHz. Channel reordering buffers 64 samples. RAM size includes double buffered channel reordering and 1 second long-term accumulation.

TABLE I

Comparison of correlator resource requirements for various SKA options

SKA option	No. of stations	No. of processing units (PU)	No. of spectral channels per PU
Aperture array	100	27	76
Luneberg lens	300	256	8
Cylindrical reflector	600	1024	2

beamforming to some extent, as the economic cost of transmitting the entire field of view back to the central processing facility, and processing it, is prohibitive. The digital beamformers provide a data format ready for digital transmission from the antenna, which also demonstrates a key SKA concept.

5. Conclusions

The SKAMP offers a valuable demonstration of SKA correlator technology through its FX architecture, large number of antennas and easy expansion ability. An SKA implementation could easily be achieved without significant alteration to the SKAMP architecture.

Acknowledgements

The authors would like to thank Duncan Campbell-Wilson and Anne Green for their assistance in preparing this paper.

References

Adams, T. J. and Bunton, J. D.: 2003, 'The SKA Molonglo Prototype Correlator,' *SKA 2003*, Geraldton, W.A., Australia.
Bunton, J. D.: 2004, 'SKA Correlator Advances', *Exp. Astron.* this issue.
Bunton, J. D.: 2001, 'A Cross Multiply Accumulate Unit', *ALMA Memo No. 92*.
Chikada, Y.: 1984, 'Digital FFT Spectro-Correlator for Radio Astronomy', in J. A. Roberts (ed.), *Indirect Imaging*, Cambridge University Press, pp. 387–404.
Ferris, R. H., Bunton, J. D. and Stuber, B.: 2001, 'The SKA: Defining the Future, Berkeley California', http://www.skatelescope.org/skaberkeley.
Hall, P. J.: 2003, The International Engineering and Management Team (IEMT) 'Report to the International SKA Steering Committee', http://www.skatelescope.org/documents/
Mills, B. Y.: 1981, *Proceedings of the Astronomical Society of Australia* **4**, 156–159.
Urry, W. L.: 2000, 'A Corner Turner Architecture', *ATA Memo*.
Wertheimer, D. and Parsons, A.: 2004, 'A Multipurpose Spectrometer for SETI and Radio Astronomy', http://seti.berkeley.edu

COTS CORRELATOR PLATFORM

KJELD VAN DER SCHAAF* and RUUD OVEREEM
ASTRON, P.O. Box 2, 7990 AA Dwingeloo, The Netherlands
(*author for correspondence, e-mail: schaaf@astron.nl*)

(Received 23 July 2004; accepted 28 February 2005)

Abstract. Moore's law is best exploited by using consumer market hardware. In particular, the gaming industry pushes the limit of processor performance thus reducing the cost per raw flop even faster than Moore's law predicts. Next to the cost benefits of Common-Of-The-Shelf (COTS) processing resources, there is a rapidly growing experience pool in cluster based processing. The typical Beowulf cluster of PC's supercomputers are well known. Multiple examples exists of specialised cluster computers based on more advanced server nodes or even gaming stations. All these cluster machines build upon the same knowledge about cluster software management, scheduling, middleware libraries and mathematical libraries. In this study, we have integrated COTS processing resources and cluster nodes into a very high performance processing platform suitable for streaming data applications, in particular to implement a correlator. The required processing power for the correlator in modern radio telescopes is in the range of the larger supercomputers, which motivates the usage of supercomputer technology. Raw processing power is provided by graphical processors and is combined with an Infiniband host bus adapter with integrated data stream handling logic. With this processing platform a scalable correlator can be built with continuously growing processing power at consumer market prices.

Keywords: correlator, COTS, streaming data, high performance computing, parallel processing

1. The environment

1.1. THE NEED FOR COMPUTER POWER

Radio astronomy started several decades ago receiving signals that were collected with a single dish and a single receiver. By building bigger dishes and better receivers, the resolution of radio astronomy images improved. When in the late 60's the physical limit of dish-sizes was reached, synthesis radio telescopes were invented in which the signals of several radio telescopes are combined (Thompson et al., 1994). This technique is so common nowadays that the radio telescopes all over the world are often combined to obtain the highest possible resolution.

As time passes the the requirements of astronomers increase, in particular with regard to sensitivity and resolution. Instead of several large dishes with single receivers, telescopes of the future will use large numbers of smaller receiving areas, often with multiple receivers. Typically these are small dishes or phased array systems. This increase in the number of receivers stresses the need for computer

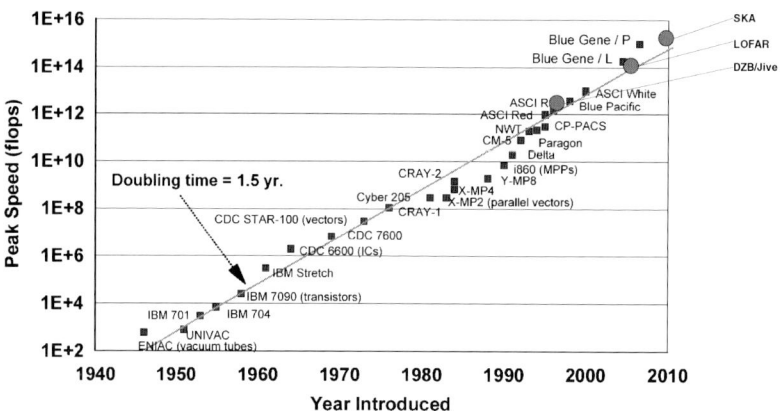

Figure 1. Comparison of processing power of radio telescope correlators with supercomputers.

power since the signals from each antenna must be correlated with signals from all other antennas.

The correlators in modern radio telescopes have processing capacities comparable to the largest computers in the world. This has been so in the past and the radio telescopes currently under construction and planned will continue this trend, as is shown in Figure 1. From this comparison between correlators and supercomputers we can conclude first of all that it should be possible to build correlator sections using supercomputer technology, as is done in the LOFAR telescope (de Vos et al., 2001; Schaaf et al., 2003).

Secondly, we can conclude that we might benefit from the developments in supercomputer technology, potentially resulting in price advantages, reduced development effort and extra flexibility and scalability. For example, we can benefit from the high bandwidth networks and associated middleware libraries, considerably reducing development costs. These arguments are valid in particular for COTS-based supercomputers. This paper presents a COTS based correlator platform prototype, thus providing good arguments to consider supercomputer based correlator platforms for future radio telescopes, in particular the Square Kilometre Array (SKA).

Table I gives more detail of the processing power needed in some radio telescopes. We can observe that a large number of complex multiplications have to be performed, ranging to the order of Peta flops. Fortunately, the calculation of the correlation of these signals is something that can easily be done in parallel. So increasing the number of CPU's eventually gives enough computing power to do the job, although power consumption becomes a major issue. Looking at the computer hardware market, it is not certain that the development of CPU's will keep up with Moore's law. The solution presented here is the use of modern graphics cards. Today, these cards are highly programmable and have several parallel pipelines optimised for streaming data processing (Figure 2). Furthermore,

TABLE I
Correlator processing power in modern radio telescopes

Telescope	No. of stations	Bandwidth (MHz)	Word size (bits)	Correlation power (T Complex 2-bit mult/s)	Technology
WSRT	14	160	2	0.06	Dedicated ASIC on custom boards
VLA	27	160	2	0.22	Dedicated ASIC on custom boards
ALMA	64	4000	2	31	Mask Prog. Logical Devices on custom boards
LOFAR	109	32	8	12	General purpose supercomputer
SKA	~100	1000	8	305	To be decided

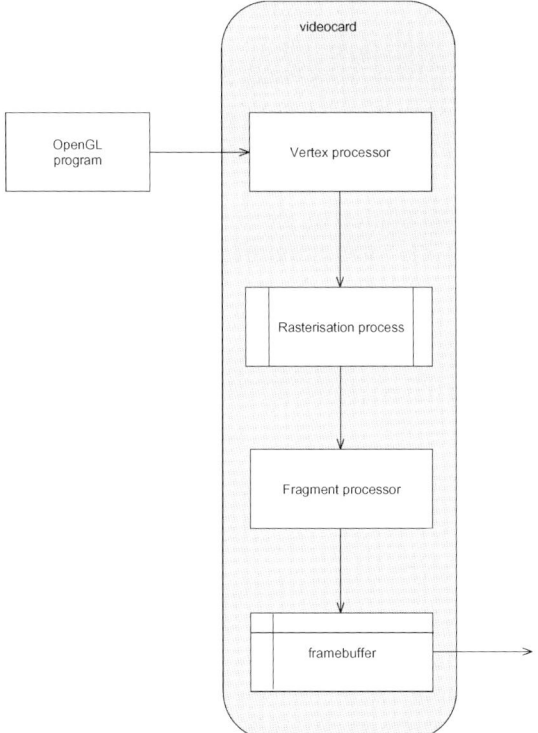

Figure 2. Simplified representation of the processing pipeline current graphics cards.

the parallelism on these cards has just started and it is likely that the development of GPU's can easily keep up with Moore's law, as is shown during the last few years. Additionally, GPU's are cheaper than CPU's and have a favourable form factor.

1.2. CORRELATING SIGNALS

The correlator is a basis function in an interferometer. The correlator produces the cross correlation of the input voltages fed to it. In modern radio telescopes cross correlations are calculated for all interferometer pairs available in the array. The inputs voltages for the correlator stem from individual antennas or from a phased array of antennas. In the common FX architecture, a Fourier transform ("F") is applied to the digitised antenna voltage time series before correlation ("X") (see Chikada, 1991). The task of the correlator is to produce all cross correlation product between corresponding samples from all antennas:

$$\forall_{i,j<i} : r_{i,j}(t) = V_i(t)V_j(t+\tau)$$

where i is the set of antennas. The delay τ of the signal V_j with respect to V_i is caused by the geometry of the source with respect to the antenna array and by instrumental delays. The correction of the geometrical effects are the fringe rotation and delay tracking functions. See e.g. Thompson et al. (1994) for more information on interferometry. In this study, we have concentrated on the correlation itself, considering the calculation and application of the time delay τ as a pre-filter function.

What remains is the calculation of the correlator output matrix resulting from a single set if input data. Time integration over series of 512 samples is applied on the output matrix in order to reduce the resulting data stream.

The spectral nature of the correlator allows one to split the correlation tasks into multiple smaller ones, each handling small a frequency window. Thus a single small correlator building block can be the basis for a large correlator.

1.3. MODERN GRAPHICS CARDS

1.3.1. *Architecture of the cards*

Before you can use a graphics card for scientific calculations you must realize that the current graphics cards are not designed with scientific calculations in mind; the design is completely focused on calculating the color of pixels as fast as possible. However, there is a growing awareness of the scientific users which will probably result in additions to the programmers interface in future graphics cards. In Figure 3 a simplified overview of the processing pipeline of a graphics card is shown, as it is applied in the current prototype. Note that the vertex processor and the fragment processor are programmable by the user.

Since graphics cards are designed to draw objects on the screen, we first describe how that process works. Later, we will describe how to use this process to perform useful calculations for us.

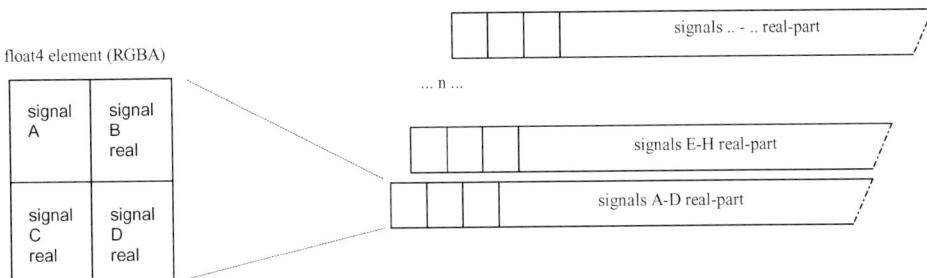

Figure 3. Optimised usage of the graphics card processing pipelines. Parallel processing in 4 pipelines is performed by mapping the input signals on a 2-dimensional texture. Shown here is the mapping of the real component is the input signals on a 2-dimensional texture; the vertical dimension n can be varied.

When you want to draw an object on your screen you need a wire-model of your object. To visualize this object to following steps are performed:

1. In your OpenGL program you send the coordinates of the junctions of the wire-model to the graphics card, specifying the color of each point.
2. This input is send to the vertex processor where you can program all kind of translation and rotations.
3. The resulting coordinates are then processed by the rasterisation processes to calculate which fragments (pixels) correspond with the area enclosed in the vertex coordinates.
4. The fragment processor is called for every fragment and should calculate the color of the fragment. For this calculation all kinds of fancy functions and mappings are available. These will be discussed later. The fragment processor has 4.16 parallel pipelines.
5. Finally the color of the pixel is written to the frame buffer that can be projected on the screen.

Looking at this pipeline we can conclude that the result of everything we do on the card will be placed in the framebuffer. To retrieve this result the framebuffer must be read or some 'screen scraping' tool must be used.

1.3.2. *The program environment*

NVIDIA has developed a special language called Cg (C for graphics) to write programs for the vertex- and fragment processor. As the name indicates, Cg is an extension to C. NVIDIA has mainly added some data types as float2, float4, float4 × 4 etc., and provided some keywords to address graphics-related issues.

Cg comes with a library that is loaded with mathematical functions for mapping and transforming vectors and matrices.

NVIDIA developed the Cg language by mutual agreement with Microsoft so that you can use Direct3D instead of OpenGL if you prefer that.

Vertex- and fragment-programs must be written in separate .Cg files and can be compiled and loaded during runtime.

2. Building the correlator

When trying to use a graphics card as a co-processor, the main question is how can we use the pipeline of the graphics card as efficiently as possible. One issue to take into account is that current graphics cards use the AGP bus for data communication. This means that data can be written to the card rapidly, but reading data from the bus is slower and may very well become a bottleneck.

The solution we describe here uses the following steps:

1. Textures are used for exchanging the data between the CPU and the GPU.
2. We project the correlation matrix on the framebuffer in the ratio 1:1. This means that one element of the matrix corresponds with one pixel in the frame buffer.
3. We draw a triangle that covers the area of the correlation matrix we have to calculate.
4. The rasterisation process is used to calculate which pixels should be 'colored' (=which correlations should be calculated).
5. The fragment-processor is loaded with a program to calculate the correlation between to data samples. The result is written as a color-component to the framebuffer.

The advantages of this method are:

- minimal data transport to and from the card. The data signals are transferred to the card only once and the result is read back after 512 integrations.
- optimised usage of the rasterisation process and parallelism. The rasterisation process addresses only the matrix-elements of the triangle we have to calculate and the parallel pipelines of the fragment processor are used as parallel as possible.

The disadvantages are:

- the vertex processor is not used. There is no useful work to do for this processor yet.
- we still have to fetch the results from the framebuffer. The results have to be downloaded from the AGP bus which is relatively slow. Using e.g. an HP SEPIA board may speed up this part of the process.

2.1. Use of videocard concepts

2.1.1. Textures
In the world of graphics textures are often used to project a surface onto an object. A texture is usually a 2 dimensional picture, but it can also be 1 or 3 dimensional. After loading the texture on the graphics card the usage of a texture is simple: before you pass the vertex coordinate to the card you first pass the corresponding texture coordinate. The rasterisation process calculates for every fragment the correct texture coordinates.

It is possible to use more than one texture at the time by passing the coordinates for each texture to the card before passing the vertex coordinates.

The correlator uses three textures: an empty input texture, the result-texture and the signal-texture.

- The input texture is 2D and has the same size as the correlation matrix and contains only 0 values. It is used to reset the calculation result on the GPU.
- The result-texture is also 2D and covers the whole correlation matrix also. It is used at the end of a series of correlation to retrieve the results.
- The signal texture is 1D and contains the data-samples to calculate the correlation of.

2.1.2. The framebuffer
When you create a window in the X-window environment a framebuffer is allocated with 3 bytes per pixel (RGB). This buffer meets the requirement for the screen but cannot be used to do the calculations. To take maximum advantage of the power of the graphics card we need a framebuffer that has 4 floats per pixel. When the program starts it creates this framebuffer on the card.

2.1.3. Triangle
To calculate all the necessary elements of the correlation matrix we simple pass the coordinates of a triangle to the card together with the coordinates of the textures. The rasterisation process will do the hard work for us.

2.1.4. Rasterisation
Since we don't do anything in the vertex process the rasterisation process receives the texture-coordinates and vertex coordinates exactly as we passed them in our OpenGL program. The rasterisation process calculates which fragments/pixels are covered by the triangle we just draw. For every pixel the program of the fragment processor is executed.

2.1.5. Fragment processor
The fragment processor code is described in two iterations. First we describe the basic model which shows the functional flow. Then we describe the enhanced model that was used in the demonstrator and has better performance.

Basic model: The fragment processor receives 4 input values each time it is executed:

1. Coordinates for the input-texture
2. Coordinates for the signal-texture
3. Pointer to the input-texture
4. Pointer to the signal-texture

The simplified algorithm is:

$$I_i = signal[input\_texture\_coord]$$
$$I_j = signal[input\_texture\_coord]$$
$$P_{i,j}red+ = \Re(I_i)\Re(I_j) - \Im(I_i)\Im(I_j)$$
$$P_{i,j}green+ = \Re(Ii)\Im(I_j) + \Im(Ii)\Re(I_j)$$

where $\Re$ and $\Im$ denote the real and imaginary components of the Fourier transforms of the digitised time series data. This algorithm is applied to all pixels i, j. Note that the coordinates for the signal-texture are not used.

This algorithm uses only half of the processing power of the graphics card because the blue and alpha channels are not used. With a small modification to the algorithm these channels can also be used, almost doubling the speed.

The correlation result is added to the existing result to make it easy to integrate over a number of samples: simply repeat passing the new data samples and drawing a triangle. After all integrations are done the frame-buffer is read back from the card and the (still empty) input-array is uploaded to the card again to start the next cycle.

Enhanced model: To make better use of the GPU architecture we should reformat the way the data is passed to the fragment processor. The fragment processor has 4 pipelines that can execute the same instruction in parallel. Moreover, we can also use the blue and alpha values of a pixel. We can make use of all this processing power by mapping of the input signal I_i on a 2-dimensional texture, as shown in Figure 3. Separate textures are used for the real and imaginary part of the input signals. Each element in a texture is filled as follows:

$$k = i/4$$
$$R\_Texture_{k,red} = \Re(I_i)$$
$$R\_Texture_{k,green} = \Re(I_i + 1)$$
$$R\_Texture_{k,blue} = \Re(I_i + 2)$$
$$R\_Texture_{k,alpha} = \Re(I_i + 3)$$

And identically for the imaginary part of the signals, a texture $I\_Texture_{k,rgba}$ is filled. After the multiplications of the texture, we can collect the part of the complex multiplications and construct the output correlation matrix.

We can use the multiple pipelines by making the a 2-dimensional texture. In this case each pipeline is calculating part of the correlation matrix. Since our target graphics card contains four pipelines, the optimal performance is reached when the two correlation sub-matrices are calculated, each utilising one pipeline for the real parts and one for the imaginary parts.

2.2. RESULTS

The tests were performed on a Dual Xeon Linux server with two different graphics cards. The first card is the NVIDIA 5900XT from the former generation (Overview of the NVIDIA 5900 http://www.nvidia.com/object/LO_20030508_6881.html), which is now priced relatively cheaply. The other card is the newest NVIDIA GeoForce 6800 Ultra card (Overview of the NVIDIA 6800 http://www.nvidia.com/object/LO_20030508_6881.html), which was just released and priced accordingly. The host PC ran RedHat 9.0 and an $8 \times$ AGP bus was used. The programs were compiled with gnu 3.2.2 with $-O3$ optimalisation.

In Table II the results for integrating 512 samples are shown. These results were measured with the openGL driver 1.4.1 for the 5900 card, and 1.5.1 for the 6800 card. The basic performance measure is the number of complex multiplications (CMAC) that are executed per second. The same processing power is also shown in equivalent floating point operations per second (Flops), counting 7 floating point operations for a complex multiplication.

The optimal size of the correlation matrix for the CPU version is 32×32 or 64×64. In these cases the cache is used most efficient. The time per correlation is 31 ns on machine 1 and 10.8 ns on machine 2.

With the graphics card the loading and retrieving of the textures is the bottleneck. The optimal matrix size for the GPUs is larger (256×256–512×512). The power consumption of the host CPU and GPU were measured. In the table, the total

TABLE II

Measured processing power for CPU and graphics cards. the quoted power consumption for the GPUs is the difference of the total system with during calculations with the idle host PC without the graphics card

	CPU 2.8 GHz	GPU 5900XT	GPU 6800Ultra
Optimal matrix size	64×64	128×128	256×256
10^6 CMAC/s	92	132	428
GFLOPS	0.6	0.9	2.9
Power consumption (Watt)	65	71	107
Price (Euro)	400	200	600
Price/GFlop	617	216	205
Power/GFlop	100	77	36

power consumption for the PC is quoted for the CPU results. For the GPU results, the difference is taken between idle mode on the PC without the graphics card and the total power consumption of the PC including the graphics card during the calculation. Note that all measurements include the power supply inefficiency.

3. Discussion

The measured processing power of 428×10^6 CMAC/s demonstrates the feasibility of GPU boards as the main building block for a modern radio telescope correlator. Based on the results with the 5900 card, we now that the OpenGL driver for the 6800 does not provide optimal performance; another 30% of performance improvement can be expected from selecting the optimal OpenGL driver. Moreover, the correlator code is still optimised for the 5900 card.

The basic building block for a distributed correlator will be a host PC with two 6800 cards attached to the PCIE16 bus, providing a total processing power of order 5×10^9 CMAC/s. The host PCs can be connected with a fast interconnect network such as Infiniband on which the required data transpose between receiver data and correlator units is performed. In an integrated system we do need additional processing elements for the fringe rotation and delay tracking tasks which are left outside of the correlator in order to allow for optimisation of the GPU code.

Some directions for improvement can be identified:

- The correlator code should be optimised for the 6800 card
- There are still some leads to optimisation of the OpenGL and fragment-processor code
- When using a HP SEPIA board the results can be read back using the fast DVI connector instead of the AGP bus.

4. Further work

To explore the full possibilities of the COTS graphics cards for data processing some further research has to be done. We are planning to do the following experiments.

With a FX5900XT card on a PCI-X or PCI-Express bus we can investigate several issues like the speed difference between the AGP and the PCI-X or PCX bus. But we can also test how using 2 graphics cards in one server scales. It is possible to gain a factor 2 in theory, but we might run into another bottleneck.

The next generation graphics cards will be tested to see if the processing power indeed grows faster than Moore's law.

To circumvent the AGP read bottleneck, and to have addition data handling and data merging facilities, two HP SEPIA 3 boards will be used. With this setup, we will investigate how multiple PC's can be coupled to form a bigger calculation platform with extra data handling possibilities. These boards also get around the AGP bus bottleneck because they capture the raw video data directly from the graphics card.

5. Conclusion

The results of the correlator platform prototype are promising. We have demonstrated the feasibility to construct a cost effective correlator based on COTS components. Moreover, we can apply the graphics card processing power for more tasks in the data processing systems of radio telescopes. This is especially interesting if we can integrate in existing cluster computers. It is a small investment to put a fast graphics card in these machines and it will more than double the available processing power.

Based on these results, we believe that the possibility to build processing systems in SKA based on COTS components should be considered as serious alternative to "hard wired" dedicated systems.

Acknowledgements

The authors would like to thank our co-operators at Hewlett Packard, and especially Steve Briggs for the extensive support in this research.

References

Chikada, Y.: 1991, 'Correlators for Interferometry Today and Tomorrow', Presented in the Astronomical Society of the Pacific Conference Series, vol. 19, in Cornwell and Perley (eds.), *Radio Interferometry: Theory, Techniques, and Applications*.

de Vos, C. M., Schaaf, K.V.D. and Bregman, J. D.: 2001, 'Cluster Computers and Grid Processing in the First Radio-Telescope of a New Generation', in Proceedings of IEEE CCGrid.

Overview of the NVIDIA 5900 http://www.nvidia.com/object/LO_20030508_6881.html.

Overview of the NVIDIA 6800 http://www.nvidia.com/object/LO_20030508_6881.html.

Schaaf, K.V.D., Bregman, J. D. and de Vos, C. M.: 2003, 'Hybrid Cluster Computing Hardware and Software in the LOFAR Radio Telescope', in Proceedings of IEEE PDPTA.

Thompson, A. R., Moran, J. M. and Swenson, G. W.: 1994, *Interferometry and Synthesis in Radio Astronomy*, ISBN 0894648594.

DSN DEEP-SPACE ARRAY-BASED NETWORK BEAMFORMER

JOHN D. BUNTON[1,*] and ROBERT NAVARRO[2]

[1]*CSIRO ICT Centre, P.O. Box 76, Epping NSW 1710 Australia;* [2]*Jet Propulsion Laboratory, Pasadena, CA 91109, U.S.A.*
(*author for correspondence, e-mail: john.bunton@csiro.au)

(Received 5 August 2004; accepted 28 February 2005)

Abstract. NASA is proposing a new receiving facility that needs to beamform broadband signals from hundreds of antennas. This is a similar problem to SKA beamforming with the added requirement that the processing should not add significant noise or distortion that would interfere with processing spacecraft telemetry data. The proposed solution is based on an FX correlator architecture and uses oversampling polyphase filterbanks to avoid aliasing. Each beamformer/correlator module processes a small part of the total bandwidth for all antennas, eliminating interconnection problems. Processing the summed frequency data with a synthesis polyphase filterbank reconstructs the time series. Choice of suitable oversampling ratio, and analysis and synthesis filters can keep aliasing below $-39\,\text{dB}$ while keeping the passband ripple low. This approach is readily integrated into the currently proposed SKA correlator architecture.

Keywords: beamformer, correlator, Deep Space Network, FX correlator, filterbank

1. Introduction

NASA proposes to increase the data rate capacity of the Deep Space Network (DSN) by arraying together hundreds of medium sized parabolic dishes ($\sim$12 m) (INPR, 2004). With up to 400 such antennas, sensitivity is equivalent to that of a parabolic dish 240 m in diameter. To achieve this all the antennas must be accurately calibrated in delay, phase and amplitude before their signals are added. Calibration data is generated by correlating antennas; for the best detection of the weak signals from 'lost' spacecraft and the minimization of calibration errors, all pairs of correlations should be formed. Although it has somewhat slower convergence the correlation between antennas and the sum of all other antennas gives similar calibration accuracy (SUMPLE, Rogstad et al., 2003). Past beamformers in the DSN have used time domain approach together with the SUMPLE algorithm for calibration. In this algorithm the full-bandwidth summed data is returned to each input node of the beamformer for correlation. This requires either a return path for the global sum or interconnections between each summing unit of the beamformer. In either case the width of the data path is proportional to the number of subarrays. If full correlation data is wanted a separate correlator is needed and the data path for the antenna data must be duplicated. In the frequency domain approach proposed here the return data paths, or interconnections and the duplication of the antenna data path for a separate correlator, are eliminated.

The proposed frequency domain approach is based on an FX correlator with summing hardware included in each correlator module. This eliminates the need for a separate antenna data paths to a beamformer and correlator. Each beamformer module handles the data for all antennas over a limited bandwidth. Thus, any interconnections or return data paths are within the module and intermodule connections are eliminated. Returning full accuracy beam data for use in SUMPLE is one of the more efficient methods. For a four subarrays system 72 connections (four 18-bit data paths) are eliminated. These connection would have been routed to every beamformer FPGA or ASIC in the beamformer module. In the proposed design antenna data is divided into a multiple sub-bands with each sub-band being processed in a single FPGA or ASIC. Full-bandwidth time data is generated a passing the beamformed data through a frequency synthesis filterbank.

In the proposed design the elimination of cross connects and return paths improve reliability as most failures affect only a small part of the total spectrum. This can then be recovered by rerouting the data to other processing modules. A second advantage of the frequency domain approach is that a full cross power spectrum (SUMPLE or full correlator) is calculated. Although a single lag cross correlator is sufficient for calibration the extra information provided by the cross power spectrum makes the correction of passband errors possible and also provides delay calibration. A search for delay calibration may be needed with a single lag correlation.

A requirement of the DSAN system is that the combining loss of the arrayed beam be less than 1 dB. This error budget is largely assigned to calibration errors. The loss due to noise and distortion added to each antenna by the beamformer processing must be much less than this, approximately 0.1 dB. As aliasing is largely coherent, aliasing must be less than -39 dB to meet this specification. Use of critically sampled polyphase filterbanks with a 10% transition bandwidth introduces noise due to aliasing at about -20 dB per antenna when reconstructing the data in the time domain. This aliasing precludes the use of critically sampled filterbanks for arraying purposes in the DSAN and can be overcome running the polyphase filterbanks in an oversampling mode.

2. Beamformer architecture

As with the SKA the DSN Array-Based Network (DSAN) beamformer faces the challenge of beamforming signals from hundreds of antennas with bandwidths of about 1 GHz. In both cases the problem is the generation of real-time calibration data which in the simplest implementation, SUMPLE (Rogstad et al., 2003) requires the correlation of each antenna against the sum of all other antenna. The more compute intensive approach is to form correlations between all pairs of antennas, which is the approach that will probably be used in the SKA. Although SUMPLE has been used alone in the past to array antennas in the DSN, the full cross-correlation pairs are desired to fully calibrate the antennas in the array and to help search for lost spacecraft by making an image of the entire sky. For the DSAN or SKA no single hardware module can process the full bandwidth data. In a time domain beamformer using SUMPLE for calibration, this leads to interconnections between modules that grow linearly with the number of beams. A better approach is to use a frequency domain beamformer which allows all antennas for a restricted bandwidth to be processed by each module. When this is done the architecture of the beamformer maps directly on to that of the FX correlator proposed for the SKA (Bunton, 2005).

A frequency domain beamformer can be based on the FFT (Williams, 1968; Rudnik, 1969), overlap–add continuous convolution (Weber and Heisler, 1984) or polyphase filterbanks (Weiss et al., 1999). The FFT approach is computationally efficient but introduces time aliasing when calibration phase and delay data or fringe stopping is applied to output of the analysis filterbanks. This can be eliminated by the use of overlap–add continuous convolution at the expense of increased data rate at the output of the FFT. However, the correlations used to calibrate the beamformer suffer from reduced signal-to-noise ratio, particularly if there are narrow band components in the signal (Okamura et al., 2001). The correlation degradation is minimized if a polyphase filterbank is used in place of the FFT (Bunton, 2000, 2005). Use of an oversampling polyphase filterbank reduces the degradation even further. Oversampling also confers the advantage that each channel can be processed in second stage of filterbank (Bunton, 2003) if high frequency resolution is required. This can be done for any part of the input bandwidth, allowing the targeting of multiple spacecraft at different frequencies and bandwidths in the DSAN and flexible spectral line observing for SKA. The second stage filterbank can be either critically sampled or oversampled. If it is critically sampled then the performance and data rates are the same as that of a very-long single-stage critically sampled filterbank with the same frequency resolution.

3. Analysis–synthesis filterbanks

The frequency domain beamformer is based on a polyphase analysis–synthesis filter bank pair (Crochiere and Rabiner, 1983). An example of a critically sampled filterbank pair is shown in Figure 1. In the analysis filter bank a segment of data kM samples long is taken and multiplied point by point by a lowpass filter impulse response. The result is broken up into k segments of length M and the segments added. The resulting M data points are then Fourier transformed to generate one set of filterbank outputs. For the critically sampled filterbank shown the next segment of data is displaced by M samples. The inverse process is used to reconstruct an approximation of the input time series.

For the critically sampled filterbank the Nyquist frequencies for two adjacent bands coincide. This is shown graphically in Figure 2. With critical sampling there

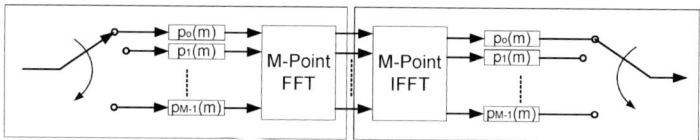

Figure 1. Critically sampled analysis polyphase filterbank (left) followed by a synthesis polyphase filterbank (right).

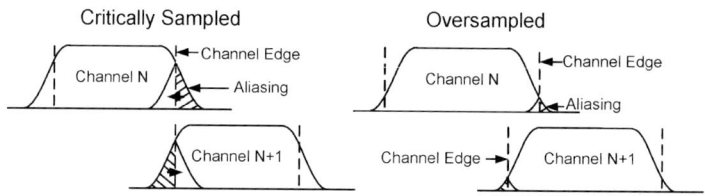

Figure 2. Critically and oversampled filterbanks – adjacent bands.

is aliasing from components beyond the channel edge. By incrementing along the data by less than M, the sample rate for each channel is increased giving an oversampled polyphase filterbank (Crochiere and Rabiner, 1983). This leads to a more complicated structure than Figure 1 but in a practical system the data is processed in multiple interleaved subsets, ~16 for a 1 GHz bandwidth. Step sizes that are multiple of the number of subsets are relatively easy to implement if there is sufficient memory for the filter coefficients. The response of the channels is shown on the right of Figure 2. The Nyquist frequency is increased but the filter response is unchanged; this allows aliasing to be arbitrarily suppressed (Weiss et al., 1999).

An example of a prototype lowpass filter and the resulting channel response that reduces aliasing to less than −60 dB can be found in Bunton (2005), this issue. For the example the reconstructed signal has a ±0.15 dB ripple which can be removed, if needed, by a post reconstruction filter. In general there is considerable flexibility in trading-off output frequency ripple and aliasing for polyphase filter length and oversampling ratio.

4. Beamformer

The architecture of the beamformer is based on a channel reordering FX correlator (Urry, 2000; Bunton, 2005), as shown within the large rectangle Figure 3. The output signal of each analysis filterbank consists of all the frequency channels for that filterbank's antenna. In the router, the data from all the filterbanks is transposed so that a data path consists of all the antennas for a fraction of the bandwidth in time serial order. Data for the same frequency band for all antennas is routed to each beamformer/correlator module. By restricting the bandwidth a single module can

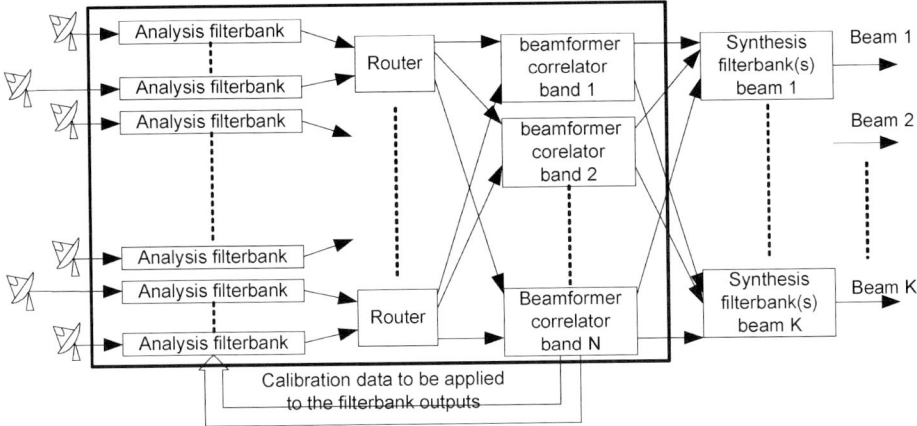

Figure 3. Combined frequency domain beamformer and FX correlator. Correlator only section shown in large rectangle.

both beamform and correlate avoiding the duplication of data paths seen in earlier designs.

For the DSAN there could be 400 antennas and each beam generated has up to 100 MHz of bandwidth. A possible solution is to have 20 routers each accepting 100 MHz signals from 20 antennas. The router would generate 20 outputs each containing the data for 20 antennas over a 5 MHz bandwidth. The interconnects denoted by the arrows in the Figure 3 would each be carrying about 100 MHz of 16 bit complex data (8 bits real, 8 bits imaginary). This would be about 1.6 Gbits s^{-1} of data which could easily fit into an OC-48 type serial link using copper for short connections or optical fibre for longer connections. At the input to the beamformer/correlator board there would be inputs from the 20 routers, thus providing 5 MHz of bandwidth for all antennas. Twenty beamformer/correlator modules are then needed to process the full 100 MHz of the beam. The board not only beamforms but also generates calibration data. The calibration data, in this example, is returned to the analysis filterbanks and applied to data on a frequency channel by frequency channel basis. In addition to this the frequency channel data in the filterbank is fringe stopped, phase gradient corrected and adjusted for magnitude and phase errors. Bulk delay corrections are applied before the filterbanks. Fringe stopping for the DSAN is accurately implemented for baselines up to 1000 km. Beyond this a separate fringe stopping circuit before the analysis filterbanks is needed or the channels have to be pass band corrected with an all pass filter (Chau et al., 2002).

Within the beamformer/correlator board it is a simple matter to form sums of subarrays of antennas. A possible circuit for this is shown in Figure 4. The Dual Register File has N registers each corresponding to a subarray. As the antenna data enters the circuit it is added to the appropriate register. One of the registers is used as location to dump data from antennas that are not part of a current beam. After all

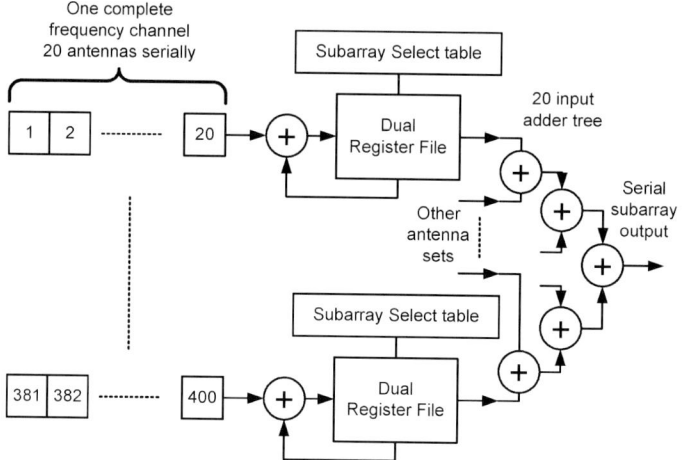

Figure 4. Circuit for beamforming multiple subarrays.

antennas have been added the register sets are swapped and the newly accumulated data is read out in subarray order. At each clock cycle all 20 data values for a single subarray appear at the inputs to the adder tree, which sums them over successive clock cycles.

The serial subarray data is then routed to synthesis filterbank modules as shown in Figure 3. The architecture of the beamformer provides great flexibility because each channel of the filterbank can be separately beamformed into a different subarray. For the DSAN there are five outputs from a filter bank each an arbitrarily chosen set of channels with a total bandwidth of 100 MHz taken from two 500 MHz input IFs. A full DSAN implementation would see all the hardware to the right of the analysis filterbanks duplicated five times giving a total processed bandwidth of 500 MHz. When beamforming, the DSAN data is (8 bit real, 8 bit imaginary) complex data. When operating in a correlator only mode, this can be reduced to $4+4$ bit complex data giving the correlator access to the full 1 GHz of bandwidth from each antenna over the hardware interconnections from the analysis filterbank through the data routers to the beamformer/correlator modules.

The system is also robust against failure as the total available bandwidth exceeds normal DSAN operational needs. Should any router, beamformer or synthesis module fail the system can be reprogrammed to route the data to units with unused capacity and full functionality restored. In the SKA the robustness leads to a graceful degradation of performance when there are system failures.

5. Conclusion

Integration of a frequency domain beamformer within an FX correlator eliminates duplication of signal transport to separate beamformer and correlator modules leading to a unified design. The design has great flexibility in forming multiple beams from arbitrary subarrays of antennas as well as considerable robustness against module failures.

Acknowledgements

The research described in this paper was carried out at the ICT Centre, CSIRO and the Jet Propulsion Laboratory, California Institute of Technology, under a contract with the National Aeronautics and Space Administration.

References

Bunton, J.D.: 2000, 'An Improved FX Correlator', ALMA memo 342, December 2000, at http://www.alma.nrao.edu/memos.
Bunton, J.D.: 2003, 'Multi-Resolution FX Correlator', ALMA Memo 447, http://www.alma.nrao.edu/memos/.
Bunton, J.D.: 2004, 'SUA Correlator Advances', *Exp. Astron.* this issue.
Chau, E., Sheikhzadeh, H., Brennan, R. and Schneider, T.: 2002, 'A Subband Beamformer On An Ultra Low-Power Miniature DSP Platform', 2002 International Conference on Acoustics Speech and Signal Processing (ICASSP'02), May 13–17, 2002 in Orlando, Florida.
Crochiere, R.E. and Rabiner, L.R.: 1983, *Multirate Digital Signal Processing*, Prentice-Hall, New Jersey.
INPR: 2004, 'Special Issue on Array Developments in the Deep Space Network", in J.H. Yuen (ed.) Vol. 42–157, May 15, 2004, at http://ipnpr.jpl.nasa.gov/progress_report/42-157/title.htm.
Okamura, S.K., Chikada, Y., Momose, M. and Iguchi, S.: 2001, 'Feasibility Study of the Enhanced Correlator for 3-way ALMA I.', *ALMA Memo 350*, February 2001, at http://www.alma.nrao.edu/memos/.
Rogstad, D., Mileant, A. and Pham, T.: 2003, *Antenna Arraying Techniques in the Deep Space Network*, Hoboken, Wiley, New Jersey.
Rudnik, P.: 1969, *J. Acoust. Soc. Am.* **46**, 1089–1090.
Weber, M.E. and Heisler, R.: 1984, *J. Acoust. Soc. Am.* **76**(4), 1132–1144.
Williams, J.R.: 1968, *J. Acoust. Soc. Am.* **44**, 1454–1455.
Weiss, S., Stewart, R.W., Schabert, M., Proudler, I.K. and Hoffman, M.W.: 1999, 'An Efficient Scheme for Broadband Adaptive Beamforming', 33nd Asilomar Conference on Signals, Systems, and Computers, Vol. I: Monterey, CA, pp. 496–500.
Urry, W.L.: 2000, 'A Corner Turner Architecture', ATA imager memo 14, Nov 2000, at http://astron.berkeley.edu/ral/ATAMemo14.pdf.

DATA PROCESSING AND SOFTWARE

SOFTWARE DEVELOPMENT FOR THE SQUARE KILOMETRE ARRAY

T. J. CORNWELL* and B. E. GLENDENNING

NRAO, Socorro, NM, U.S.A.

(*author for correspondence, e-mail: tcornwel@nrao.edu)

(Received 12 August 2004; accepted 1 November 2004)

Abstract. Software development costs for the Square Kilometre Array are likely to be very large – in the range of 1000–2000 person-year total. This level of software effort is unprecedented in radio astronomy. Consequently the risk associated with software development is very large. This is common to many large science projects and so we can learn from such projects how to best mitigate against the risk. We present a shopping list of suggestions drawn from the experience in other projects.

Keywords: radio interferometry, software, Square Kilometre Array

1. Introduction

Radio telescopes have grown larger both in ambition and in necessary complexity. As a result, software has come to lie at the heart of radio telescopes, bringing computational power and flexibility. Although it was once true that a small group led by a single expert could assemble all the software required, now that era is gone. For example, the ALMA project has software development costs of over 400 person-year. A recent estimate for SKA software development derived by scaling the ALMA budget yields numbers of well over 1000 person-year (Kemball and Cornwell, 2004). If the construction of the software is to be done in 5–10 years, then a team consisting of perhaps 100 or 200 people is needed. The team must have a broad range of skills and knowledge – from programming methodologies to arcane details of interferometry, from parallel computing to user interface design, from code management to people management. It is highly unlikely that the team can reside in one location and geographical separation will be one of the challenges.

To only add to the challenges, it seems probable that development of SKA will occur at a time of large change in the computing industry, compared to, for example, the mainframe and deck-of-cards era of the 1960's and 1970's. Paradigms for computing are changing constantly and the rate of change will probably only increase. Software methodologies continue to evolve in response to sometimes-painful experience: current (or slightly expired) buzzwords are Extreme Programming and Aspect-Oriented Programming, both of which aim at improving programmer productivity. On the hardware side, there is a push towards Grid computing. Finally, we share the commonly held view that computing in general and astronomy in particular are moving towards more service-orientation, as in the numerous national Virtual Observatories.

How then can all this activity be managed? The possible failure modes are budget overrun, broken timelines, incomplete software, and drawn-out commissioning, to name but a few. One logical possibility is to radically de-scope – reduce the functionality expected of the telescope. This is often advocated, at least implicitly in statements that SKA software development should not, *a priori*, cost more than X person per year where X is a number drawn from personal experience. We reject this line of argument as being defeatist and unnecessary. Instead we advocate the usual approach to a fresh challenge – learning from our past experience and borrowing best practices, both from the radio astronomy community and from other disciplines.

This paper is first and foremost an opinion piece, representing the opinions of the authors. We recognize that others have quite different views and hope that countervailing arguments will be written down, perhaps in response to this paper. We choose to write a collection of suggestions, a good fraction of which are backed up with references. In addition to our own experience and those of people at NRAO[1] and in ALMA that we talk to a lot, we've found the following documents very helpful:

- LCK: An excellent paper by Lewis et al. (2002) on *"Do larger telescopes need larger hardware?"*
- PS: Presentations from the NSF-funded series of workshops on Project Science. See http://131.215.125.172/
- BPS: the presentation in PS by Kent Blackburn on the LIGO data analysis system.
- RL: A paper by Robert Lupton and others (2001) on the SDSS imaging pipelines. This is notable for Lupton's view on the challenges of software development for astronomy. See http://arxiv.org/pdf/astro-ph/0101420
- CSNAP: A presentation by Bill Carithers on the SNAP software. See http://snap.lbl.gov/review/2003_11/Carithers_Computing.ppt
- LCG: The site for the Large Hadron Collider Grid Computing Project. See http://lcg.web.cern.ch/LCG/Overview.htm
- Jim Gray's web site (http://research.microsoft.com/~Gray) provides highly insightful commentary on many aspects of computing, but most particularly the economics of large data and distributed computing (Gray, 2003).
- The entire series of ADASS proceedings at http://www.adass.org/. These contain much interesting and relevant experience.

---

[1] The National Radio Astronomy Observatory is operated by Associated Universities, Inc., under cooperative agreement with the National Science Foundation.

We assume that the SKA will have a strong foundation in project management and so we will mostly confine our suggestions to computing (but see PS for real experience by managers of large science projects, and LCG for an example of a well constructed oversight and review structure).

Many of the things that make working in current projects hard and unappealing are not specific to computing. Examples are:

- Inoperative management and/or management structures,
- Quantity rather than quality of oversight,
- Cultural differences at the team and organizational levels,
- Politicization of technical designs,
- Drastically insufficient funding.

We hope that none of these will occur in the happy world that is SKA.

2. Things not to do

Don't omit software and data analysis – Surprisingly this does happen, though we refrain from pointing fingers.

Don't write unrealistic requirements – Committees are, of course, prone to add more and more features as meetings continue. Setting priorities (e.g. vital, desired, useful) helps but really just amounts to a culling since it is common experience that only the highest rated priorities are actually implemented. Iterating requirements writing with even first-order cost estimation would help.

Don't underestimate the cost of complexity – A rule of thumb (Glass, 2002, fact #21) is that a 25% increase in requirements complexity, realized as the number of distinct requirements, doubles the development costs.

Don't let the hardware design define the software – We believe that software is more difficult than hardware so why do the hardware engineers get to define the playing surface? Often hardware, such as a correlator, is designed and built before the software is even started. Hardware choices often end up constraining the software in very troublesome ways, such as in the lack of real hardware testing capabilities. Let's turn that around and have the software constrain the hardware. If changes must be made as the construction continues, consider changing the hardware instead of the software.

Don't allow "Not Invented Here" thinking – Re-use if done properly can significantly cut costs. Re-use of known packages should be the default, not the exception.

Don't forget some class of users – The users of the software will be astronomers, engineers, technicians, operators, the public, etc. It's easy to focus on the needs of the astronomers and forget that tools are needed for operations. Software support will also be needed for testing during construction and that need can come to dominate the schedule unless properly planned.

Don't separately manage control and other software – Although separation of concerns helps the construction of complex systems, separating the "control" software from the rest of the system can result in arbitrary interfaces and complexity in the software.

Don't claim accurate cost estimates too early – Early cost estimates will be top–down and as such, poorly founded and inaccurate. If requirements accepted by the project in all areas (technical, operational, scientific) are not available to base a bottom-up cost estimate on, use a rule of thumb based on total capital cost (10–20% are typically suggested). Bottom-up costs should only be based on requirements for the current project requirements and software re-use scenarios.

Don't divide the work before the software architecture is defined – While a geographical distribution of the development team might be inevitable, the work should not be divided until the architecture is defined, and then work should be distributed to minimize the interfaces between groups.

3. Things to do

Do make software a key part of the project – software will be vital to making the vision of SKA a reality. The core management team (Manager, Engineer, Scientist) should have deep experience in software. Most likely this will require two Project Engineers and perhaps two Project Scientists (as the EVLA has) with different skills.

Do assemble a first rate team – Most of the productivity comes from a small fraction of a typical team, so it really pays to get the best people, even though it may not be possible to collocate them in one place. The team must be diverse – architects, designers, developers, testers, and managers all have different skill sets, perspectives, and roles. There is a real concern that the talent pool of people working in this area is too small. Lupton (RL) has many interesting things to say about the challenges of staffing a project with good people, especially given the dearth of established career paths for such people. Maintaining the team for construction period and beyond requires some careful choices – form the core from people who are likely to stay around, and give them a good working environment.

Do hire a dedicated management team – There is plenty of experience to show that at least 10% of the budgeted effort should go in management, partly by team members but mostly by experienced software managers. If the current trend continues, there will be a lot of oversight for the managers to respond to, in addition to their internal management responsibilities.

Do track risk – Software will inevitably be an area of high risk. SKA should analyze and track risk at the project level. Some organizations are using a Risk Manager or Risk Office for this activity. The LHC Computing Grid Project has a nice example of risk tracking: see http://lcg.web.cern.ch/LCG/PEB/risk/. Such risk estimates can be used in allocating contingency.

Do plan for logistical complexity – Blackburn (BPS) shows that in a typical year only about 30% of developer time went to writing new code. The rest went to activities like code maintenance (14%), testing (10%), documentation (6%), lab support (22%), meetings (5%), and system administration (10%). For a geographically distributed project, the overhead of meetings and travel is likely to be larger.

Do engage domain experts – Domain experts (i.e. in the case of SKA, people who know the nuts and bolts of building and operating radio synthesis arrays) are vital to bring perspective and insight to the software development. People with deep knowledge of and experience in running current interferometers wrote the ALMA Science Software Requirements. A substantial number of those people have continued to be involved in the software development. This will be particularly important for SKA since it will combine techniques – synthesis imaging, mosaicing, wide-field, connected VLBI, pulsar observing, SETI. It also helps to have a few genuine visionaries since, like all big steps, SKA will place us in unfamiliar territory.

Do demand operational model early on – It's a curious fact that the software team is often the first to ask what the operational model is. Definition of the model seems to take 1–2 years of committee work so it's best to start on it early. Include commissioning in the operational model since it will probably continue for quite a few years, and it is a big source of stress and possible disruption on the software development teams.

Do manage and document change – Any substantial changes should be reviewed and accepted (or not) by an appropriate review body. Changes in scope should require approval all the way up to the stakeholders and funders. Changes in interfaces should require approval by an internal review board.

Do prototype, test, and simulate – SKA constitutes a large jump in capabilities. Such jumps are hard to execute and require new technology (see, for example, Ekers, 2002). How can we be sure that it will all work? There are three types of checks available: testing, prototypes, and simulations. All three will be important for SKA, and all three provide ways to validate the software. Prototypes and demonstrators will be built prior to SKA – these provide a place to learn about the software issues. Testing is now an intrinsic, and some would say primary, part of software development methods. Simulators should be built alongside the software development, providing a means to test the software and algorithms via a continuing series of challenges of escalating complexity.

Do establish and debug a software process early – Debug a process with the early design and development team rather than after you staff up the group. This process will probably have to be somewhat formal given the size of the development team and the oversight the SKA project will probably be subject to.

Do build a complete system early and often – Carithers (CSNAP) describes a process in which a complete but stubbed system is built early, thus establishing interfaces first and implementation later. ALMA is also following the same approach, focusing on biannual builds that integrate all subsystems. The early freezing of interfaces allows the entire team to focus on implementation confident that interfaces they use will not change. This evolves into an approach where testing is always the highest priority. This also prevents the common problem where substantial design mistakes are found too late.

Do integrate early and often – integration can be the most difficult part of a project. Do not establish a development model in which separate systems are only brought together in a final commissioning stage.

Do invest in infrastructure – LCK advocate substantial investment in well conceived and executed infrastructure that is relevant to the job at hand. The "Buy–Build–Borrow" question is key to making a project work. ALMA is reusing three major packages – AIPS++ (pipeline and offline data reduction), NGAST (archiving), and a component-container package built on top of CORBA (used in the ALMA Common Software). The net savings are at least at the 25–50% level. If it is to be useful, infrastructure must be in place early and well executed, otherwise it can act as a shared problem.

Do review appropriately – The traditional NASA approach of PDR, CDR has too much rigidity for the typical software development process. ALMA is using multiple, annual CDRs in recognition of the continuing need to evaluate many aspects of the design as they inevitably evolve. To conclude the design phase, Blackburn (BPS) advocates data challenges in place of a Final Design Review. The review panels should be chosen carefully – the skills to review science, operations, electronics, and software are not often found in one person.

Do establish a separate testing group – Getting software tested by knowledgeable people is always difficult. Budget for a small group of core testers, and use external scientists to test targeted areas.

Do get help – SKA will be one of the most impressive scientific projects in history. We should be able to get the interest of internationally known experts in computing. SDSS benefited immensely from the addition to the team of Jim Gray of Microsoft (a Turing Prize winner for his work on databases). The recent partnership between ASTRON and IBM on LOFAR computing is another example.

4. Other related suggestions

Algorithm development will be vital for SKA – It is not the same as computing and should not be done by the same people. Hire experts and support them with a small dedicated computing group.

Plan for continuing software development in operations – Software systems need maintenance and new capabilities for the lifetime of the project. Haggouchi et al. (2004) describe how the Very Large Telescope Data Flow System has evolved since initial deployment, and how the activities of the DFS group has been structure to support that need.

Acknowledgements

Both authors have been involved in radio astronomical software development for many years. TJC was Project Manager of the AIPS++ Project from 1995 to 2001 and head of Data Management for the NRAO from 2001 to 2003. Brian Glendenning was Technical Leader of the AIPS++ Project from 1992 to 1998, and has been the head of the ALMA Computing Integrated Project Team since 1998. We thank Neil Killeen and Wim Brouw for insightful comments and suggestions on an early draft of this paper. We also thank the referees for helpful comments.

References

Ekers, R. D.: 2003, 'Astronomical Data Analysis Software and Systems XII,' in H. E. Payne, R. I. Jedrzejewski and R. N. Hook (eds.), *ASP Conference Series*, Vol. 295, ASP, San Francisco, p. 125.
Glass, R.L.: 2002, *Facts and Fallacies of Software Engineering*, Addison-Wesley, Reading, MA.
Gray, J.: 2003, *Distributed Computing Economics,* MSR-TR-2003-24, Microsoft Research.
Haggouchi, K. et al.: 2003, 'The VLT Data Flow System, A Flexible Science Support Engine,' *ADASS 2003 Conference*, Strasbourg, France, 13–15 October 2003.
Lewis, H., Conrad, A. and Kilbrick, B.: 2002, 'Do Bigger Telescopes Need Bigger Software?' in Lewis, H. (ed.), *Advanced Telescope and Instrumentation Control Software II.* Proceedings of the SPIE, Vol. 4848, pp. 167–174.
Kemball, A. J., and Cornwell, T. J.: 2004, 'A Simple Model of Software Costs for the Square Kilometre Array,' *Exp. Astronom*, this issue.

A SIMPLE MODEL OF SOFTWARE COSTS FOR THE SQUARE KILOMETRE ARRAY

A. J. KEMBALL[1] and T. J. CORNWELL[2,*]

[1]*NCSA/UIUC;* [2]*NRAO, Socorro, NM, U.S.A.*
(*author for correspondence, e-mail: tcornwel@nrao.edu)

(Received 12 August 2004; accepted 17 November 2004)

Abstract. Software costs for radio telescopes have nearly always been underestimated. Since the Square Kilometre Array is often called a software telescope, repeating the usual error would be particularly egregious. We estimate software costs by scaling from the reasonably well-known costs for the Atacama Large Millimeter Array. The resulting model has sharp dependency on the complexity of the SKA, suggesting the obvious – that software costs can most easily be limited by constraining the scientific and operational requirements. A bottom-up costing will not be possible until SKA is much more clearly defined. For the moment, we recommend that 20% of the SKA budget be allocated to software development.

Keywords: radio interferometry, software, Square Kilometre Array

1. Introduction

The science goals of the Square Kilometre Array (SKA) pose significant computing challenges, particularly in the areas of complexity, scale, and throughput. In this note, we consider the costs of the software required to meet those challenges. Software is an integral part of the SKA and should be considered a capital cost on the same footing as, for example, the antenna or correlator costs; these software costs needs to be factored in to the early design decisions and costing. In addition to these considerations, software is inevitably associated with high risk. According to various studies, only about 10% (Royce, 1998) of software projects meet their goals on time and within budget. Without going into details, it is probably safe to say that this rule is not violated in astronomy projects. Thus possible non-completion of the SKA software should feature prominently in the SKA risk analysis.

The SKA is representative of a new class of modern astronomical facilities that relies on the recent geometric advances in computing power and data throughput rates in order to meet new science challenges. The computing advances are best known in the form of Moore's Law, which predicts exponential increases in chip density as a function of time. Exponential increases in processor performance, network bandwidth and data storage are also expected, and observed, albeit with separate, individual exponent scaling factors. These advances in computing are having a profound impact across the physical sciences, and observational astronomy is no exception. The computing advances offer major new scientific opportunities, but require commensurate increases in the relative resources assigned to the computational science and engineering of these instruments, and in the effectiveness of the use of those resources.

The Moore's Law increases in hardware computing power are required for the success of the SKA, but do not address the costing and risk associated with developing and deploying the software for the SKA. There is unfortunately no Moore's Law in software engineering; in fact, the opposite holds true. As there are $O(N^2)$ interactions between N software components, in the worst-case, software costs increase with the size of the overall development. Broadly speaking, software costs can be parameterized as (Royce, 1998):

$$\text{Effort} = (\text{Personnel}) (\text{Environment}) (\text{Quality}) (\text{Size}^{\text{Process}})$$

The *diseconomy of scale* (Royce, 1998) results from the fact that the *Process* exponent is greater than unity. This exponent is a scaling driver, in the terminology of a common costing model, COCOMO II, and is increased for projects which have fewer technical precedents, more limited development flexibility, greater architectural risk, lower team cohesion or less maturity in the software development process employed (Boehm et al., 1998). The overall relation for effort clearly has a sharp, and unavoidable, dependence on complexity; we consider means by which this complexity can be qualitatively estimated for the SKA in the remainder of this memorandum. This basic relation can be re-expressed in a simpler form as Effort $= A$ Size^{process}; we know from existing software projects in astronomy that the factor A is approximately one person-year per 10000–15000 lines of debugged and tested code, incorporating all associated development costs in user support, design and operations implicit in all software development projects. We are taking some liberties here in averaging over multiple project development phases, where the balance of tasks may differ in individual phases. However, the basic dependencies in the complexity relationship are well established in software engineering. The scaling factors differ between industry and academic research projects due to differing quality goals, economic drivers and available resources.

In some senses, this memo is clearly premature – the SKA concept has not been chosen and so we do not know whether SKA will have 300 × 300m elements or 10,000 × 10m elements. These choices are obviously key drivers in the complexity of the resulting software development tasks. There are no formal requirements on operations, scientific capabilities, hardware monitor and control, testing, engineering support, etc. There is no construction timetable to set deliverables and timescales. In face of these uncertainties, we must find parameterized software costs. Our approach is inspired by that of Lewis et al. (2003) for costing of a putative 30-m optical telescope. We first find plausible software costs for a system design close to the types of arrays now in operation or construction, and second examine how this equation scales with known differences between SKA concepts, such as number of antennas or stations.

Before discussing software costs, we describe the source of the problem – the ever-increasing complexity of telescopes.

2. The cost of complexity

The instrumental models for modern radio telescopes have become steadily more detailed, reflecting increases in the complexity of the underlying hardware as well as increasingly stringent calibration requirements imposed by the more challenging science goals posed by modern instruments. This can be illustrated in a number of ways – from the scientific and operational requirements, the block diagram(s), the number of monitor points, the telescope proposal form, the data reduction user manual, etc. The scientific and operational requirements would be particularly interesting. Unfortunately, documentation of such requirements is a fairly recent trend with the first substantial examples dating back no more than 10 or 15 years. Therefore for our purposes, we choose to capture the first-order complexity from the data formats used to hold the observational data that are, by definition, closely coupled to the instrumental complexity. In Table I and Figure 1, we show the number of tables and sub-tables in the schema for various telescopes and software packages going back about 30 years. This ignores the complexity of each table (such as the number of keywords and columns) but does capture the overall structure of the data. The trend is obvious – at the turn of the millennium, 2–3 sub-tables were being added per year. The complexity of the data schema increases to allow more faithful modeling of the observation process, thus allowing (or tracking) improvements in processing methods. The upside is continually improving scientific capabilities,

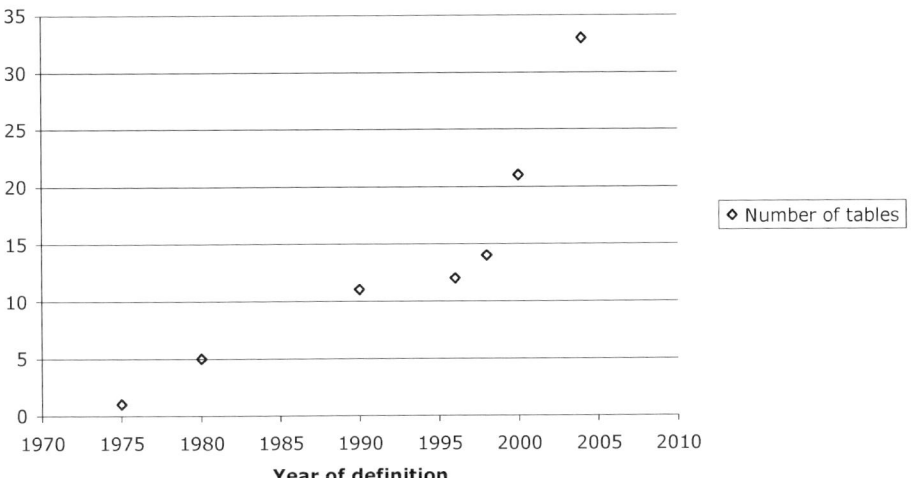

Figure 1. Growth of interferometry data schema.

TABLE I
Number of tables and sub-tables for various software packages

System	Date	Number of tables	Table structure
Caltech MERGE format (VLBI)	1975	1	Single table
VLA (DEC-10)	1980	5	VIS, INX, GAI, CAL, ANT
AIPS	1990	11	AN, BL, BP, CL, CH, CS, FG, FQ, NX, SN, SU
AIPS++ MeasurementSet Version 1	1996	12	MAIN, ANTENNA, FEED, FIELD, FLAG_CMD, HISTORY, OBSERVATION, POLARIZATION, SOURCE, SPECTRAL_WINDOW, SYSCAL, WEATHER
FITS-IDI (export format from VLBA correlator)	1998	14	ANTENNA, ARRAY_GEOMETRY, FLAG, FREQUENCY, GAIN_CURVE, INTERFEROMETER-MODEL, PHASE-CAL, SOURCE, SYSTEM_TEMPERATURE, UV_DATA, BANDPASS, BASELINE, CALIBRATION, WEATHER
AIPS++ MeasurementSet Version 2	2000	21	MAIN, ANTENNA, DATA_DESCRIPTION, DOPPLER, FEED, FIELD, FLAG_CMD, FREQ_OFFSET, HISTORY, OBSERVATION, POINTING, POLARIZATION, PULSAR_GATE, PROCESSOR, SOURCE, SPECTRAL_WINDOW, SAMPLER, STATE, SYSCAL, TRACKING_STN, WEATHER
ALMA Export Data Format (still under discussion)	2004	33	MASTER, ALMA_CORRELATOR_MODE, ANTENNA, CALDEVICE, CONFIG_DESCRIPTION, DATA_DESCRIPTION, DOPPLER, EXECUTE_SUMMARY, FEED, FIELD, FLAG_CMD, FOCUS, FREQ_OFFSET, GAIN_TRACKING, HISTORY, OBS_SUMMARY, OBS_UNIT, POINTING, POLARIZATION, PROCESSOR, RECEIVER, SCAN_SUMMARY, SCHEDULE, SEEING, SOURCE, SOURCE_PARAMETER, SPECTRAL_WINDOW, STATE, SWITCH_CYCLE, SYSCAL, TOTAL_POWER_MONITORING, WEATHER, WVMCAL

such as improved calibration, and queue observing. The downside is increasing software and operational complexity.

In some senses, the growth of complexity is inevitable. Telescopes are becoming more complex both in construction and in the scientific and operational requirements. However, we note the following points in coping with complexity:

- Accurate costing is vital – too often the software costs of implementing new scientific or operational capabilities are not known accurately at the time of system definition, resulting in either the fiction of capabilities "not designed out" or lower priority requirements that are never implemented. It is the purpose of this memo to propose a first, simple model for software costs.
- System partitioning limits the visibility of sub-systems. The data schema reflect a desire to know everything about the system so that it can be corrected after the fact. A relevant example from SKA where partitioning could be helpful would be the functioning of the stations. The internal workings of the stations should not necessarily be recorded in the export data schema.
- Appointment of a Software System Engineer who has overall responsibility for information for the flow throughout the telescope of information, control, and data.

3. The role of the operational model

The operational model for any telescope is key in determining the overall software costs. The style of operation of the telescope drives the type of software capabilities that must be delivered. To a first approximation, the two competing styles are expert (EXPERT) and automated (AUTO). In the first style, only experts can interact with the telescopes, and in the second, any person with a graduate-level education (or better) can do so. For AUTO, a lot of effort must be spent on specifying, analyzing, designing, developing, and testing tools that provide an intelligible, navigable, powerful interface to the telescope. In EXPERT, the user (an expert) is expected to work with simpler, less powerful tools. For example, in AUTO, the observer (after receiving a time allocation) would only have to provide a minimal description of the source to be observed and the observing characteristics, whereas in EXPERT, the observer would be expected to provide a detailed schedule complete with frequency settings, dwell and drive times, calibration source, etc.

Although the recent tendency has been to move from EXPERT (VLA, for example) to AUTO (ALMA), the choice between EXPERT and AUTO is by no means obvious. From the perspective of a truly external user, both can be made to look roughly the same, and thus the telescope can have equal scientific impact in both models. In EXPERT, the telescope is tended by people, and in AUTO by computers and software. Thus the EXPERT/AUTO choice may be seen as a choice between operational (people) and developmental costs (software).

For a single-purpose telescope with a known and limited user community, EXPERT is almost certainly the best choice. A small number of people can master the operation of such a telescope, and there is little to be gained by moving on AUTO. However, for the SKA, the user community will presumably be very diverse, drawing from many different fields of astronomy (and even beyond). Ultimately the choice of operational model is to be determined by the SKA community. We urge that the definition of the operational model be completed early on in the development of SKA. In this document, we explore the software costs for the AUTO model.

4. ALMA software costs

For our base cost, we must choose a telescope that has or will have operations close to that of SKA, and that has a comparable level of complexity. The best fit is ALMA – it has a large number of antennas, ambitious scientific and operational goals (AUTO), a remote site, and an international construction team.

The ALMA computing budget is split into two parts – before and after first scientific operations. In some ways, this is an artifact of the ALMA budget, and we regard the sum of these as the important number. However, it also reflects the difficulty of single-pass software planning – the detailed planning of the second part has yet to be done. To help control costs, ALMA has reused a number of existing systems: the principal ones being ACS (system framework, see Chiozzi et al., 2002), AIPS++ (calibration and imaging), and NGAST (archiving). Without reuse, the costs would be substantially higher.

The overall budget is 430 FTE-years, split 270/110 in pre- and post-first science operations, plus another 50 for further development of AIPS++ by NRAO[1] and NGAST by ESO. As a matter of policy, it was decided that ALMA computing should be equally split between North America and Europe. Furthermore, some of the subsystem divisions were created to apportion work to a number of organizations on a finer basis. This reflects the reality of international collaborations and one might expect that SKA would be split similarly. Thus the subsystems do not necessarily reflect real problem-domain boundaries. In Table II, we list the subsystems arranged in groups according to our own classification.

While this budget may seem high to those accustomed to the typical radio astronomy funding of software, the real question is whether the budget is large enough? More exactly, does the budget match the requirements? The ALMA Computing Preliminary Design Review characterized it as lean compared to the scientific requirements. A more exact statement is notoriously hard to make for software. Without disrespecting the capabilities of the ALMA team, we can easily imagine that the cost to meet **all** requirements regardless of priority could be a factor of two higher, reaching close to 1000 FTE-years, or roughly $100M.

---

[1]The National Radio Astronomy Observatory is operated by Associated Universities Inc., under cooperative agreement with National Science Foundation.

TABLE II
ALMA-computing sub-systems

Subsystem	Purpose
Process	
Management	Management of computing development
Science software requirements	Development of scientific requirements, scientific oversight, testing, etc.
Analysis and design	High level analysis and design of the entire ALMA computer system
Software engineering	SE methodology, tools, and testing
Integration, test, and support	Integration of subsystems, testing, and support
Infrastructure	
Common software	Development and maintenance of ALMA Common Software
Preparation	
Observation preparation	Development of ALMA observation preparation software, including tools for use by astronomers for preparing proposals and observations
Scheduling	Development of ALMA scheduling software
Observing	
Control software	Development of array monitor and control software, including antenna control
Executive software	Development of executive subsystem that coordinates the computing system
Correlator software	Development of ALMA correlator software, including embedded software
Telescope calibration	Development of synchronous calibration software
Post Observing	
Pipeline	Development of pipeline facility, including pipeline system and pipeline heuristics
Archiving	Development of archive software and system, including hardware, APIs.
Off-line data reduction	Development of software for standard off-line processing of ALMA observations, including engines to be used in pipeline

Note. The grouping is of our devising.

When telescope computing budgets are too low, there are two non-exclusive outcomes – commissioning is long drawn out, thus transferring unexpended computing construction costs to operational losses, and the delivered system is less capable than desired, again leading to operational problems and deferred or unrealized scientific capabilities.

5. SKA-computing costs

The complexity of SKA has two parts; one part due to the intrinsic complexity of building the telescope independent of design concept, and one part due to the specific aspects of a particular design concept.

Independent of design, the SKA will be an amalgam of different types of telescope – a standard connected element interferometer like the VLA or WSRT, a low frequency array like LOFAR, a mosaicing array like ALMA, and a real-time VLBI network like the eEVN. As such, it will have challenges arising from all these aspects. Signal transmission will take place over longer distances and perhaps more complex, commodity networks. Correlation will be on a much larger scale than currently done. The data processing will be larger in volume and more demanding in technique. The data processing will certainly require both high-performance computing resources and techniques, including the use of grid computing methods dominant in high-end computing for the physical sciences. There will almost certainly be two separate systems for high and low frequency, doubling the cost of some software sub-systems. In addition, multiple experiments may be run simultaneously (e.g. synthesis imaging, pulsar timing, and SETI searches) requiring sophisticated operational capabilities.

Depending on the design, the array stations may be substantially more complex than a simple antenna. For example, the Large Aperture Reflector (LAR) antennas are more complicated than a VLA antenna, say, while at the other extreme, the Large-Number-Small-Diameter (LNSD) antennas are simpler but there will be vastly more.

We can capture these dependencies in some crude parameterizations. In Table III, we list the numerical factors affecting software costs, and some typical values. In Table IV, we connect these factors to a set of subsystems mirroring those in the ALMA design.

Some of these dependencies are additive, some are multiplicative. Some come linearly and some as some power (only slightly above zero). Our information from ALMA does not warrant too much interpretation so we have limited our model to including the dominant multiplicative terms only.

An example of the application of our model is given in Table V. The budget numbers given are our estimates, not official project numbers.

The number of staff has been determined from the requirement that the software be completed in 10 years.

There is much to argue with in this necessarily simplistic analysis. We have ignored all but the grossest scaling behaviors. However, it does give a model with which to evaluate the incremental cost of various changes. Some conclusions are as follows:

- The cost of the hybrid is roughly a factor of two in computing costs.
- Halving the number of observing modes (from the perhaps high number of 10 down to 5) halves the cost.

TABLE III
Parameters affecting software costs

Parameter	Description	Typical values
N_{Hybrid}	Number of separate telescopes needed to meet scientific requirements	1 or 2
N_{obs}	Number of major observing modes	5–10
$N_{concurrent}$	Number of concurrent observations	1–10
N_{ant}	Number of antennas	30–10000
$M_{antenna}$	Number of distinct antenna types	1–3
$N_{stations}$	Number of stations	30–1000
$N_{correlators}$	Number of correlators	1–1000
$M_{correlators}$	Number of distinct correlators	1 or 2
$N_{noncorrelation}$	Number of non-correlation backends	~1–5
N_{staff}	Number of computing staff	50–200
$N_{sub-systems}$	Number of sub-systems in computing design	~10–20
$M_{embedded}$	Number of distinct embedded systems that need programming	~10–20

Note. We separate out costs for the absolute number (N) and the number of distinct types (M).

TABLE IV
Sub-system cost dependencies

Sub-system	Dependencies
Management	N_{staff}
Science software requirements	N_{Hybrid}, N_{obs}
Analysis and design	N_{Hybrid}, N_{obs}, $N_{sub-systems}$
Software engineering	N_{staff}, $N_{sub-systems}$
Common software	$N_{sub-systems}$
Control software	N_{Hybrid}, $M_{antenna}$, $N_{concurrent}$, $N_{noncorrelation}$, $M_{embedded}$
Executive software	N_{Hybrid}, $N_{concurrent}$, $N_{sub-systems}$
Correlator software	N_{obs}, $N_{concurrent}$
Pipeline	N_{Hybrid}, N_{obs}, $N_{concurrent}$
Archiving	N_{Hybrid}, N_{obs}
Scheduling	N_{Hybrid}, N_{obs}, $N_{concurrent}$
Observation preparation	N_{Hybrid}, N_{obs}, $N_{concurrent}$
Off-line data reduction	N_{Hybrid}, N_{obs}
Telescope calibration	N_{Hybrid}, N_{obs}, $N_{concurrent}$
Integration, test, and support	N_{Hybrid}, N_{obs}, $N_{concurrent}$

TABLE V
Software cost model

	Cost (FTE-years)	System independent	Staff	Hybrid	Obs	Concurrent	Subsys	Cost multiplier	Cost (FTE-years)
ALMA			50	1	5	3	15		
SKA			200	2	10	5	16		
	ALMA								SKA
Process	107							5.1	548
Management	14		4.0					1.1	4.3	60
Science software requirements	19			2.0	2.0			4.0	76
Analysis and design	10			2.0	2.0	1.7		1.1	7.1	71
Software engineering	32		4.0					4.0	128
Integration, test, and support	32			2.0	2.0	1.7		6.7	213
Infrastructure	54							1.0	54
Common software	54	1.0						1.0	54
Preparation	67							5.0	335
Scheduling	25			2.0	2.0	1.7		6.7	167
Observation preparation	42			2.0	2.0			4.0	168
Observing	127							4.5	810
Control software	20			2.0	2.0	1.7		6.7	133
Executive software	11			2.0		1.7		3.3	37
Correlator software	60			2.0	2.0	1.7		6.7	400
Telescope calibration	36			2.0	2.0	1.7		6.7	240
Postobserving	81							2.7	216
Pipeline	20			2.0	2.0			4.0	80
Archiving	36	1.0						1.0	36
Off-line data reduction	25			2.0	2.0			4.0	100
Total	436								1963

Note. Only dominant, linearly scaling terms have been included. The ALMA costs per sub-system have been used but are not given. The left hand column of numbers shows the ALMA costs, and the right hand column shows the predicted SKA costs. The columns between hold the scaling for each factor (left hand group), and the total scaling (next column) – the product of all factors.

- Adding another concurrent mode costs about 300 FTE-years.
- Increasing the number of computing subsystems has negligible effect on the cost.
- The number of antennas or stations has little effect on the overall software cost.

For the configuration we have chosen the cost is about 2000 FTE-years, or 5 times the cost of ALMA software. This amounts to a significant fraction of the SKA cost – for example, in the US or Europe, an FTE-year costs about $100K, so the total software budget would be about $200M, or 20% of the target capital cost.

The cost could increase if reuse of a scale similar to that of ALMA is not possible. ALMA reuses AIPS++, NGAST, and parts of ACS. We estimate the total (replacement) costs as between 100 and 200 FTE-years. Thus reuse is a 10–20% factor in this estimate, and is thus much less than other uncertainties.

The cost can be reduced in a number of ways: reducing the mode space, limiting the number of simultaneous observing sessions, moving to a single telescope design, reusing existing software. Most important would be moving the operational style along the curve from AUTO (the ALMA model) to EXPERT. In addition, the software could be developed in a country where an FTE-year costs less than in Europe or the US. Ultimately, the best use of this model is to put a cost to complexity in the SKA, and thus ensure that only scientifically important modes are designed into the software. Most importantly, the SKA project would be best served by incorporating software cost estimates as early as possible in the funding and design phases of the project as well as the project plan. *At this stage of development of SKA, the best approach to software costs is probably to allocate a large fixed fraction. Based on our analysis a number like 20% is reasonable.*

Finally, our modeling can be checked against software costs for other projects, especially those with the "Automated" operational style. We plan to do so in future work.

Acknowledgement

We thank the referees for helpful suggestions.

References

Boehm, B. et al.: 1998, COCOMO II Model Manual, http://sunset.usc.edu/research/COCOMOII/index.html.

Chiozzi, G. et al.: 2002, 'CORBA-Based Common Software for the ALMA Project', in H. Lewis (ed), *SPIE Advanced Telescope and Instrumentation Control Software II*, Proceedings of the SPIE, Vol. 4848, pp. 43–54.

Lewis, H., Conrad, A., and Kilbrick, B.: 2002, 'Do Bigger Telescopes Need Bigger Software?' in H. Lewis (ed), *Advanced Telescope and Instrumentation Control Software II*, Proceedings of the SPIE, Vol. 4848, pp. 167–174.

Royce, W.: 1998, *Software Project Management*, Addison-Wesley, Reading, MA.

SKA AND EVLA COMPUTING COSTS FOR WIDE FIELD IMAGING

T. J. CORNWELL
NRAO[1], Socorro, NM, U.S.A.
(e-mail: tcornwel@nrao.edu)

(Received 12 August 2004; accepted 8 December 2004)

Abstract. I investigate the problem of high dynamic range continuum synthesis imaging in the presence of confusing sources, using scaling arguments and simulations. I derive a quantified cost equation for the computer hardware needed to support such observations for the EVLA and the SKA. This cost has two main components – from the data volume, scaling as D^{-6} (where D is the antenna diameter), and from the non-coplanar baselines effect, scaling as D^{-2}, for a total scaling of D^{-8}. A factor of two in antenna diameter thus corresponds to 12 years of Moore's law (18 month doubling time) cost reduction in computing hardware. For a SKA built with 12.5 m antennas observing with 1 arcsecond at 1.4 GHz, I find the computing load to be about 150 Petaflops (costing about $500 million in 2015). For 25 m antennas, the load is about 256 times lower, costing $2 million in 2015. This new cost equation differs from that of Perley and Clark (2003), which has scaling as D^{-6}. This is because I find that the excellent Fourier plane coverage of the small antenna design does not significantly change the convergence rate of the Clean algorithm, which is already satisfactory in this regime.

Keywords: interferometry, radio astronomy, Square Kilometre Array, wide field imaging

1. Introduction

The Square Kilometre Array (SKA) will provide a large step forward in sensitivity. It is still unclear how this improvement in sensitivity is to be realized in steel, and less so on the silicon, computing side. The combination of many antennas, long baselines, large bandwidths will inevitably drive the computing needs upwards. Just how large the computing requirements are has had relatively little discussion, largely due to the focus being on the prime problem of making cheap collecting area. In this paper, I make a start by estimating the computing costs for sensitivity limited continuum imaging at decimeter wavelengths. While this is only one type of experiment, the experience with current arrays tells us that it is often one of the most demanding. My intention is not to claim a definitive analysis but instead to start the discussion of this topic with the hope that others with different ideas will contribute. Earlier versions of this paper have been published in the EVLA and SKA memo series. This version incorporates comments arising at the Penticon 2004 meeting, and supersedes the earlier memos.

[1]The National Radio Astronomy Observatory is operated by Associated Universities, Inc., under cooperative agreement with the National Science Foundation.

My work follows on from that of Perley and Clark (2003), who have recently derived a cost equation for synthesis arrays that includes the computing costs to counteract the non-coplanar baselines aberration. One conclusion from their work is that the cost equation should include a cubic term in the number of antennas. Consequently the minimum cost antenna diameter for fixed collecting area is increased over that derived, while ignoring the costs of non-coplanar baselines. To determine how much the diameter is increased, the actual scaling coefficient must be known. In this paper, I estimate the scaling relationships using analysis of the processing algorithms and large simulations performed in AIPS++.

The layout of this paper is as follows: in Section 2, I describe the basic operations required for full field imaging, and the ways in which these operations scale. In Section 3, I check these arguments by simulations and also determine the absolute costs, measured in Teraflops and scaled to dollars using Moore's Law. In Section 4, I summarize the scaling laws, and in Sections 5 and 6, I investigate the implications for the expand very large array project (EVLA) and the SKA. In Section 7, I describe some possible ways to decrease the computing costs.

2. Scaling behavior

It is important to emphasize that the case being considered here is sensitivity-limited imaging of a region much smaller than the primary beam, in the presence of background sources spread across the entire primary beam. If the primary beam were otherwise empty of emission, we would need only to construct an image of that limited region. However, the sidelobes from the background sources will limit the noise level achievable in the region of interest, and hence these sources must be estimated and their sidelobes removed from the region of interest. There is thus no advantage in increasing the field-of-view (by making the antenna diameter smaller). The case of surveying a region larger than a primary beam is not considered here.

Wide-field imaging with synthesis radio arrays is potentially limited by a number of aberrations. One of the most important is the "non-coplanar baselines" effect (see e.g. Cornwell and Perley, 1992). Perley and Clark (2003) analyzed the time taken to clean an image afflicted by non-coplanar baselines smearing using the facet based algorithms (see Cornwell et al., 2005 for more on the taxonomy of wide field imaging algorithms). The w-projection algorithm in AIPS++ outperforms the facet-based algorithms in AIPS and AIPS++ by about an order of magnitude (Cornwell et al., 2003, 2005), and so I choose to focus on it in what follows.

Calculation of the work required to make a dirty image using w-projection is straightforward, but for clarity, I first consider the case where the non-coplanar baselines effect can be ignored. Let there be N antennas of diameter D on baselines up to B. To minimize smearing of sources far out in the primary beam, the number of channels and the integration time (ignoring the atmospheric time scale) must both be chosen to scale as B/D. The number of baselines goes as N^2 and so the data rate goes as $N^2 B^2/D^2$. For a constant collecting area, this is B^2/D^6. The data rate can be higher than this, of course, if the atmospheric coherence time is very short or radio frequency interference must be sampled and excised.

In constructing an image, the cost of gridding the data nearly always dominates over that for the fast Fourier transform. The data are gridded onto the $(u, v, w = 0)$ plane using a convolution function of fixed size, typically 7×7 or 9×9 pixels. Hence the number of operations required to grid the data goes as the data rate $N^2 B^2/D^2$.

If the non-coplanar baselines effect is important, either w-projection or facet-based imaging must be used. For the former, one image is made and the area of the gridding function in pixels goes as $\lambda B/D^2$, and for the latter, the gridding function is constant but the number of images goes as $\lambda B/D^2$. Hence the number of operations required to grid the data goes asymptotically as $\lambda N^2 B^3/D^4$, which for a constant collecting area goes as $\lambda B^3/D^8$.

For the cleaning, there are two main operations – the minor cycle clean and the major cycle calculation of the residuals. The latter nearly always dominates, and so the total cost goes as the number of major cycle times the gridding cost. The number of cycles is driven by the maximum sidelobe, p_b, exterior to the beam patch in the Clark minor cycle. Each major cycle lowers the noise floor by roughly the sidelobe level, and so the dynamic range achieved after N_c major cycles is roughly:

$$\Lambda \sim \left(\frac{1}{2 p_b}\right)^{N_c} \tag{1}$$

The factor of 2 occurs because for stability the clean in a major cycle typically terminates somewhat above the maximum sidelobe. The number of major cycles is:

$$N_c \sim -\frac{\log(\Lambda)}{\log(2 p_b)} \tag{2}$$

I show in appendix the familiar result that the rms sidelobe level goes as the inverse of the square root of the number of visibility samples. If the field-of-view is held fixed, then the number of visibilities goes as D^{-4} and the sidelobe level as D^2, which is the scaling assumed by Perley and Clark. However, the field-of-view must expand as the antenna diameter decreases. Taking this into account, the number of visibilities goes as D^{-6}, and the rms sidelobe as D^3.

If the peak sidelobe level is five times the rms, the number of major cycles needed to clean for a point spread function constructed from N_s samples is:

$$N_c \sim \frac{\log(\Lambda)}{\log(\sqrt{N_s}/10)} \tag{3}$$

TABLE I

Expected scaling of imaging time with antenna diameter

	Number of baselines	Time and frequency sampling	Non-coplanar baselines	Cleaning		Total	
General	N^2	$\dfrac{B^2}{D^2}$	$\dfrac{\lambda B}{D^2}$	$\dfrac{\log(\Lambda)}{\log(\sqrt{N_s}/10)}$		$\dfrac{N^2 B^3 \lambda}{D^4}$	$\dfrac{\log(\Lambda)}{\log(\sqrt{N_s}/10)}$
Fixed collecting area D^{-4}		$\dfrac{B^2}{D^2}$	$\dfrac{\lambda B}{D^2}$	$\dfrac{\log(\Lambda)}{C - 3\log(D)}$	D^{-8}		$\dfrac{\log(\Lambda)}{C - 3\log(D)}$

This includes the costs of gridding, Fourier transform, and deconvolution; C is a constant.

This scaling behavior holds as long as the work in the minor cycle can be ignored, which may not be true for imaging of complex sources.

Also note that the source spectral index enters only via the logarithm of the dynamic range. Hence that source of wavelength dependency can be ignored.

Taking all of these factors into account, I obtain scalings as shown in Table I. In comparison, Perley and Clark found that the computing scales as:

$$\frac{N_{ag} B^3 \lambda^{1.7}}{D^6} \qquad (4)$$

The term N_{ag} is the number of antennas in a station. The discrepancy between the two scalings occurs because of the differences in the cleaning behavior. There is surprisingly little gain in speed for improvements in sidelobe level. I take special note that this is mostly due to the excellent characteristics of the Clark Clean algorithm.

I agree with Perley and Clark that there is a substantial penalty attached to using small antennas. This analysis is somewhat unfair to small antennas in that the benefits of excellent Fourier coverage for imaging complex sources are not incorporated but it does represent well the cost of removing confusing sources. This may change the numbers by 2 or 4 but will not compensate for the power of eight scaling.

3. Simulations

There are many subtle points to get right in the above analysis. Hence simulation is essential to check the scaling, and to determine the scaling coefficient.

Simulation of SKA observing on current computers is barely possible. I therefore choose to simulate only a short period of observing: 50 s of time spread over 3000 s of hour angle, looking at a source that was close but not at the zenith. The maximum baseline length was chosen to be only 10 km. The antenna locations were chosen using a random process designed to give approximately Gaussian Fourier plane coverage. I performed three sets of simulations, with the same baselines and antennas but with wavelengths separated by factors of ten to separate out the influence of the non-coplanar baselines effect (this does not mean that I expect that a 12.5 m antenna would be used to observe at 2.1 m – just that it is convenient to scale the simulations in such a way).

These simulations have been constructed to scale appropriately with antenna diameter – hence as the antenna gets smaller, the integration time and bandwidth decrease linearly, and so the data volume does indeed scale as D^{-6}.

The angular density of sources was chosen to be approximately correct for 20 cm imaging down to about 1 mJy, but weaker sources are underrepresented. This is in line with our intention of investigating the cleaning of bright sources spread across the primary beam.

The simulations were performed using the AIPS++ (version 1.9, build 549) simulator and qimager tools, running on a Dell 650 Workstation (dual processor Xeon 3.06 Ghz processors, 3 GB memory, Redhat Linux 7.2, special large memory kernel). The SPEC2000 floating point benchmark (CFP2000) is 13.8. Only one processor was used (Table II).

The quantitative simulation results are given in Table III. Some notes are:

- The Fresnel number is $D^2/\lambda B$.
- The width of the w-projection gridding function is determined from the numerically calculated form, and is closely related to the inverse of the Fresnel number.
- Times shown are noted from a wall clock.
- The image properties shown are the minimum (affected by cleaning errors around bright sources), and the median absolute deviation from the median (a robust statistic showing the off source error level.
- I was not able to complete the most time-consuming case – 12.5 m antennas observing at 2.1 m. My estimate is that it would have taken about 4 days.

4. The scaling laws

The rms sidelobe scales with the cube of the antenna diameter as expected (Figure 1), for both natural and uniform weighting. These are very low by the usual standards in radio synthesis but there is no qualitative change in behavior for small antennas.

I find in the simulations that the scaling index for imaging time with antenna diameter varies with the Fresnel number as shown in Figures 2 and 3 and as summarized in Table IV.

The scaling is steeper than -8 at the extreme ends but I believe this is most probably due to an onset of moderate paging. Hence for the low Fresnel number case, the scaling power can be taken to be -8. The required throughput therefore

TABLE II
Details of simulation

Total collecting area	Equivalent to 1600 twelve-meter antennas within 10 km
Antenna diameter	12.5, 15, 17.5, 20, 22.5, 25, 27.5, 30, 32.5, 35, 37.5, 40 m
Number of antennas	Set by antenna diameter to achieve fixed collecting area
Array configuration	Random antenna locations
Frequency	14 GHz, 500 MHz bandwidth; 1.4 GHz, 50 MHz bandwidth; 140 MHz, 5 MHz bandwidth
Observing pattern	50 s at transit, integration time 10 s, scaling as antenna diameter, with gaps of 600 s
Number of spectral channels	Eight channels maximum, scaling inversely with antenna diameter
Array latitude	34°N
Source declination	45°
Source details	250 point sources per primary beam with source count index −0.7. Peak strength: 1 Jy (but two sources may be in same pixel).
Antenna illumination pattern	Unblocked, uniformly illuminated
Synthesis imaging details	0.15, 1.5, 15 arcsec pixels, uniform weighting, with 0.6, 6, 60 arcsec taper. The image size scales inversely as antenna diameter. For the AIPS ++ FFT, the image size must be chosen to be the next largest composite of 2, 3, and 5.
Number of w planes in w-projection algorithm	128
Clean details	Cotton-Schwab algorithm, loop gain 0.1, maximum 100 000 iterations, stopping threshold 0.1 mJy for 40 m scaling as $1/\sqrt{N_s}$, beam patch 51 by 51 pixels
Resolution	0.6, 6, and 60 arcsec

obeys the scaling law:

$$T_{SKA} = T_{12.5\,m} \left(\frac{0.1}{\eta}\right) \left(\frac{f}{0.5}\right)^2 \left(\frac{B}{5\,\text{km}}\right)^3 \left(\frac{N}{1600}\right)^2 \left(\frac{D}{12.5\,\text{m}}\right)^{-4} \left(\frac{\lambda}{0.2\,\text{m}}\right)$$
$$\times \left(\frac{\Delta \nu}{500\,\text{MHz}}\right) \quad (5)$$

Assuming Moore's Law (with 18 months halving time) for the cost of processing, the scaling law is:

$$C_{SKA} \approx C_{12.5\,m} \left(\frac{0.1}{\eta}\right) \left(\frac{f}{0.5}\right)^2 \left(\frac{B}{5\,\text{km}}\right)^3 \left(\frac{N}{1600}\right)^2 \left(\frac{D}{12.5\,\text{m}}\right)^{-4} \left(\frac{\lambda}{0.2\,\text{m}}\right)$$
$$\times \left(\frac{\Delta \nu}{500\,\text{MHz}}\right) 2^{\frac{2(2010-t)}{3}} \quad (6)$$

TABLE III
Simulation results

Antenna Diameter (m)	Fresnel number	GCF Width (pixels)	Ant	Int	Chan	Sources	Sizes Image (pixels)	Vis Records	MS GB	To construct	To predict	To clean	Threshold (mJy)	Clean Comps	Cycles	PSF Min	Outer	Image properties Minimum	Robust
15.0	0.011	129	1111	8	13	694	4320	64126920	3.917	750.3	8865.5	73767.2	1.9	16516	4	−0.0015	0.0015	−2.97E-05	2.99E-06
17.5	0.015	128	816	7	11	510	3600	25604040	1.612	224.0	1848.4	15858.8	1.2	14517	4	−0.0011	0.0011	−7.66E-05	2.61E-06
20.0	0.019	109	625	6	10	390	3200	11700000	0.691	85.1	574.5	4924.7	0.8	11791	4	−0.0013	0.0015	−7.87E-05	2.29E-06
22.5	0.024	90	493	5	8	308	2880	4851120	0.297	43.2	183.4	1582.2	0.5	10951	4	−0.0017	0.0023	−4.19E-04	2.97E-06
25.0	0.030	71	400	5	8	250	2500	3192000	0.196	28.8	85.1	738.5	0.4	10457	4	−0.0023	0.0030	−7.74E-05	2.75E-06
27.5	0.036	66	330	4	7	206	2304	1519980	0.096	14.5	41.4	358.4	0.3	9357	4	−0.0033	0.0059	−3.39E-05	2.34E-06
30.0	0.043	54	277	4	6	173	2160	917424	0.06	10.1	22.3	202.7	0.2	7533	4	−0.0052	0.0077	−2.65E-05	1.79E-06
32.5	0.050	48	236	3	6	147	1944	499140	0.035	5.8	12.2	134.6	0.2	7187	5	−0.0097	0.0157	−5.80E-05	2.74E-06
35.0	0.058	39	204	3	5	127	1800	310590	0.022	4.2	5.8	71.2	0.1	9524	5	−0.0150	0.0149	−6.47E-05	4.79E-06
37.5	0.067	39	177	3	5	111	1728	233640	0.017	3.5	4.7	69.0	0.1	10556	6	−0.0193	0.0180	−8.40E-05	5.08E-06
40.0	0.076	32	156	3	5	97	1600	181350	0.014	2.8	5.4	73.6	0.1	8218	6	−0.0238	0.0241	−9.03E-05	3.63E-06
Wavelength 0.21 m																			
12.5	0.074	32	1600	10	16	1000	5000	2.05E+08	11.46	2288.3	2547.9	21671.3	3.4	20631	4	−0.0003	0.0003	−2.83E-05	3.43E-06
15.0	0.107	25	1111	8	13	694	4320	64126920	3.917	677.0	660.9	5678.9	1.9	16353	4	−0.0005	0.0007	−2.28E-05	2.72E-06
17.5	0.146	19	816	7	11	510	3600	25604040	1.612	203.7	217.6	2007.5	1.2	14270	4	−0.0007	0.0011	−4.01E-05	2.37E-06
20.0	0.190	16	625	6	10	390	3200	11700000	0.691	85.9	95.3	867.6	0.8	11619	4	−0.0011	0.0018	−8.83E-05	2.15E-06
22.5	0.241	14	493	5	8	308	2880	4851120	0.297	43.5	44.5	416.7	0.5	10468	4	−0.0019	0.0034	−4.09E-04	2.42E-06
25.0	0.298	11	400	5	8	250	2500	3192000	0.196	29.0	26.4	254.0	0.4	10196	4	−0.0021	0.0038	−1.99E-04	2.54E-06
27.5	0.360	11	330	4	7	206	2304	1519980	0.096	14.4	17.4	217.4	0.3	8099	5	−0.0048	0.0072	−2.60E-05	1.68E-06

(Continued on next page)

TABLE III
(Continued)

Antenna Diameter (m)	Fresnel number	GCF Width (pixels)	Ant	Int	Chan	Sources	Image (pixels)	Vis Records	MS GB	To construct	To predict	To clean	Threshold (mJy)	Comps	Cycles	PSF Min	PSF Outer	Minimum	Robust
30.0	0.429	10	277	4	6	173	2160	917424	0.06	10.1	12.4	123.7	0.2	7576	4	−0.0057	0.0078	−5.60E-05	1.76E-06
32.5	0.503	9	236	3	6	147	1944	499140	0.035	5.8	8.4	113.7	0.2	7376	5	−0.0097	0.0169	−5.51E-05	2.99E-06
35.0	0.583	8	204	3	5	127	1800	310590	0.022	4.2	3.6	59.7	0.1	10777	5	−0.0150	0.0152	−1.06E-04	5.40E-06
37.5	0.670	8	177	3	5	111	1728	233640	0.017	3.4	3.3	65.5	0.1	12111	6	−0.0193	0.0171	−8.50E-05	5.80E-06
40.0	0.762	7	156	3	5	97	1600	181350	0.014	2.9	4.6	67.7	0.1	7847	6	−0.0237	0.0243	−4.46E-05	3.31E-06
Wavelength 0.021 m																			
12.5	0.744	7	1600	10	16	1000	5000	2.05E+08	3.202	2372.8	1195.5	10339.1	3.4	20621	4	−0.0003	0.0003	−2.75E-05	3.38E-06
15.0	1.071	6	1111	8	13	694	4320	64126920	3.917	727.9	380.6	3354.4	1.9	16333	4	−0.0004	0.0006	−2.11E-05	2.70E-06
17.5	1.458	6	816	7	11	510	3600	25604040	1.612	201.2	141.3	1374.2	1.2	14209	4	−0.0007	0.0011	−6.48E-05	2.37E-06
20.0	1.905	5	625	6	10	390	3200	11700000	0.691	87.2	71.0	659.9	0.8	11632	4	−0.0012	0.0018	−4.33E-05	2.15E-06
22.5	2.411	5	493	5	8	308	2880	4851120	0.297	43.1	36.6	351.1	0.5	10697	4	−0.0021	0.0032	−1.06E-04	2.60E-06
25.0	2.976	5	400	5	8	250	2500	3192000	0.196	29.6	23.0	222.2	0.4	10169	4	−0.0021	0.0040	−6.26E-05	2.53E-06
27.5	3.601	5	330	4	7	206	2304	1519980	0.096	14.7	15.8	201.3	0.3	8051	5	−0.0044	0.0072	−2.58E-05	1.69E-06
30.0	4.286	5	277	4	6	173	2160	917424	0.06	10.1	12.8	118.7	0.2	7597	4	−0.0055	0.0078	−3.00E-05	1.75E-06
32.5	5.030	4	236	3	6	147	1944	499140	0.035	6.2	8.3	111.5	0.2	7744	5	−0.0097	0.0168	−3.33E-05	3.28E-06
35.0	5.833	4	204	3	5	127	1800	310590	0.022	4.2	3.5	58.4	0.1	10727	5	−0.0150	0.0151	−1.16E-04	5.36E-06
37.5	6.696	4	177	3	5	111	1728	233640	0.017	3.5	3.0	64.3	0.1	11896	6	−0.0193	0.0171	−1.12E-04	5.74E-06
40.0	7.619	4	156	3	5	97	1600	181350	0.014	5.6	4.2	68.3	0.1	7633	6	−0.0237	0.0235	−2.68E-05	3.27E-06

SKA AND EVLA COMPUTING COSTS FOR WIDE FIELD IMAGING 337

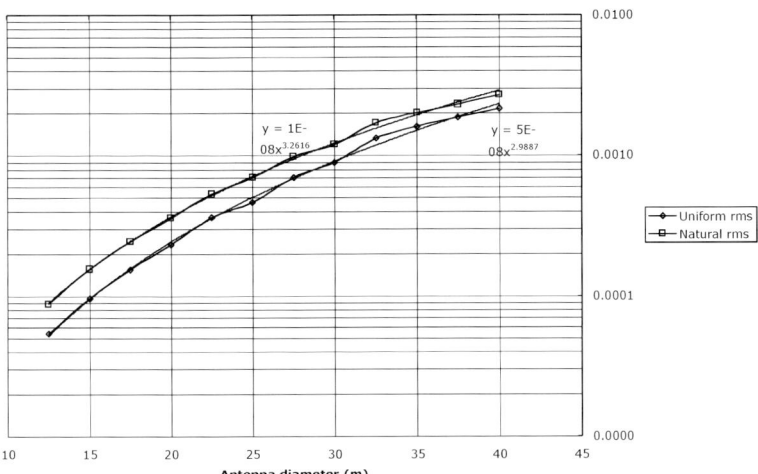

Figure 1. Sidelobe levels as a function of antenna diameter, showing scaling as the cube for both uniform and natural weighting.

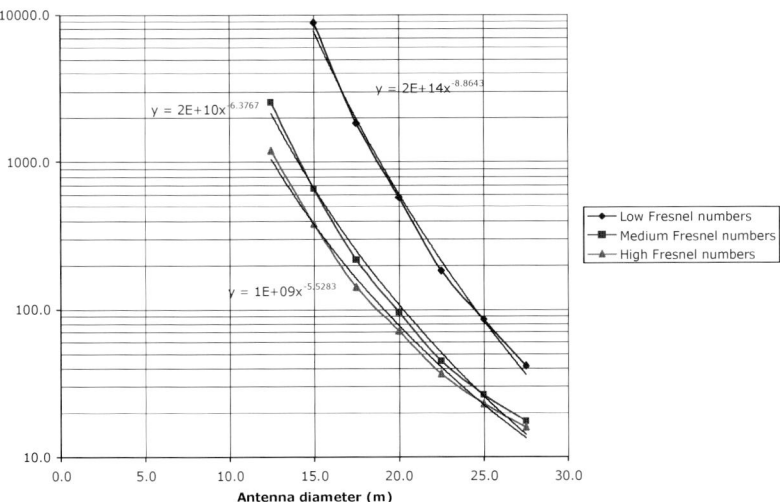

Figure 2. Model prediction times as a function of antenna diameter for low ($\ll 1$), medium (~ 1), and high ($\gg 1$) Fresnel numbers. Antennas >27.5 m have been excluded.

or, for a constant collecting area:

$$C_{\text{SKA}} \approx C_{12.5\,\text{m}} \left(\frac{0.1}{\eta}\right) \left(\frac{f}{0.5}\right)^2 \left(\frac{B}{5\,\text{km}}\right)^3 \left(\frac{D}{12.5\,\text{m}}\right)^{-8} \left(\frac{\lambda}{0.2\,\text{m}}\right)$$
$$\times \left(\frac{\Delta \nu}{500\,\text{MHz}}\right) 2^{\frac{2(2010-t)}{3}} \quad (7)$$

TABLE IV
Observed scaling index for cleaning time as a function of antenna diameter

Fresnel number (approx.)	Model prediction (approx.)	Clean (approx.)
0.001–0.1	−8.9	−8.8
0.1–1	−6.4	−6.0
1–10	−5.5	−5.2

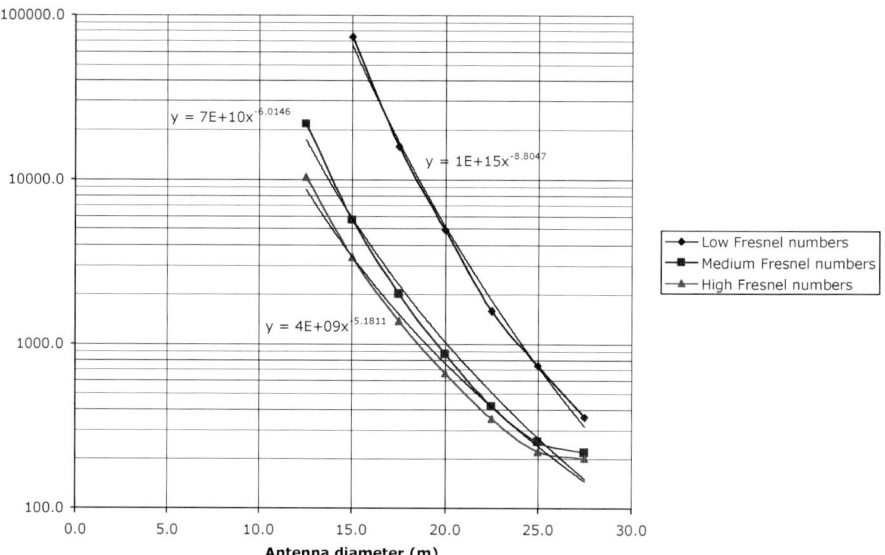

Figure 3. Clean times as a function of antenna diameter.

The filling factor f is the fraction of collecting area within the baseline B. For the SKA, the scientific specification on collecting area is 50% within 5 km. The efficiency of processing, η, is both very important and as yet unknown. It includes, for example, the cost of correcting for source spectral effects, and antenna primary beams, and the efficiency of parallel processing. A reasonable value for this efficiency is about 10%.

For a 17.5 m antenna design, the ratio between observing time and real time in our simulation is roughly 3000, so the efficiency is about 0.03% (for 25 m, the ratio is ~100, efficiency is ~1%). The computer used in the simulations were of the of cost about \$8 K in 2003. On solving, I found that the coefficient $C_{12.5m}$ is about \$7 million (2010) equivalent to about $T_{12.5\,m} \approx 200 T$ flops.

Since the antenna size for the EVLA has been chosen, I write the EVLA cost equation as:

$$C_{\text{EVLA}} \approx C_A \left(\frac{0.1}{\eta}\right) f^2 \left(\frac{B}{35 \text{ km}}\right)^3 \left(\frac{\lambda}{0.2 \text{ m}}\right) \left(\frac{\Delta \nu}{500 \text{ MHz}}\right) 2^{\frac{2(2010-t)}{3}} \qquad (8)$$

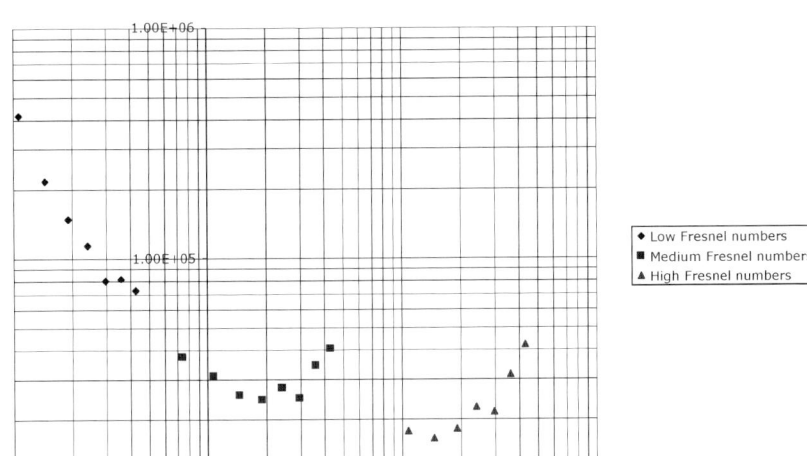

Figure 4. Operations per data point versus Fresnel number, calculated by scaling time by clock frequency, dividing by number of data points. The left end of each curve is biased upwards by constant cost terms.

Scaling appropriately from C_{SKA}, I find that C_A is \$170 K ($T_{12.5\,m} \approx 5$ T flops).

The number of operations required per data point can be estimated by scaling by the CPU clock rate. The curves shown in Figure 4 reach a minimum at about 20 000 floating pointing operations per data point. This should be taken as correct in order of magnitude only but it does reflect the scale of processing per data point.

5. Implications for the EVLA

For the EVLA A configuration (baselines up to 35 km), the computing load for wide-field processing is 5 Tflops (costing 170 K in 2010, 17 K in 2015). This is quite modest and not dissimilar to previous estimates. In phase II, the EVLA will have baselines up to 350 km baselines, and the required throughput would be 500 Tflops (\$17 and \$1.7 m). Such observations would be fairly rare and so the actual required duty cycle would be low.

Algorithm improvements help. The advent of w-projection brings the cost down by about an order of magnitude, which is equivalent to five years of Moore's law gains. Poor symmetry and stability of e.g. primary beams and pointing will hurt a lot by decreasing the efficiency (see e.g. Cornwell, 2003; Bhatnagar et al., 2004).

In addition, there remains a lot of software development to be done. It is clear that parallel processing using tens or hundreds of processors will be required to handle

EVLA data. There has been relatively little work on parallelization of synthesis imaging algorithms.

Finally, operational models of the EVLA will affect the cost estimates. If the most demanding observations occur infrequently and turnaround can be a few days or weeks (as is now often the case) then the computing costs can be reduced proportionately.

6. Implications for the SKA

The canonical case of SKA imaging with 12.5 m antennas on the 5 km baselines at 20 cm would require only 360 Tflops ($12 million in 2010, and $1.2 million in 2015). However, for the more interesting case of the 35 km baselines, the costs rise to 16 Pflops ($540 and $54 million). Increasing the antenna diameter to 25 m brings the costs down to 63 Tflops ($2.1 million and $210 K). For 350 km baselines, the cost increases to 63 Pflops ($2100 and $210 million), even with 25 m antennas!

A key point is that the scaling behavior is very dramatic, as the cube of the baseline and the inverse eighth power of the antenna diameter. In comparison, the effects of more bandwidth and longer wavelength are linear. Thus the SKA computing budget will be determined by the scientific emphasis placed on baselines in the range of 10 km and longer. The primary conclusion is that computing hardware is a major cost driver for the SKA, and much more attention is required before the concept cost estimates can be viewed as accurate. In addition, simulations should start to include the non-coplanar baselines effect, so as to raise awareness of the importance of the effect for SKA. In the specific case of the LNSD concept, the cost minimization with respect to antenna diameter should be repeated with these more accurate computing costs included.

7. Possible remedies

Are there ways to avoid this large cost penalty for small antennas?

- *Invent an algorithm for non-coplanar baselines with better scaling:* this is a good idea, of course. The newest algorithm, *w*-projection, is much faster than the old algorithm (faceted transforms), but has fundamentally the same D^{-8} scaling behavior, arising from the data volume, D^{-6}, and the physics of Fresnel diffraction, D^{-2}. More algorithm development is probably needed, but we should not cash in breakthroughs not yet made.
- *Reweight the data taking account of the superb Fourier plane coverage of LNSD:* this has been done in these simulations. Tapered uniform weighting brings the sidelobes down by some factor but the basic scaling with the inverse square root of the number of visibility sample still applies. In any event, I have shown that the number of major cycles is only weakly determined by the sidelobe level.

- *Average data per facet:* it has been suggested (Lonsdale, personal communication) that in a faceted transform, the data can be averaged in time and frequency before gridding, thus removing a factor of D^{-2}. While this is correct, it must be done appropriately for each facet, and therefore the factor is immediately lost. There may be a slightly lower scaling coefficient, bringing faceted processing closer to w-projection in speed.
- *Ignore the sources outside the minimum field-of-view needed for science:* rely on averaging in time and frequency to suppress the sidelobes from these sources. Lonsdale et al. (2004) have investigated this in some detail. They find that it works well for calibrated data but is likely to have deleterious effects on the efficacy of self-calibration.
- *Use a pre-existing model of the sky:* this is a good idea but not very relevant for early observations. Imaging the entire sky at the required sensitivity level would take many months of observing.
- *Form stations from clusters of antennas:* this would help a lot and certainly seems necessary on the longest SKA baselines. Whether it is acceptable for the shorter, 35 km, spacings needs more study. It would undermine the strength of LNSD – the superb Fourier plane coverage.
- *Only observe snapshots:* for snapshots, the effect of non-coplanar baselines is less. At the dynamic range required, the integration time would have to be very short (~minutes or less).
- *Only do hard cases infrequently:* the VLA followed this path in the early eighties when spectral line observing was almost impossible. As technology improves, the duty cycle can be changed. This requires continuing investment in computing, and deferred gratification.
- *Mandate an efficiency of 100%:* the costs scale inversely as the efficiency of processing, η. We could mandate a "one-shot" policy. This seems to be counter-productive – why build a $1G telescope and then not reduce the data correctly? Efficiency of 100% is unlikely, anyway, since self-calibration will almost always be needed.
- *Build special purpose hardware for imaging:* this is quite plausible and should be investigated. The best approach would probably be to build a special (digital) processor to do the w-projection part of the imaging, and keep the rest of the processing in general purpose computers. Most of the work in w-projection arises from convolving the measured visibilities onto a regular grid, using a convolution function that varies in width as $\sqrt{w}$ pixels to reflect the Fresnel diffraction effect (Cornwell et al., 2003, 2004). The half-width of the convolution kernel is given in Table III. For the most difficult cases, this can be in the range 100–300. This type of processing is also well suited to graphical processing units (GPUs), which are now a commodity item, and some investigation of their use would be worthwhile.

Moving to very large antennas might seem the best way to solve this problem. However, the simulations with 40 m antennas were marginally stable, tending to diverge for fewer hour angles. This should be understood in more detail before concluding that very large antennas are acceptable.

8. Summary

I found that for the specific problem of imaging in the presence of confusing sources, the use of small antennas comes at huge computing cost, as the inverse eighth power of the antenna diameter or the fourth power of the number of antennas. The cost has two main components – from the data volume, scaling as D^{-6}, and from the non-coplanar baselines effect, scaling as D^{-2}.

Continued algorithm research in this area is vital. We should investigate deploying existing algorithms on parallel machines, and possibly GPUs and special purpose hardware.

Finding a way to avoid this cost should be a high priority for SKA concepts that use relatively small antennas, such as the LNSD and the aperture array, as should be justifying the use of such small antennas.

Appendix

For the scaling relations, we need approximate relations for the typical sidelobe level outside the center region of the point spread function (PSF). This number determines how deeply any one major cycle of the clean algorithm can go.

The PSF is the Fourier transform of the weights attached to the visibility samples. For a gridded transform, the weight per grid cell is a product of the sum of the weights in that cell, optionally divided by some uniform weighting correction, and optionally multiplied by a taper function. The weighting correction is chosen to minimize the noise level (natural weighting), the sidelobe level (uniform weighting), or some compromise (robust weighting). See Briggs (1995) for more details.

We can use a random model for the distribution of Fourier plane samples. The PSF is simply a linear combination of the N_g grid weights. Thus the variance per grid cell adds:

$$\sigma_{\text{psf}}^2 = N_g(\sigma_w^2 - \langle w \rangle^2)$$

We will limit our considerations to the case where all samples have the same intrinsic weight (before gridding). Normalizing the PSF then amounts to dividing by the total number of samples N_s.

For natural weighting, we can use a Poisson model of mean N_s/N_g. The variance about the mean is N_s/N_g and so the rms sidelobe level is:

$$\sigma_{\text{psf,nat}} = \sqrt{\frac{1}{N_s}}$$

For uniform weighting, the weight per grid cell is either 0 or 1. If the number of empty cells is N_e, then the rms sidelobe level is:

$$\sigma_{\text{psf,uni}} = \frac{\sqrt{N_g - N_e}}{N_g - N_e}$$

We can apply the Poisson model again, but this time we are interested in the number of empty cells given that I distributed N_s samples at random. The probability that a given cell is empty is:

$$e^{-\frac{N_s}{N_g}}$$

Hence the rms sidelobe goes as:

$$\sigma_{\text{psf,uni}} = \frac{1}{\sqrt{N_g\left(1 - e^{-\frac{N_s}{N_g}}\right)}}$$

For small average sample density much less than 1, this can be approximated by the natural weighting result. For high density, the number of grid points plays the role of the number of samples.

Acknowledgements

I thank Sanjay Bhatnagar for help thinking about sidelobe levels, and Rick Perley for comments on an early version of this memo. The referees also provided helpful comments.

References

Briggs, D. S.: 1995, 'High fidelity deconvolution of moderately resolved sources', Ph.D. Thesis, New Mexico Tech, http://www.nrao.edu/~tcornwel/danthesis.pdf.
Bhatnagar, S., Cornwell, T. J. and Golap, K.: 2004, EVLA memo 84, see http://www.nrao.edu/evla.
Cornwell, T. J. and Perley, R. A.: 1992, *Astron. Astrophys.* **261**, 353.
Cornwell, T. J.: 2003, EVLA memo 62, see http://www.nrao.edu/evla.
Cornwell, T. J., Golap, K. and Bhatnagar, S.: 2003, EVLA memo 67, see http://www.nrao.edu/evla.
Cornwell, T.J., Golap, K. and Bhatnagar, S.: 2005, *Astron. Astrophys.*, submitted.
Lonsdale, C. J., Doeleman, S. S. and Oberoi, D.: 2004, SKA memo 54, see http://www.skatelescope.org.
Perley, R. A. and Clark, B. G.: 2003, EVLA memo 63, see http://www.nrao.edu/evla.

EFFICIENT IMAGING STRATEGIES FOR NEXT-GENERATION RADIO ARRAYS

COLIN J. LONSDALE*, SHEPERD S. DOELEMAN and DIVYA OBEROI

Massachusetts Institute of Technology, Haystack Observatory, Westford, MA 01886, U.S.A.
(*author for correspondence, e-mail: cjl@haystack.mit.edu)

(Received 7 September 2004; accepted 21 February 2005)

Abstract. The performance goals of the Square Kilometre Array (SKA) are such that major departures from prior practice for imaging interferometer arrays are required. One class of solutions involves the construction of large numbers of stations, each composed of one or more small antennas. The advantages of such a "large-N" approach are already documented, but attention has recently been drawn to scaling relationships for SKA data processing that imply excessive computing costs associated with the use of small antennas. In this paper we examine the assumptions that lead to such scaling laws, and argue that in general they are unlikely to apply to the SKA situation. A variety of strategies for SKA imaging which exhibit better scaling behaviour are discussed. Particular attention is drawn to field-of-view issues, and the possibility of using weighting functions within an advanced correlator system to precisely control the field-of-view.

Keywords: imaging, radio-astronomy, radio-telescopes

1. Introduction

The performance goals for SKA are ambitious. In addition to a very wide frequency range, the instrument is expected to cover wide fields-of-view, and also provide VLBI-like angular resolution. The high sensitivity of SKA coupled with wide fields-of-view implies a need for very high imaging dynamic range, in very complex fields, far in excess of dynamic ranges routinely achievable today. It is unlikely that all of the main SKA goals can be met with a single design, or even with a single instrument – in particular, optimal approaches for low and high frequencies are likely to be different. Because of the areal density of astrophysically interesting sources, wide fields-of-view are most sought after at the lower frequencies, and a key SKA specification is 1 deg^2 at 1.4 GHz with 0.1" resolution. Much wider fields-of-view are considered desirable.

These performance goals cannot be met by simply scaling up current instruments, which were developed with far less demanding requirements in mind. The same applies to data handing and reduction techniques. At present day sensitivities, with narrow fields-of-view and small fractional bandwidths, many potentially troublesome effects can be, and are, ignored. SKA design must be approached from the ground up, in such a way that hardware capabilities and software algorithmic approaches work together to permit correction for a variety of subtle effects. An interferometer with a much larger number of stations than current instruments has, among many advantages, the property of generating large numbers of independent measurements with which to drive advanced algorithms. This is one of the most compelling reasons for the development of large-N designs within the international SKA community.

However, concerns exist (Cornwell, 2004; Perley and Clarke, 2003,) that large-N SKA designs generate excessive data volumes, and place unsupportable burdens on computing resources for imaging and deconvolution operations. We contend that poor scaling of algorithms and approaches developed for present-day, small-N arrays is not surprising, and that data handling and reduction approaches need to be redesigned. In this paper we seek to identify some key issues and provide pointers to innovative solutions, to be further developed in future work. We discuss in more detail one specific proposed technique that can sharply reduce the above-described imaging computational problem. The present discussion focuses on imaging and deconvolution, and as in the above referenced EVLA and SKA memos, we assume that the data are perfectly calibrated. In the summary we comment on the calibration problem, which poses significant risk for any SKA design.

2. General considerations

2.1. Data rates and volumes

It has been shown (Cornwell, 2004) that based on certain assumptions regarding array performance and postprocessing algorithms, the computational load of creating high dynamic range images from calibrated SKA data scales as D^{-8}, where D is the diameter of the constituent antennas. Using the derived scaling relationship, plus an estimate of the cost coefficient based on simulations, projected computing costs for the year 2015 to meet SKA specifications with a design based on 12.5 m antennas with correlation of all antenna pairs would be of order $0.5 billion. This figure is much higher than can reasonably be accommodated.

This conclusion is, however, not surprising and follows straightforwardly from correlator output data rate considerations. Following Lonsdale (2003), we can examine the case for imaging 1 deg^2 with 1″ resolution at 1.4 GHz, with a fractional bandwidth of 0.25. Expression (3) of the Lonsdale memo describes the data rate in somewhat more general terms, allowing for the possibility of covering the individual antenna field-of-view using multiple beams formed at stations composed of multiple antennas. For the case in question with 12.5 m antennas, SKA specifications on the radial distribution of collecting area combined with the 4400-antenna USSKA design yield ∼2800 antennas within a 40 km diameter. The corresponding data rate is[1] $C_{out} \equiv 550/(n_a)^2$ GB/s, where n_a is the number of antennas per station. For $n_a = 1$ (the full cross-correlation case considered by Cornwell), the rate of 550 GB/s is problematic regardless of the efficiency or otherwise of imaging algorithms. A value of $n_a > 1$ (i.e. using beamformed stations outside some diameter substantially smaller than 40 km) appears necessary, and we will argue below that concerns about high fidelity imaging in the face of high station sidelobes are misplaced.

We also note that because $n_t \propto D^{-2}$, where n_t is the total number of antennas, for a fixed total array collecting area, expression (3) of Lonsdale (2003) confirms the D^{-6} data volume scaling relationship found by Perley and Clarke (2003) and Cornwell (2004). This scaling arises if the goal is to achieve a given point source sensitivity, and if full field-of-view (FOV) imaging is needed to reach that sensitivity. In later sections we critically examine the latter assertion, and propose methods both for reducing data volume, and for generating images in a more computationally efficient manner.

This D^{-6} scaling is misleading, however, because point source sensitivity is not the only valid measure of SKA sensitivity. Many SKA applications, including multiple key science projects, instead require survey speed, which scales linearly with the field-of-view in square degrees. The data volume for constant survey speed then scales only as D^{-4}.

It may nevertheless be asked, given these scaling laws, why large-N SKA designs are attractive. While it is not the purpose of the current paper to answer this question, we note that large-N and small-D are not synonymous. The effective aperture diameter of a SKA station is $\sim 1000\, N^{-1/2}$ m, so for 50 m dishes, N would be of order 500, certainly considered large. Small diameter dishes make it easier to achieve large fields-of-view, while a large number of stations increases the number of independent measurements with which to solve the challenging SKA calibration problem. The point is that postprocessing computing loads are only one of multiple considerations for SKA performance and feasibility, and addressing one problem by increasing D exacerbates other problems. Many researchers are of the opinion that overall, small-D and/or large-N solutions are superior.

It is also worth pointing out that expression (3) of Lonsdale (2003) can be used for the case of large antennas, for which focal plane arrays are employed to meet the SKA field-of-view specification. In this case, each antenna can be treated as a station of n_a smaller antennas, each with the required field-of-view, and a station filling factor f of unity. Each focal plane array feed then corresponds to a station beam, and the data rate expression applies directly. The data rate issue, and the scaling, remains the same. For high angular resolution, stations, or equivalently large dishes, must be employed.

---

[1] It is interesting to note that this rate is independent of the areal filling factor f of the station. This factor increases the correlator *input* data rate (Bunton, 2004), but does not affect the output rate. This is because as f decreases, the station beams shrink and more of them are required, but for each beam the correlator output data can be averaged over a greater range in time and frequency.

Finally, a comment about the current SKA specifications, as described in SKA memo 45, item 10. In the example discussed above, we derived a correlator output data rate of $550/(n_a)^2$ GB/s. The requirement of $0.1''$ angular resolution at 1.4 GHz increases this already troublesome data rate by a factor of 100, or equivalently requires an increase in n_a by a factor of 10, and a decrease in the number of stations N by a similar factor of 10. The specification of 0.2 arcsec over 200 $\deg^2$ at 0.7 GHz is 50 times more stringent still in terms of data rate. These specifications appear unrealistic, and fields-of-view must be allowed to decrease as the angular resolution increases.

2.2. FIELDS-OF-VIEW

For various reasons, a large-N SKA with a baseline length dynamic range approaching 10^5 will not be used for imaging studies in a manner that uses all the collecting area at the same time, for the same observation. In practice, SKA will be used in low, medium and high angular resolution modes, perhaps simultaneously, but never generating monolithic data sets covering the whole range. Low, medium and high resolution types of observation will naturally feature wide, medium and narrow fields-of-view respectively. In this subsection we briefly point out the various reasons for and consequences of this.

The reasons that observations must be segmented by resolution are as follows:

- SKA specifications call for full FOV observations at moderate resolution, but such observations at full resolution yield unsupportable data rates (Lonsdale, 2003).
- In order to serve a wide range of science goals, SKA must have substantial sensitivity on a wide variety of spatial scales, which mandates an array configuration with heavy central concentration of collecting area. A naturally weighted PSF for such an array would yield a point spread function (PSF) with very poor imaging properties, and roughly uniform weighting is instead required. For a given target resolution, this downweights the shortest baselines to the point of irrelevance, so they might as well be used for different science, enhancing overall instrument efficiency (we note that the ability to maintain superb imaging of simultaneous low and high resolution subarrays is a particular strength of the large-N approach).
- On long baselines, practical matters mandate the collection of antennas into clusters, or stations. Data rate considerations both before and after correlation strongly encourage station-based beamforming. The station beam presents a natural field-of-view for higher resolution observations. Tiling a wider field-of-view with multiple station beams carries significant complications, and there will be a strong natural tendency to shrink the field-of-view as the resolution increases. Full antenna FOV observations are better served by full cross-correlation, which is attractive, and likely feasible, only at lower resolutions.

EFFICIENT IMAGING STRATEGIES FOR NEXT-GENERATION RADIO ARRAYS 349

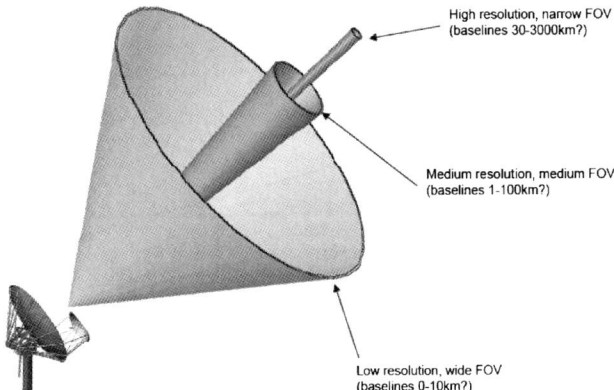

Figure 1. Different fields of view for different target resolutions. In practice a given experiment is unlikely to use more than a 100:1 baseline length range for producing a single image.

- Among the techniques for effective FOV restriction and associated reduction in computational load is tailoring of the correlator FOV, as discussed at length in Section 5 below. This potentially powerful technique is fundamentally restricted to a limited range of baseline lengths, perhaps covering a range of order 100:1. The 10^5 baseline length dynamic range of a large-N SKA thus requires of order 3 separate FOV domains in which the technique can be applied.

In essence, one can have wide fields or high angular resolution, but not both at the same time. SKA observing will be conducted separately for different target angular resolutions, probably simultaneously, by using different ranges of baseline length. This is illustrated in cartoon fashion by Figure 1.

3. Layered attenuation

The response of an interferometer array to sources far from the pointing direction depends on several factors. In present-day arrays, the most important factor is typically the primary beam of the antennas. Very high sensitivity observations can become limited by our ability to locate and subtract emission far from the field center. Such emission often lies in regions where instrumental performance is poorly known, yet signal attenuation is modest; extrapolation to SKA sensitivities then causes concern. In this section we look more closely at the attenuation that one can expect from various factors in a large-N SKA array, and illustrate that appropriate array design can readily compensate for the more rigorous demands of SKA sensitivity levels. Further, we point out that imaging the full antenna field-of-view is not necessarily a requirement, which bears directly and strongly on postprocessing computing loads.

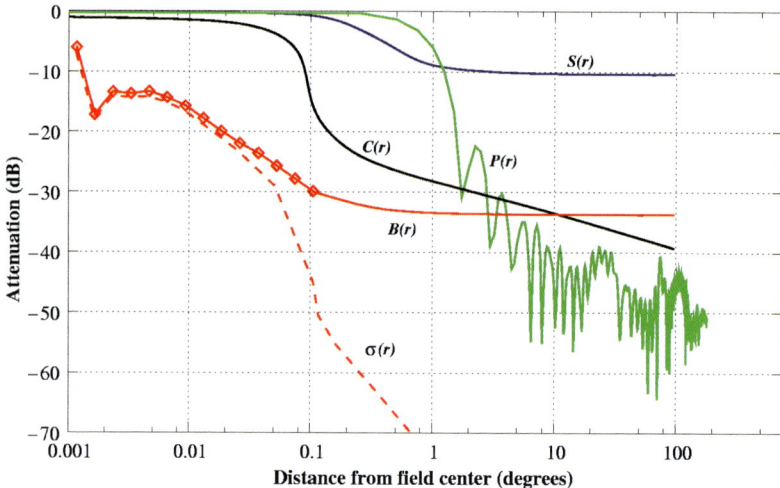

Figure 2. The solid and dashed red lines represent RMS sidelobe confusion levels at the phase center due to a source at distance r. $B(r)$ is the "dirty beam" RMS, governed by the sampling function, while $\sigma(r)$ is the RMS after various layers of signal attenuation are taken into account. The green line is $P(r)$, the primary beam of the antennas, the blue line is $S(r)$, the mean attenuation of the station beam. The black line is $C(r)$, the correlator attenuation factor (Section 5). Details are given in the text.

3.1. ATTENUATION EXAMPLE

Figure 2 is a graphical illustration of the concept of layered attenuation, that will be central to affordable computation for SKA.

Below are some comments on the detailed behavior of the functions illustrated in Figure 2. The curves are approximately what can be expected for a 1-h SKA observation at 1.4 GHz with 6% bandwidth, using 12 m paraboloids clustered into stations of ~10 antennas, and a maximum baseline of ~35 km. For this example, we have assumed that only 100 stations comprising a total of 1000 antennas participate in the observation with full weight.

- *Dirty beam RMS B(r).* The inner part of this PSF RMS curve is based on simulations using the MIT array performance simulator (MAPS), using uniform weighting. The baseline length range has been restricted in order to justify the use of a tailored correlator field-of-view $C(r)$ (see below).
- *Primary beam attenuation P(r).* This curve is a scaled version of the power pattern of the 6 m offset paraboloid ATA antenna, and is roughly appropriate to a 12 m paraboloid of similar unblocked design at 1.4 GHz. The far sidelobe levels scale roughly as the wavelength squared. We are grateful to Dave deBoer of the ATA project for providing these data.
- *Station beam attenuation S(r).* Here, we have assumed a minimum antenna separation of 2 times the antenna diameter, and a station area that is 50 times the

antenna area (station areal filling factor $f = 0.2$). The station beam main lobe is ~7 times smaller than the primary antenna beam. The sidelobe levels asymptote to ~0.1 beyond angular scales corresponding to approximately twice the antenna diameter.
- *Correlator attenuation factor C(r).* For conventional correlators, this is a complicated function that depends on baseline lengths and orientations. In this paper we propose a method of tailoring and matching the function for each visibility point. At small r, the function is flat, and at large r it asymptotes to a value and slope determined by the granularity of the data in time and frequency before correlator weighting. The detailed shape of this curve is only estimated here, and will be the subject of a future memo. Correlator field-of-view tailoring allows us to implement a sharp increase in attenuation as illustrated. Details are discussed in Section 5.
- *Effective dirty beam RMS $\sigma(r)$.* This is simply $B(r)$, multiplied by the various attenuation factors.

3.2. ATTENUATION FACTORS AND SOURCE COUNTS

The concepts can be clarified by expressing the situation in more mathematical terms. Consider the PSF of the array as determined solely by the weighted distribution of visibility samples on the *uv* plane. For a large-N array, at large distances r from the center, the PSF sidelobes represent the sum of contributions from many different baselines and spatial frequencies, with essentially random phases. As such, the sidelobe amplitude distribution can be expected to be Gaussian, with RMS given by a function $B(r)$. For an observation pointed at a particular location on the sky, a source of flux density s at a distance r from the region of interest will contribute sidelobe-related noise at the field center of RMS $\sigma_{s,r} = s.B(r)$. Contributions from many sources of different flux densities and at different distances will add in quadrature at the field center, to generate an aggregate noiselike fluctuation with RMS σ_t. Ignoring attenuation as a function of r, the number of sources contributing sidelobe noise, for given values of r and s, will be given by $rN(s)\,dr$. The differential source count $N(s)$ at 1.4 GHz is well described by $N(s) = ks^{-1.65}$ where k is a constant (e.g. White et al., 1997).

We now consider the fact that attenuation by the antenna primary beam $P(r)$, by the station beam $S(r)$ if multiple antennas are beamformed at a station, and by the correlator field-of-view $C(r)$ will reduce the effect of a source of flux density s by a factor $A(r)$

$$A(r) = P(r)S(r)C(r) \tag{1}$$

Here, we ignore dependencies on baseline length and on position angle due to various effects, and consider only the mean radial attenuation due to each effect. We can now express the aggregate sidelobe noise with RMS σ_t at the field center as follows

$$\sigma_t^2 = k \int_{r_0}^{\infty} \int_0^{s_{\max}} [sA(r)B(r)]^2 rs^{-1.65} ds\, dr \qquad (2)$$

The distance r_0 is the radius of the region at the field center within which high fidelity imaging with accurate deconvolution is performed as needed, so that sidelobes from sources within that radius are removed. If σ_t is comparable to or exceeds the expected RMS thermal noise in an observation, it is necessary to reduce the sidelobe contribution by some means. Both Cornwell (2004) and Perley and Clarke (2003) assume that deconvolution will always be necessary for all r out to the edge of the antenna primary beam, based on a simplified consideration of sidelobe levels. They further assume that this deconvolution will be done using the existing CLEAN algorithm.

There are some useful observations we can make about the above analysis of attenuation layers and their consequences. The sidelobe noise σ_t can be reduced by multiple, independent methods.

1. *Increase r_0*. If it is increased to the radius of the primary antenna pattern, we return to the Cornwell assumption that the entire field-of-view must be deconvolved, though perhaps not to the depth that Cornwell assumes. This is a brute-force approach.
2. *Make the attenuation function $A(r)$ a steeper function of r*. This can be accomplished by tailoring the correlator attenuation function $C(r)$, as discussed at length in Section 5.
3. *Design the array to minimize $B(r)$ for a wide range of observing parameters*. This places a premium on large-N solutions and good array configuration design.
4. *Reduce $s_{\max}$*. This implies the removal of bright sources from the dataset, with the threshold flux density likely to be a strong function of r. The sources need be removed only with sufficient accuracy so that the residual flux density does not significantly distort the source counts and elevate σ_t. In the next section, we discuss how this might be accomplished at a small fraction of the computational cost of the iterative CLEAN algorithm.

A quantitative assessment of achievable sidelobe confusion levels σ_t is needed for a range of SKA observational and design parameters. This can be accomplished by estimation and simulation of $A(r)$ and $B(r)$, followed by numerical integration of Equation (2). This will be the subject of a future memo.

3.3. Station Beamforming versus Full Cross-Correlation

One other interesting observation can be made, with relevance to the debate about the need for and advisability of a full cross-correlation architecture for an array based on small paraboloids. The instantaneous monochromatic PSF of an array has asymptotic RMS sidelobes at the $1/N$ level, where N is the number of stations (i.e. $B(r) \to 1/N$). The product $A(r)B(r)$ describes the net effect of a distant source in terms of sidelobe confusion. Assuming for the moment that $P(r) = C(r) = 1$, noting that the asymptotic value of $S(r)$ is $1/n_a$, and also noting that $N = N_t/n_a$ where N_t is the total number of antennas in the array, we see that $A(r)B(r) = 1/N \times 1/n_a = 1/N_t$, which is independent of n_a. This lack of dependence is unchanged by realistic $P(r)$ and $C(r)$ functions.

In other words, the degree of protection one gets from sidelobes generated by sources at large distances is unaffected by the choice of station beamforming versus full cross-correlation. Perceptions of increased problems in data reduction due to high station beam sidelobe levels are based on concerns about poorly behaved sidelobes that are hard to calibrate. In reality, such sidelobes are a concern only within the antenna primary beam, and SKA sensitivity is such that their relatively smooth pattern in that limited area can be readily determined using calibration sources. This is an instance where extrapolation from current experience to the SKA case is not well justified, and is not a valid argument against a station-based architecture on medium and long baselines.

3.4. CONSEQUENCES OF ATTENUATION LAYERS

As is apparent from Figure 2, sources that are distant from the phase center of an observation, even if they fall within the primary antenna beam, do not necessarily contribute significantly to the sidelobe confusion noise at the field center. This calls into question the basic assumption in earlier work that the full antenna field of view must be imaged in order to reach sensitivity goals. The steep slope of $\sigma(r)$ indicates that the flux density of sources about which we must be concerned increases quickly with distance from the center. The slope is so steep in our example that the total number of sources that must be subtracted from the visibility data in order to suppress sidelobe confusion noise to a given level is sharply reduced. This reduction is equivalent to a reduction in the sky area across which sources must be subtracted, provided a method of subtraction is used which depends primarily on the number of sources, not their locations. Direct subtraction from visibility data, instead of subtraction in the image plane, is such a method.

4. Alternatives to CLEAN based deconvolution

For the problem it is designed to solve, namely the simultaneous identification and removal of sources from a mostly empty dirty image, CLEAN is highly successful and efficient in its modern form, as noted by Cornwell (2004). For the problem facing a large-N SKA, however, it is an inefficient method of suppressing sidelobe confusion noise, which is the issue at hand. This inefficiency stems from the fact that CLEAN is an iterative blind search for emission, despite the fact that in the SKA era, the great majority of the sky will be known *a priori*, down to a given flux density level. Instead of a blind search, direct subtraction of a previously generated global sky model (GSM) from the visibilities is preferable. Conceptually this is equivalent to a single "major cycle" in the CLEAN algorithm, and considerable savings in computational load.

One may ask whether a chicken-and-egg problem exists for the GSM – do we need deep SKA surveys to generate it in the first place? The answer is no, because over the vast majority of the sky area over which the GSM is needed for a given imaging problem, strong attenuation applies. A typical attenuation factor, given by the combination of the synthesized beam sidelobe level, the station beam, and the correlator field-of-view, might be 10^3. This implies that generation of a GSM sufficiently deep to contain sources whose sidelobes fall at or below the target noise level of a SKA scientific observation can be done using a SKA observation with a duration 10^6 times shorter. In other words, the GSM does not need to be very deep, provided the array PSF is of very high quality (as is the case for large-N arrays).

Subtraction can still be expensive, but there are two considerations which can lead to further reductions in cost. First, depending on applicable levels of attenuation and sidelobe protection (see Figure 2 and Section 3), the required precision with which GSM sources must be subtracted will vary strongly. Most of the sources are relatively weak, and need only be approximately subtracted, which allows the use of various shortcut methods. Second, in principle, it is possible to precompute the visibilities due to a GSM for all locations in a (u, v, w) volume that will ever be sampled by the SKA. Subtraction of the GSM would then consist of little more than a lookup operation in this volume for each visibility point. This also eliminates the penalty associated with non-negligible w-terms in conventional approaches. In practice, application of calibration information is necessary, and significant computation per visibility point would be required. Nevertheless, the concept of using a precomputed Fourier representation of the GSM is interesting, and offers the potential for computational gains provided massive data storage and access become practical on SKA timescales.

We emphasize that these considerations are for perfectly calibrated data. Computational load implications of these approaches including calibration tasks need to be carefully analyzed.

5. Correlator field-of-view

The phenomena of bandwidth smearing and time-average smearing are well understood in the practice of imaging interferometry. Both effects result in loss of coherence and a consequent reduction of the interferometer response to flux density far from the phase center. As such, these effects can in principle be employed to advantage in suppressing troublesome sidelobe responses from distant sources, a central issue for high dynamic range SKA imaging. In this section, we examine this possibility in detail.

Correlators perform a sample-based cross-multiplication operation, followed by accumulation of results over some time interval often referred to as an accumulation period. The data also have a spectral resolution that is typically adjustable, and is set by the parameters of the "F" Fourier transform operation in either an "XF" or "FX" architecture. The correlator field-of-view arises because sky positions away from the interferometer phase center are associated with phase gradients in the (u, v) plane, which map into phase variations in time and frequency. When the correlator averages over finite spans of time and frequency, these phase variations cause decorrelation, and a reduction in the contribution of off-center flux density to a given visibility measurement.

More generally, the field-of-view to which a single visibility measurement is sensitive is given by the Fourier transform of the (u, v) plane sampling, which is in turn determined by the correlator time and frequency averaging operations. The response of an individual visibility point to the sky can be tailored to a degree that is limited only by the range of available times and frequencies that can be integrated into that point, and by the granularity of the time and frequency samples with which to perform the integration. In principle, arbitrarily high levels of signal rejection can be achieved away from the phase center. Tailoring the correlator field-of-view is potentially an attractive option for suppressing troublesome sidelobes of distant sources in SKA imaging.

5.1. LIMITATIONS AND PROBLEMS

Cornwell (2004) notes that the use of time and frequency averaging to suppress sidelobes does not work well at the VLA. This is not surprising. The VLA correlator applies the same time and frequency averaging to all baselines, regardless of baseline geometry (this fact incidentally results in a simplification of the application of self-calibration algorithms, which is relied upon to a significant extent by current calibration software). The transformation from (f, t) to (u, v) is highly variable between baselines, so that for constant (f, t) extent, sensitivity to distant sources is also highly variable between baselines. The effective array PSF then becomes a strong function of position – long baselines become increasingly down-weighted with increasing distance from the phase center, with significant fluctuations superimposed on this general trend. No facility exists in postprocessing software to compute or account for such PSF variations. Existing hardware and software systems are not designed to exploit this technique, and the poor VLA experience is not of general applicability.

A more fundamental problem with tailoring the correlator field-of-view stems from the very wide dynamic range in baseline lengths that is a scientifically essential feature of large-N SKA designs. This dynamic range is $\sim$50 dB for the USSKA design, as compared to $\sim$15 dB for a given VLA configuration. Matching the correlator FOV on the longest and shortest SKA baselines would require weighting functions differing in time extent and frequency extent by factors of order 10^5, which is impractical. We estimate that the technique is likely to be most useful for datasets with baseline length dynamic ranges of 100 or so, but not much more. We do not see this as a problem in practice, and refer to Section 2.2 which presents multiple arguments for limiting particular imaging problems to such baseline length ranges.

5.2. SIMULATIONS

We have quantitatively explored the concept of tailoring the correlator field-of-view by implementing different weighting schemes in the (u, v) plane for simulated data using MAPS (MIT Array Performance Simulator). In this software package, a brightness distribution consisting of a test pattern of point sources was Fourier inverted to the (u, v) plane. A snapshot data set was created using a VLA-like array configuration with baseline lengths ranging from 2 km to 36 km. This restricted range of baselines was chosen purely in order to illustrate the technique using modest map sizes, and does not reflect a fundamental limitation on the viable range. True limits will be set by calibration variations with time and frequency, and as mentioned above will probably result in baseline length ranges of order 100:1.

For each baseline, the corresponding visibilities on the (u, v) plane are numerically integrated over a frequency and time range that maps to a non-rectangular (u, v) patch, whose size is a function of baseline geometry, integration time and observing bandwidth. In this way, the simulator implements a model of operations inside a real correlator. It is in this integration step that we impose various weighting schemes in a hypothetical advanced correlator, and thence the (u, v) plane. Since weighting at this stage is equivalent to convolution, the effect is a multiplication in the image plane by the Fourier inversion of the weighting function. The main results of these simulations are described in the subsections below.

5.2.1. Time and frequency averaging
Simple averaging in both time and frequency can be used to restrict the field-of-view, but the same averaging time and bandwidth will produce different effective fields-of-view for different baselines. For the VLA-like array used, the baseline length dynamic range is ~20:1, so averaging times appropriate for the smallest baselines are 20 times too big for the longest baseline. With such mismatched fields-of-view, imaging breaks down as shown in Figure 3.

5.2.2. Use of shaped weighting functions to tailor field-of-view
Applying a single weighting function in the (u, v) plane imposes the same field-of-view restrictions on all baselines and avoids the image distortion shown in Figure 3. A simple windowing function in the (u, v) plane multiplies the image plane by a

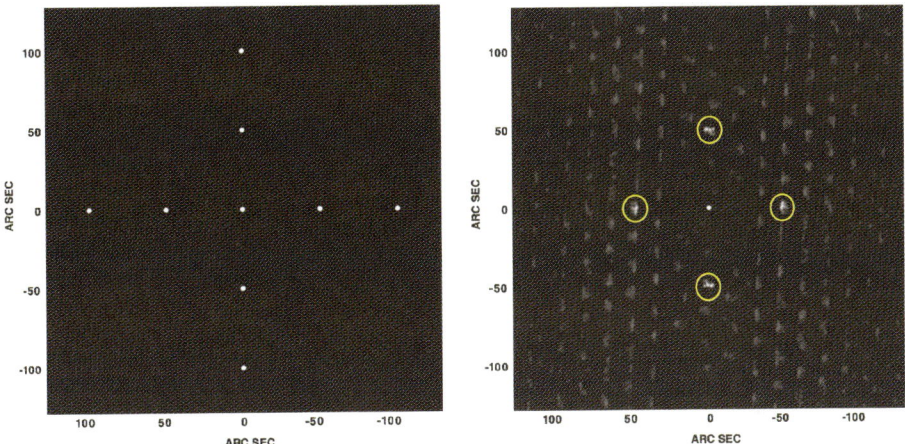

Figure 3. The effects of using simple averaging in frequency and time to restrict field-of-view. Left panel shows CLEANED image of grid of point sources with no averaging. The right panel shows the same data set averaged over constant time and frequency, with identical transfer functions. The central point source is unaffected by the averaging. The next innermost four sources (○) are severely distorted due to the mismatch of averaging extent in the (u, v) plane for different baselines. Strong, unsubtractable sidelobe confusion noise is evident across the image.

Jinc function (J_1 is a Bessel function of order 1):

$$W\left(\frac{r}{2a}\right) = \begin{cases} 1, & r < a \\ 0, & r > a \end{cases} \Leftrightarrow \tilde{W}(q) = \frac{J_1(2\pi q a)}{\pi q a} \qquad (3)$$

And a Gaussian weighting function in the (u, v) plane imposes a Gaussian in the image plane:

$$W(r) = e^{-\pi a^2 r^2} \Leftrightarrow \tilde{W}(q) = e^{-\pi q^2/a^2} \qquad (4)$$

For both of these functions, r is the radial distance on the (u, v) plane, and q is the radial distance on the image plane. We note that a disadvantage of the windowing function is that the ringing of the Jinc function can map positive sources on the sky into negative features on the image. The ringing also falls off slowly at large distances. An example image created with a Gaussian weighting function is shown in Figure 4. It also shows the measured and predicted fall off in point source flux density with distance from the image phase center, for a Gaussian weighting function.

5.2.3. Removing PSF sidelobes of distant sources

The main objective of averaging in this way is to attenuate the contribution of distant sources within a desired field of view. To gauge the effectiveness of this method, we simulated a data set with six sources of flux density 1 Jy, all ∼232″

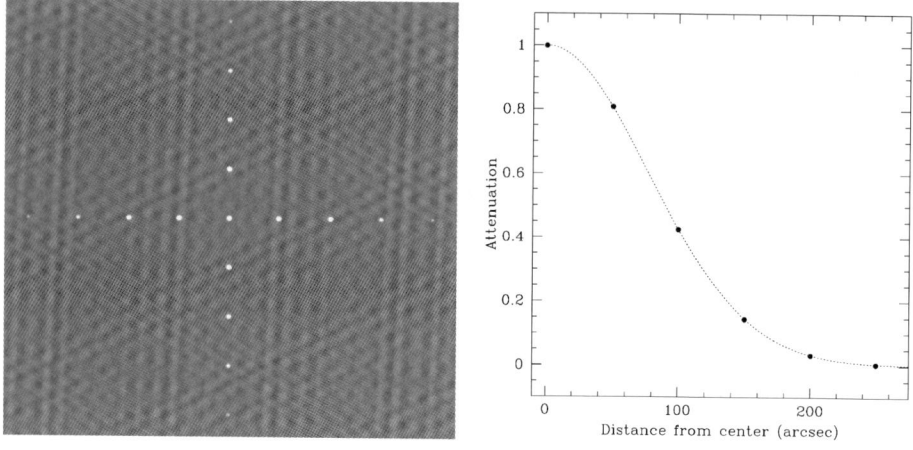

Figure 4. This figure shows results of using a Gaussian weighting function. The left panel shows the CLEANED image for a test grid of 1 Jy point sources as in Figure 3, but with a slightly wider field-of-view. The size of the Gaussian in the (u, v) plane has been adjusted to give an image plane Gaussian with a 90″ 1/2 power point. The right panel shows the measured flux (•) of the point sources as a function of radial distance from the phase center. The dotted curve is the expected flux for the Gaussian weighting used. This clearly demonstrates the ability of this technique to strongly suppress sources away from the phase center without compromising the imaging performance (compare with severe sidelobes in Figure 3 – the gray scale in the above image has been adjusted to show lower sidelobe levels).

from the field center in random directions. We then examined four different methods of averaging: no averaging, averaging over a fixed time and bandwidth, using a windowing function in the (u, v) plane, and using a Gaussian weighting function in the (u, v) plane. For each method we produced a "dirty" image with no deconvolution and measured the map RMS at the phase center. Table I shows the results with comments.

The RMS of the array PSF at a distance of 232 arcseconds was measured to be 0.048. In the case with no averaging we expect that all six sources will contribute to the RMS level at the phase center in quadrature. The RMS should then be $\sqrt{6}$ times 0.048 Jy/beam, or 0.118 Jy/beam, which is very close to the RMS value we measure.

The suppression of the distant sources by simple but aggressive time/bandwidth averaging as in present-day correlators is effective, but as discussed above, the imaging characteristics within the 1/2 power point of the field-of-view will be poor. The top-hat windowing function provides relatively poor protection from the effects of distant sources, primarily because of the ringing of the corresponding Jinc function in the image plane. The Gaussian weighting function provides the best protection of those tested here, but we note that there are other choices that fall off faster and have better characteristics in the image plane. The distant sources, along with their sidelobes, were attenuated by 20 dB.

TABLE I

Results of 4 different methods of averaging to restrict FOV and limit contributions from distant sources

Method	1/2 Power point of field-of-view	RMS at phase center (Jy/beam)		Comments
		Measured	Expected	
No averaging	Much larger than distance of sources	0.115	0.118	Noise from all distant sources adds in quadrature.
f, t Averaging	~45" for shortest baselines	0.0014	Hard to predict	Bad imaging characteristics within FOV.
Window in (u, v)	~90" for all baselines	0.0144	0.014	Ringing of Jinc function produces distant sidelobes.
Gaussian in (u, v)	~90" for all baselines	0.0012	0.0012	Good choice, but there are other functions that fall off faster.

5.3. COMPUTATIONAL COST

There is a computational cost incurred by the application of a weighting function to the visibilities during the accumulation stage. Here, we argue that it is a small fraction of the correlation cost, and thus if included in the design of the correlator system from the start, should not be problematic.

The dominant computational cost in large-N digital correlation is the CMAC, or complex multiply and accumulate. This operation proceeds at the Nyquist sample rate with accumulation of some large number of correlated samples to reduce the output data rate. For the proposed strategy of applying weighting functions in time and frequency, we need to preserve a higher time and frequency resolution within the correlator before weighting. This higher resolution does not by itself increase the processing load of the basic CMAC operations. The additional load is incurred only in the application of weighting functions, and is significant only if it must be done at close to the Nyquist sample rate.

Lonsdale (2003) gives approximate expressions for the accumulation period τ_{int} and channel bandwidth $\Delta \nu$ necessary to preserve the field-of-view of an observation (these expressions are equivalent to those derived by Perley (2003), with $k = 4$):

$$\tau_{int} \sim 4000 \frac{d_a}{D} \left(\frac{n_a}{f}\right)^{1/2}; \quad \Delta \nu \sim \frac{\nu}{4} \frac{d_a}{D} \left(\frac{n_a}{f}\right)^{1/2}; \quad (5)$$

where n_a is the number of antennas per station, and f is the areal filling factor of the station.

Here, D is the array diameter and the antenna diameter is d_a. For the moment, consider only the full correlation case, with $n_a = f = 1$.

For effective weighting with good control of the correlator FOV, we must subdivide the frequency and time range into bins. Let the required number of bins in each dimension be n_{bin}. The subchannel sample rate will then be $2\Delta\nu/n_{bin}$, while the accumulation period will be τ_{int}/n_{bin}. The product of these two quantities describes the number of sample periods n_s per accumulation period:

$$n_s = \frac{2\Delta\nu\tau_{int}}{n_{bin}^2} \sim \frac{10^3 \nu d_a^2}{n_{bin}^2 D^2} \qquad (6)$$

If $n_s \gg 1$, many samples can be accumulated into bins before application of weighting coefficients, and the additional computational burden of weighting is small. Taking some sample numbers, for $n_{bin} = 10$, $D = 350$ km, and $\nu = 1.4$ GHz, we find $n_s \sim 20$. Problems arise in principle for longer baselines and lower frequencies. However, we note that there are already compelling arguments for using station beamforming on longer baselines, and not employing a full cross-correlation architecture. This increases n_s by a factor n_a/f.

There will be some extra cost in providing high spectral resolution antenna data to the cross-multipliers (assuming a FX architecture). The extra resolution required to facilitate accurate weighting represents, however, only a modest increase in FFT size over that needed in a more conventional design, and the aggregate cost increases only as the logarithm of the FFT size. Particularly for a large-N correlator that is dominated by the baseline-based cross multiplications, the added station-based FFT burden is small.

A potentially difficult issue for design of a correlator system that implements such weighting is how to present the weights to the high-speed arithmetic sections in a timely fashion. Fortunately, FOV tailoring does not demand high accuracy, and baseline geometries change slowly, so relatively infrequent updates of table data will suffice. We do not anticipate a bandwidth issue relating to weight computation and delivery.

We conclude that for the great majority of the observational parameter space of a large-N SKA, the computational burden of weighting in time and frequency to tailor the correlator field-of-view is not a significant concern.

6. Conclusions and the calibration issue

The main points covered in this work are as follows:

- The data rates and volumes from large-N designs are problematic for a combination of wide FOV and high angular resolution. The problem is significantly reduced if survey speed is heavily weighted in the telescope sensitivity figure of merit.

- For multiple reasons, SKA fields-of-view will become narrower as the resolution increases. SKA will not be used to cover the full range of resolutions in single observations, and realistically, baseline length ranges of order 100:1 will be typical.
- There are multiple layers of attenuation that progressively reduce the sidelobe contamination levels due to sources far from the phase center. These attenuations are strong for a widefield large-N SKA, and the assumption that the full primary beam must be imaged to high fidelity is not well justified. This can have a large impact on computational requirements.
- In an architecture based on station beamforming, high station beam sidelobes due to a small number of antennas per station are unlikely to compromise performance.
- There are likely to be much more efficient methods than CLEAN to suppress sidelobe confusion noise in a large-N SKA image. The use of a predetermined global model of the static sky will reduce computational costs by a factor of at least several, even with no new model subtraction algorithms.
- The use of shaped weighting functions during correlator accumulation is shown to be an effective and promising method for controlling the field-of-view, and limiting downstream computing loads. As illustrated by simulations, this should be an effective method for suppressing sidelobe confusion from distant sources in medium to high angular resolution SKA observations in which imaging of the full antenna FOV is either not needed, or is impractical. Another useful application of this technique will be to match the FOVs across the large fractional bandwidth of SKA.

However, this paper like those of Perley and Clarke (2003) and Cornwell (2004), ignores the calibration problem. Our analyses assume perfectly calibrated data. It is certain that existing calibration algorithms will not meet the needs of SKA. For any given observation, many instrumental and environmental parameters will need to be determined as functions of time, frequency and position. Algorithms to determine these parameters will doubtless be iterative, and will need to work with the full data volumes discussed by Lonsdale (2003). The potential for unmanageably large computational burdens due to calibration needs clearly exists.

In our view, creation of high dynamic range images from large-N SKA data that has been accurately calibrated is unlikely to present a serious computational obstacle. We fear the same cannot be said of calibration for such an array. Innovative approaches to efficient calibration of next-generation radio arrays have been explored for the LOFAR project (Noordam, 2002), and it is recommended that these approaches be critically examined for potential application to the SKA case.

References

Bunton, J.: 2004, *SKA correlator input data rate*, SKA Memo 43.

Cornwell, T. J.: 2004, *SKA and EVLA Computing Costs for Wide Field Imaging (Revised)*, SKA Memo 49 (EVLA Memo 77).

Lonsdale, C. J.: 2003, *Data rate and processing load considerations for the LNSD SKA design*, SKA Memo 32.

Noordam, J.: 2002, *LOFAR Calibration Strategy*, LOFAR Memo series LOFAR-ASTRON-MEM-059 (LOFAR_ASTRON-DOC-005 1.0).

Perley, R. and Clarke, B.: 2003, *Scaling Relations for Interferometric Post-Processing*, EVLA Memo 63.

White, R. L., Becker, R. H., Helfand, D. J., and Gregg, M. D.: 1997, A catalogue of 1.4 GHz sources from the FIRST survey, *Astrophys. J.* **475**, 479.

SYSTEM ISSUES AND MISCELLANEOUS

SYSTEM OPTIMISATION OF MULTI-BEAM APERTURE SYNTHESIS ARRAYS FOR SURVEY PERFORMANCE

JAAP D. BREGMAN
ASTRON, P.O. Box 2, 7990 AA, Dwingeloo, The Netherlands
(e-mail: bregman@astron.au)

(Received 14 September 2004; accepted 28 February 2005)

Abstract. Investigating the requirements for an aperture synthesis array that optimise the performance for surveying shows that, next to collecting area and system temperature, the field-of-view (FOV) is key parameter. However, the effective sensitivity not only depends on bandwidth and integration time but could be seriously limited by side lobe confusion and by gain errors that determine the effective dynamic range. From the basic sensitivity equation for a radiometric system we derive an optimum cost ratio between collecting area and processing electronics, where the latter should be less than a third of the total cost. For an instrument that has to cover a fraction of sky larger than its field-of-view, the FOV enters the equation for survey sensitivity and we identify the number of independent feed systems per unit collecting area as a key parameter. Then the optimum cost distribution allows the electronics to account for up to half the total cost. Further analysis of the sensitivity formula shows that there is an optimum design frequency for a survey instrument below 1 GHz. We discuss the impact of station size and array configuration on self-calibration, side lobe confusion and effective sensitivity and conclude that a minimum station size of 20 m diameter is required at 0.3 GHz as long as multi-patch self-calibration procedures need, per baseline, a signal-to-noise ratio of more than two for each ionospheric coherence patch.

Keywords: aperture synthesis, multi-beam, self-calibratability, side lobe confusion, survey performance, system optimisation

1. Introduction

The SKA science case indicates four key science applications for frequencies below 1.4 GHz, each of which requires imaging of a substantial fraction of the sky with the full sensitivity of an SKA (Jones, 2004). The Epoch of Reionization concentrates on $7 < z < 13$, Cosmic Magnetism at $z < 5$, Dark Energy at $1 < z < 2$, and Pulsar Surveys for the Strong Field Gravity at $z < 1$. This means that the whole frequency range from 0.1 to 1.4 GHz needs an array design optimised for surveying.

Bunton (2003) proposed a figure of merit for survey speed where the field-of-view (FOV) and relative bandwidth are introduced and then compared various SKA white paper proposals. The main conclusion was that from a survey perspective most systems are poorly designed in allocating RF and digital signal processing resources. On the other hand neither the imaging quality required for reaching a certain survey sensitivity, nor the optimum size of the digital signal processing chain for a given survey task were discussed.

The purpose of the present analysis is to investigate the limitations in survey sensitivity of multi-beam radio telescopes, and in particular of aperture synthesis instruments as a function of instrumental design parameters. We identify two key parameters: the number of receiver chains associated with the number of independent feeds per unit collecting area that allows appropriate FOV, and the relative bandwidth that determines potential image quality as limited by the U,V-coverage following from the array configuration.

We basically follow the system analysis given by Bregman (2000). Starting with the conventional sensitivity formula for a single dish we show that the equivalence between a multi-beam single dish and a synthesis array also holds in terms of signal processing requirements. Then we derive the sensitivity for surveying including parameters such as the number of receiver chains per unit collecting area, the relative bandwidth and the fraction of sky to be observed in a given total survey time. Inspecting the resulting formula we show the existence of an optimum frequency for surveying. We derive the optimum cost distribution over collecting area, receiver chain electronics, digital cross-correlation and post-processing power. It will be shown that these ratios are different for situations where the field to be observed is smaller or larger than the instrumental FOV.

The effective sensitivity in a synthesised image depends on the image quality as determined by the U,V-coverage and on the dynamic range as determined by the calibratability and we show that there exists a critical range of telescope sizes and a minimum number of stations in a synthesis array.

Before the final summary of conclusions, we discuss mosaicing as an alternative to cover a large fraction of the sky in a time comparable with the time needed for azimuthally complete U,V-coverage and its consequences for station requirements.

2. Field-of-view, processing power and sensitivity

In this section we summarise the formulae needed to derive the sensitivity for surveying. We give a formula for the FOV related to design parameters such that it can be combined with the standard sensitivity formula to provide a formula for survey sensitivity. We compare the formula for the FOV with the one for beam solid angle and relate it with aperture efficiency. Then we discuss the multi-beam dish and compare it with the multi-dish array and show their equivalence in FOV and required processing power.

2.1. FIELD-OF-VIEW

The half power beam width of a uniformly illuminated circular aperture with diameter D at wavelength λ is given by $\theta_{1/2} = b\lambda/D$. For uniform illumination the broadening factor b equals 1.02. When the field illumination is tapered off with the square of the radius to zero at the rim we find $b = 1.28$; see for instance appendix A-10 in Kraus (1988). This value is representative for practical telescopes and will be used in this paper. Then, the FOV Ω_t, defined as the solid angle within the half power beam width, is given by

$$\Omega_t = \frac{a_t \lambda^2}{A_p} \quad [\text{sr}]$$

where A_p is the physical area of the aperture that may be elliptical and where a_t is an aperture tapering factor, which equals $(\pi b/4)^2$. The illumination taper leads to a reduced aperture efficiency given by the ratio of the squared modulus of the average aperture field strength to the aperture average of the squared modulus of the field strength. The taper mentioned above results in 75% aperture efficiency see figure 12–30 in Kraus (1988). We could absorb this factor in the aperture efficiency η_a defined as the ratio of effective over physical aperture A_e/A_p and express the FOV in effective aperture just as for the beam solid angle. Indeed, there are factors like foreshortening in aperture arrays that have an inverse relation between FOV and effective aperture. However, aperture tapering results in a different relation as shown by the example where $\eta_a \sim b^{-1}$. Centre blockage leads to a narrower beam, and aperture phase errors mainly lead to increased side lobes with little effect on the width of the main beam. Obviously there is not a universal relation between FOV and either physical or effective aperture, such as there is between beam solid angle and effective aperture.

We need to realise that the observed signal intensity of an object varies by a factor 2 within the FOV, and is on average reduced by a factor 0.75

2.2. MULTI-BEAM DISH

Illuminating the same effective area by N_f independent feed systems each with their own receiver and detector chain leads to a FOV Ω_f, defined as the solid angle within the half power beam width, which is given by

$$\Omega_f = N_f \Omega_t \quad [\text{sr}]$$

It should be noted that a focal plane array (FPA) needs to combine the signals of a number of FPA elements to form a feed that feeds a receiver chain. Independence of feeds means that groups of FPA elements are chosen such that their associated beams have a separation larger than the half power beam width.

Full polarisation processing needs two receiver chains per feed. Detection of all four Stokes parameters requires the full correlation matrix between the two complex channels, which needs eight real multiplications (one pair for each auto-correlation and two pairs for only one cross-correlation). For a signal bandwidth B

we need complex sampling at rate B for each of the two polarisation channels. The total processing power R_{FOV} for this N_f fold increase of the FOV expressed in real multiplication operations per second equals

$$R_{FOV} = 8BN_f \quad [\text{ops s}^{-1}]$$

In principle we could make also cross-correlations between all complex feed signals (upper triangle only with four times two pairs per feed receptor pair). This would allow creating nulls in the side lobe pattern of the dish and would reduce the sensitivity for radio frequency interference (RFI) signals. The required processing power R_{RFI} for this option is

$$R_{RFI} = 8BN_f^2 \quad [\text{ops s}^{-1}]$$

The power flux density of the noise for each Stokes parameter of each field with beam size Ω_t observed with effective collecting area A_e and system temperature T_s reached after integration time τ_o with bandwidth B is given by

$$\Delta S_o = 2^{1/2} k T_s A_e^{-1} (B\tau_o)^{-1/2} \quad [\text{W m}^{-2} \text{Hz}^{-1} \text{beam}^{-1}]$$

where k is Boltzmann's constant, see for instance Perley et al. (1994).

2.3. APERTURE SYNTHESIS ARRAY

The total area A_p of an aperture synthesis array can be divided over N_s stations, which leads at a wavelength λ to an FOV equal to the beam size Ω_s of a station, defined as the solid angle within the half power beam width, and is given by

$$\Omega_s = \frac{\lambda^2}{A_p/N_s} = N_s \Omega_t \quad [\text{sr}]$$

This N_s fold increase of the FOV relative to a single contiguous collecting area requires for full polarisation detection the full correlation matrix of all $2N_s$ complex signals, since each signal chain has collected only a fraction $1/N_s$ of the total available signal power.

For a signal bandwidth B we need complex sampling at rate B for each of the two polarisation channels and the total processing power R_{inf} expressed (diagonal and upper triangle of the covariance matrix) in real multiplications per second equals

$$R_{inf} = 8BN_s^2 \quad [\text{ops s}^{-1}]$$

The power flux density of the noise in such a synthesis observation is given by

$$\Delta S_o = 2^{1/2} k T_s D_c A_e^{-1} (B\tau_o)^{-1/2} \quad [\text{W m}^{-2} \text{Hz}^{-1} \text{beam}^{-1}]$$

with cross-correlation degradation factor D_c and where beam is here each synthesised beam in the FOV given by Ω_s.

Compared with the single dish situation there is no difference in total FOV when the same number of feed systems is used. Also there is no difference in required signal processing power if the same RFI rejection capabilities with spatial nulling are required. The cross-correlation degradation factor can be found in literature, for instance in Perley et al. (1994). The single dish application needs a processing degradation factor as well, which depends on how the system noise is subtracted to obtain the sky signal. This means in practice that there is also no sensitivity difference between a multi-beam single dish application and a multi-station synthesis array.

3. Surveying

Bunton (2003) presented a figure of merit for survey speed which is the product of FOV, bandwidth and A_e/T_{sys} squared. This relation does not take into account that the FOV is a function of wavelength, physical aperture A_p and aperture taper. If we want to analyse the performance of telescope systems as function of frequency and compare systems with different FOV, we need to work with a performance formula that has independent parameters that relate instrumental design to astronomical performance. We have identified the number of independent feeds per unit physical aperture as the key design parameter for the FOV of a diffraction limited instrument.

We start from the basic sensitivity formula and then include relevant telescope design parameters and observational parameters that include the FOV implicitly. Analysis of the formula for survey sensitivity reveals hitherto non-obvious aspects. It has be shown in Section 2.1 that aperture tapering affects both FOV and aperture efficiency but it turns out that this cancels to first order in the survey sensitivity.

3.1. SENSITIVITY FOR SURVEYING WITH A MULTI-BEAM APERTURE SYNTHESIS ARRAY

If we want to complete a survey over a fraction η_{sky} of the sky within a survey time τ_{surv} we need N_o observations given by

$$N_o = \frac{4\pi \eta_{sky}}{\Omega_{ms}}$$

where Ω_{ms} is the FOV of the multi-beam synthesis array. From the analysis of the FOV for a multi-beam dish and a synthesis array we have seen that the FOV of an instrument is only determined by the total number of independent feed systems N_{if} that are cross-correlated. So

$$\Omega_{ms} = \frac{a_t N_{if} \lambda^2}{A_p} \quad [\text{sr}]$$

We identify the ratio $n_{if} = N_{if}/A_p$ [m^{-2}] as the key design parameter for the FOV of a survey instrument. Then our sensitivity per field is limited by the effective time per observation

$$\tau_o = \frac{\tau_{surv}}{N_o}$$
$$= \frac{\tau_{surv} a_t n_{if} \lambda^2}{4\pi \eta_{sky}}$$

The sensitivity for surveying ΔS_{surv} is calculated by inserting into the sensitivity formula the expression for the effective observing time. We express the bandwidth as fraction η_ν of the observing frequency ν_o to get an observational parameter that relates to image quality as will be discussed in a later section.

$$\Delta S_{surv} = 2^{1/2} D_c c^{-1} k T_s A_p^{-1} \left(\frac{n_{if} a_t \eta_a^2 \eta_\nu \tau_{surv}}{4\pi \eta_{sky} \nu_o} \right)^{-1/2} \quad [\text{W m}^{-2} \text{Hz}^{-1}]$$

where c is the velocity of light.

We have not included the average loss in sensitivity of objects in the FOV, and also ignored the improved sensitivity arising from the overlap of adjacent fields. Both effects relate to details of the surveying program that are not relevant for the following discussions.

Quite surprisingly all observational and most instrumental design parameters appear now on equal footing under the square root except for T_s/A_p, our basic system specification. It is worth noting that the sensitivity loss due to aperture tapering is to first order compensated by the larger FOV, making the taper issue a second-order effect that can be used to improve imaging quality.

The parameter n_{if} is a key parameter in the design of multi-beam stations. For a single dish system with a single feed, this term causes A_p to move under the square root, demonstrating the well-known poor survey performance of large single feed single dish systems.

The $\nu_o^{+1/2}$ term is the result of a bandwidth and of a FOV effect and indicates that for a given system temperature the highest survey sensitivity is reached at low frequencies.

3.2. OPTIMUM SURVEYING FREQUENCY

The system noise temperature T_s has a receiver component T_r and a sky component, which results for frequencies below 1 GHz in

$$T_s = T_r + 3\nu_0^{-2.5} \quad [\text{K, for } \nu_0 \text{ in GHz}]$$

Inserting this into the formula for the survey sensitivity allows us to find a minimum value for ΔS_{surv} by differentiation with respect to ν_0 and finding an optimum survey frequency

$$\nu_{\text{opt}} = \left(\frac{12}{T_r}\right)^{0.4} \quad [\text{GHz, for } T_r \text{ in K}]$$

Uncooled receiver systems have below 1 GHz have a typical receiver temperature of about 30 K, which leads to $\nu_{\text{opt}} \sim 0.7$ GHz. Quite fortunately this is close to where three key science programs want extensive surveys. This means that a radio instrument optimised for surveying is indeed an efficient instrument to serve that science case.

For a station with a focal plane array we must realise that the FOV is to the first order defined by the size of the FPA. Then the figure of merit for survey speed as introduced by Bunton (2003) is a useful analysis tool. Since the area of the focal spot increases proportionally with wavelength squared the number of independent beams becomes proportional with frequency squared. This frequency dependence should be included before the survey sensitivity formula can be differentiated to find the optimum operating frequency of an array with stations using FPAs. Applying this procedure identifies the highest frequency supported by the FPA as the optimum frequency. This shows that a dish with FPA is not an optimised instrument for wide-band surveying, since the investment in receiver chains is not used at lower frequencies.

Nevertheless the derived frequency of 0.7 GHz is the optimum frequency for which any survey array should be designed, be it dishes with FPA, or aperture arrays or cylinders with line feeds. This optimum follows from increased survey sensitivity at lower frequencies by the larger FOV for a given number of independent feeds per unit physical aperture and is limited by the increasing sky noise.

We have optimised the instrumental system sensitivity, while we should optimise the signal-to-noise ratio for the class of objects that is searched for. Including the frequency dependence of the object intensity will lead to a different optimum survey frequency, but does not influence our basic conclusion that there is an optimum design frequency for a survey instrument determined by the technology of antenna receptors with integrated low noise amplifiers providing a receiver noise temperature that matches the sky noise temperature.

4. System optimisation

The sensitivity formulae show that there are components that contribute proportionally to the sensitivity such as collecting area and system temperature, and components that contribute proportionally with the square root of their processing bandwidth such as the receivers, digitizers and filters, or by increasing effective integration time, such as larger FOV to cover more relevant objects in a given observing time.

From the point of view of optimum system design we want to maximise value for money. For an observing instrument the sensitivity is the key value parameter and then we need the sensitivity reached in the final images that have passed all the necessary processing steps. Flexibility of the instrument to support a range of observing modes that allow scientific discoveries to be made is essential and is a major cost driver in software development, but will not be addressed.

4.1. GENERIC ARCHITECTURE

In Figure 1 we present a basic architecture that allows us to compare aperture synthesis arrays with stations in multi-beam collector technology or as aperture arrays. The latter sample the incoming wave front with receptors that have a separation of about half a wavelength at the highest frequency and cover a total physical collecting area A_p Every receptor needs a low noise amplifier with effective system temperature T_s. Sets of receptor signals are combined in an RF beam former and each set forms a beam that is electronically pointed to the same position on the sky. The collector with a total physical area A_p being a dish or a cylinder, parabolic or spherical, has a set of receptors in its focal plane and forms partial overlapping beams on the sky. For each station there is a set of receivers that selects the band

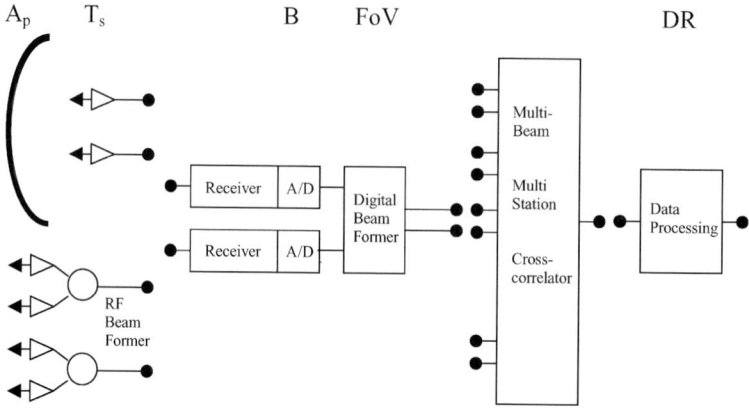

Figure 1. Generic architecture for an aperture synthesis array.

of interest with bandwidth B and digitises the signal. Then a digital beam-former creates a proper set of independent station beams that form together the total FOV of the station. All signals are transported to a central location where the correlator cross-correlates the signals from all the stations for each beam separately. Finally the correlator output is self-calibrated and imaged by a data processor.

We make explicit distinction between the LNA and the receiver chain. The LNA is either in a cryostat or integrated with the antenna structure. In both cases the dominating cost is mechanical and driven by the connectivity between receptor, LNA and receiver where the band limiting is done. Therefore the LNA and its input and output connections that drive the design for low system temperature belong, cost-wise, to the aperture. By 2015 we expect that transistor technology would have reached its physical limits and that system noise temperature would be just a matter of design trade-off between the optimum locations of elements versus the cost involved in the mechanical complexity of those optimum locations.

The receiver chain comprises the whole chain after the LNA output cable till the input of the digital cross-correlation modules and includes band selection, down-conversion, digitisation, sub-band filtering, digital beam forming, spectral channel formation and data transport. RF beam forming is excluded since its function is to point a beam and belongs to the aperture just as for a steerable collector. Practical receiver systems for aperture synthesis arrays replicate most parts to create multiple RF beams, multiple digital beams per RF beam and sufficient total bandwidth that needs to be cross-correlated. This replication makes the total cost for bandwidth elements proportional to the total processed bandwidth per station.

The effective dynamic range (DR) is determined by the effectiveness of the self-calibration executed on the data processing platform and will be discussed in later sections.

4.2. Optimum Cost Distribution over Collecting Area and Electronics

From the sensitivity equation follows that for e.g., a 10% increase in aperture or LNA performance gives a 10% increase in sensitivity, while a 10% increase in system bandwidth gives only 5% increase in sensitivity. An optimum distribution of the total available budget that maximises the sensitivity would then be two to one, or two-thirds in aperture and low noise amplifiers (including cooling if applicable) and one third in multi-beam receivers, correlator and post-processing electronics. This follows from differentiating the following performance equation,

$$(1 - x)x^{1/2}$$

where x is the fraction of the total budget spent on receiver, correlator and post-processing electronics, assuming that budget is spent on system elements using cost-effective state-of-the-art technology. A further assumption is that the performance of the particular parameter such as sensitivity or processing bandwidth is linearly related to its cost, and that this performance lies within a range that allows it to be increased or decreased without encountering a limit. This is apparently true for multi-beam aperture synthesis arrays where sensitivity is proportional with the total number of stations, and where we can trade a multi-beam station for more beams on all the remaining stations. Although a single digital receiver chain might have low cost elements that provide sufficient bandwidth for the 0.3–1.4 GHz range in 2015, we still assume that the total cost of a complete multi-beam digital receiver system with fibre optical output to the central processor has a cost that is proportional with bandwidth per beam.

Optimum distribution means that not more than one third of the system budget should be spent on subsystems that contribute to the sensitivity by virtue of their offered correlation bandwidth. If less than one third is spent it might indicate that more post processing is needed to support the end-user community in absorbing the potential data.

4.3. Optimum cost distribution over receivers, cross-correlator and post processing

We call a system optimised when the subsystem costs are balanced such that each part has the same marginal performance to cost ratio. In other words the cost distribution should be such that spending an additional cost unit on subsystem one gives the same system performance increase as spending that same cost unit on subsystem two.

From the time multiplex over more fields we have seen that FOV and integration time are equivalent contributions to the survey speed at a given sensitivity. So we are free to use an investment in processing power for additional beams or for additional bandwidth if that reduces the integration time for a required sensitivity (as for continuum observations). More beams support either a larger FOV per single user or longer observing times in a multi-user mode, providing higher sensitivity even for line observations.

From the FOV analysis we have seen that e.g., a 10% increase in FOV requires either a 10% increase in multi-beam electronics (receiver chains plus correlator), or a 10% increase in number of stations (that get 10% smaller collecting area), which would increase the receiver chain cost also by 10%, but the correlator cost by 20% since the processing power is proportional with the square of the number of inputs. In terms of marginal performance a correlator is only half as effective in producing FOV in cross-correlation mode as in multi-beam mode. To satisfy the optimisation criterion the cost of the correlation subsystem should be half that of the receiver chain subsystem. Spending an additional cost unit on the correlator subsystem then gives a double relative increment multiplied with half the marginal performance factor and results in the same absolute performance increment as would be provided by the receiver chain subsystem.

This result suggests that an optimum distribution of the total cost for electronics would be a two to one ratio or two-third in multi-beam receiver chains with digitization and pre-processing systems and one-third in the digital cross-correlator with post-processing electronics. However this follows from FOV perspective only and we continue with some alternative perspectives.

The optimum cost distribution discussed so far is valid for thermal noise limited observations on objects smaller than the FOV. However, when dynamic range limits the system performance we need self-calibration processing power to find correction factors per receiver chain. Then we need processing power proportional to the number of correlator output channels as well as proportional to the number of sources to be subtracted per channel, where the latter scales with the FOV and roughly inversely proportional with the sensitivity that needs to be obtained. This argues for post-processing components competing on equal footing with aperture, with receivers and with correlator cost, i.e. cost exponents of one, a half and a quarter respectively. Anyhow the investment in post-processing power should be larger then in pure cross-correlation power. This will lead to a larger fraction than the 5% of the system cost, as demonstrated by LOFAR, which is the first telescope designed with focus on this processing cost.

An important electronic cost item in aperture arrays is the cost of the RF beam-former, which provides the electronic pointing and should from that perspective be compared with the cost of making a collector pointable on the sky. In practice these cost dominate in collector as well as in aperture array realisations, which argues for absorbing these costs in the aperture and not in the receiver electronics. On the other hand it could be argued that pointing is used for tracking objects to increase their effective observing time, which puts it on equal footing with bandwidth and multi-beaming.

Conclusion is that for maximum-thermal noise-sensitivity per beam, an optimum investment distribution is given by at least 67% in collecting area including (cooled) LNA's (and RF beam forming for aperture and focal plane arrays), at most 22% in multi-beam receiver chains and digital pre-processing electronics, and about 11% in digital cross-correlator and post-processing electronics. These numbers should reflect the discounted value including the maintenance and upgrades over the lifetime of SKA.

4.4. OPTIMUM COST DISTRIBUTION FOR A SURVEY INSTRUMENT

However, for survey application where we want to maximise sensitivity over an area much larger than the FOV, the effective integration time depends on the FOV size. For diffraction limited collecting aperture the total FOV size is proportional to the number of independent feeds that produce beams on the sky that overlap at their half power point. A specified FOV leads to a fixed ratio of collecting area to the receiver chains. Putting this constraint in the equation for survey sensitivity brings the aperture under the square root on equal footing with bandwidth and integration time. The consequence is then that up to 50% could be spent on receiver and processing electronics. The system temperature keeps its exponent of unity making it the most critical item of the design of a survey telescope. For a general-purpose telescope that needs to serve a wide user community, we could argue that each individual user needs only a limited FOV, but considering the period that spans the lifetime of the instrument, we image a considerable fraction of the sky and in fact conduct a survey that needs an instrument designed for that purpose right from the beginning.

As an example we give the cost distribution for LOFAR as projected at the time of the preliminary design review. Antenna including land acquisition one-fourth, digital receiver chains one-fourth and correlator and calibration processing pipeline half of the total budget. Apparently half of the "optimum" antenna budget as defined for a survey instrument has moved to the calibration processing, which is indeed instrumental in reaching the thermal noise limit defined by collecting area and sky noise at the LOFAR frequencies.

5. Effective sensitivity and minimum number of stations in a synthesis array

Apart from the thermal noise discussed so far, we have to deal with main lobe confusion noise, with side lobe confusion noise and with dynamic range noise, which are dominant for continuum observations with the SKA.

To avoid main lobe or classical confusion we need an array with sufficient resolving power. Formulated in the spatial frequency domain we require a number of independent U,V-samples in a synthesis image that is significantly larger than the number of point sources that has to be imaged in the FOV. Basically we need a longest baseline that is long enough. For a SKA sensitivity of 20,000 m^2/K and 5 h integration with 20 MHz bandwidth we have a r.m.s noise floor of $0.16\,\mu$Jy and need at 0.7 GHz a maximum baseline of at least 150 km and 500 km at 0.3 GHz (Bregman, 1999) assuming 10 resolution elements per background source. It seems more appropriate to take fifty resolution elements, which increases the baselines by a factor 2.24 leading to a maximum baseline of at least 130 km at 1.4 GHz. Reducing the effective integration time by a factor ten reduces the required maximum baseline by a factor 2.

At an empty spot in a synthesis image we still find some intensity equal to the sum of the side lobe intensities of all the objects in the field at that spot. The side lobe level can be reduced to any level when the U,V-coverage is made uniform and then tapered appropriately. The array configuration needs to support such an appropriate basic coverage and we give two examples showing the feasibility of such complete U,V-coverage.

The first example is a randomised distribution of stations in exponential shells that could provide complete U,V-coverage for continuum observations. The required relative bandwidth depends on the relative width of the shells, and the synthesis time is to first order 24 h divided by the number of stations per shell. With fifty shells ranging from 0.1 to 250 km radius and five stations per shell an almost uniform U,V-coverage can be obtained in less than five hours tracking with a few percent relative bandwidth and provides good image quality (Noordam, 2001b). The expo-shell configuration satisfies automatically a basic SKA requirement to have a large fraction of the collecting area in a so-called compact core to provide sufficient brightness sensitivity especially for line observations. Instead of a continuously increasing telescope density towards the centre of the array, we could use a combination of a dense and a coarse regular rectangular grid, where the size of the dense grid is equal to the increment of the coarse grid (Bunton et al., 2002). With only 81 stations separated 0.1 km in the dense grid and 81 stations in the coarse grid, we could obtain a completely regular U,V-grid with baselines ranging from 0.1 to 8.1 km. Depending on shape, orientation and small randomised position deviations of the stations with an equivalent diameter of 50 m, full U,V-coverage could be obtained with very small additional bandwidth and or rotation. Such a core configuration needs to be complemented by for instance expo-shells to get proper filling of the U,V-plane with the longer baselines.

Line observations have a typical bandwidth less than 0.2%, resulting into less radial U,V-filling, but we do not suffer from various types of side-lobe confusion since we can apply differential image processing methods.

Atmospheric and ionospheric phase disturbances cause side lobes that cannot be cleaned away using a nominal point spread function and leads to a noise floor proportional to the strongest objects in the field. The ratio between this noise floor and the strongest object intensity is called the dynamic range in a synthesis image. This range could be increased by self-calibration, which provides for every station appropriate amplitude and phase calibration per coherence area per atmospheric coherence time. At frequencies around 1 GHz this coherence volume is determined by the tropospheric coherence time and telescope FOV, leading to a required minimum system temperature. Below 0.3 GHz the system temperature is determined by the sky and we need a telescope aperture large enough to provide sufficient sensitivity per ionospheric coherence time for the strongest source in a sky area defined by the ionospheric patch size.

Practical self-calibration algorithms, e.g. Boonstra and van der Veen (2003), need at least one point source per coherence area that has a signal-to-noise ratio per baseline per coherence time of order two This limitation is present in those algorithms that perform some form of phase unwrapping. According to the Cramer–Rao lower bound criterion the sensitivity constraint for an ideal algorithm could be relaxed by a factor equal to the square root of the number of stations.

6. Minimum station size

First we verify whether the sensitivity criterion for practical algorithms could be satisfied by an array with stations that have the lowest possible system temperature and look then into the consequences for alternative design choices for a synthesis array.

The best system temperature with cooled receiver systems is about 20 K and requires an effective aperture of 0.4 km^2 to satisfy the SKA sensitivity goal. Let us take 300 stations as a reference figure; that is, the geometrical mean of the most extreme SKA white paper proposals. Then each station would have 1333 m^2 effective area, which results at 70% efficiency in a 50 m diameter. Such a station would have a beam of 0.3 degrees at 1.4 GHz or 0.07 square degrees beam solid angle. The basic sensitivity of 4 mJy after ten seconds of integration time and 10 MHz bandwidth on a single baseline is then just sufficient to detect in a single station beam at 1.4 GHz on average one source with a signal-to-noise ratio of two, which is indeed sufficient to allow self-calibration. The tropospheric coherence time is typically of order minutes, and related to size and movement speed of water vapour structures.

At lower frequencies the intensity of the average 21 cm source increases roughly proportionally with wavelength to the power 0.7, gaining a factor of two at 0.46 GHz. However, at this frequency the galactic sky brightness is 20 K and has doubled the system temperature, which just offsets the increased calibrator intensity, but still allows calibration. At lower frequencies the sky noise increases with wavelength to the power ~ 2.5, and at 0.3 GHz the total system temperature has increased by a factor four and offsets the spectral increase in source intensity.

On the other hand at lower frequencies, the FOV increases and given the power law exponent of the integrated source count, the 21 cm flux of the strongest source increases with the FOV to the power 1.4 (Bregman, 1999). This means that a smaller telescope with reduced sensitivity still finds in its larger FOV a source that is stronger and more than compensates for the loss in sensitivity. With un-cooled receivers and 45 K system temperature we need a telescope with 38 m diameter to provide sufficient sensitivity and FOV to find a source strong enough for self-calibration. For a larger station with less FOV in a single beam we need multi-beaming and apply the atmospheric phase correction found in one beam to the adjacent beams.

Considerably smaller antenna stations down to 12 m diameter with 20 K system temperature have a factor seventeen lower sensitivity per baseline, but the proportionally increased FOV has still one source that is a factor 53 stronger, enough for self-calibration at 1.4 GHz and just strong enough at 0.3 GHz. However at 0.3 GHz the FOV has got a diameter of six degrees and contains of order seven ionospheric patches, each needing a separate self-cal (Noordam, 2001a). In at least one patch there is a source that is just strong enough, but in the other six patches the sources are too weak by a factor up to 12. We could increase the sensitivity by raising the bandwidth and the integration time to ~ 5 min as limited by the ionospheric coherence time for travelling ionospheric disturbances. This will not be enough and we need to increase the station to a diameter of 20 m or larger to observe at frequencies as low as 0.3 GHz.

In summary:

- The peculiar behaviour of the integrated source count and its spectral properties allows proper self-calibration of tropospheric phase errors in the frequency range 0.3–1.4 GHz for antenna stations with a diameter smaller than 38 m assuming 45 K system temperature and 70% aperture efficiency.
- Larger stations need multi-beaming to cover sufficient FOV to find an object that is strong enough and transfer the tropospheric phase error to the other beams.
- At 0.3 GHz, antenna stations need to have at least 20 m diameter to limit their FOV to a single ionospheric patch.
- Only when more advanced multi-patch self-calibration procedures have been proven to work satisfactorily, could we use smaller stations.

7. Conclusions

We investigated the constraints for an optimised aperture synthesis array with a collecting area of order one square kilometre with emphasis on large scale surveying at frequencies below 1.4 GHz.

Analysis of the thermal sensitivity formula of a collecting aperture shows that an aperture synthesis instrument optimised for sensitive imaging of objects smaller than the (multi-beam) FOV should spend at least 67% of its total cost on collecting area and (cooled) low noise amplifiers, at most 22% on receiver chains including digital pre-processing, and about 11% on cross-correlation and calibration processing. When we become dynamic range limited a larger budget fraction has to go to the calibration processing.

The basic sensitivity formula can be turned into a sensitivity for surveying by expressing the effective integration time per-beam-field as function of the total survey time, the sky fraction that need to be observed and the total FOV. Since the FOV is a function of aperture, frequency and number of receiver chains we introduce scale free observational and instrumental design parameters such as the relative bandwidth and the number of independent feeds per unit collecting area. Analysis of the final formula shows that there is an optimum frequency for surveying, which is a shallow function of the receiver noise temperature. For a typical SKA system temperature of 30 K the optimum frequency is 0.7 GHz, which happens to be the target frequency for three key science programs.

Further analysis of the performance expression shows that an instrument optimised for large-scale surveying could spend up to 50% of its total cost on receiver chains, multi-beam correlation and post-processing electronics, still with emphasis on minimised system noise temperature. This supports the application of aperture arrays with their large fraction of electronics cost for large scale surveying at frequencies below a gigahertz.

For frequencies below 1.4 GHz the effective sensitivity of continuum observations is dynamic range limited when observed with a SKA. Self-calibration can remove the atmospheric and ionospheric phase errors provided that the signal-to-noise ratio per baseline is more than two on the strongest point source in a single beam-field. For system temperatures of order 30 K we can find at least one such source in the field of a multi-beam telescope with a diameter between 12 and 100 m for frequencies at 1.4 GHz. For lower frequencies the field size of a small dish contains more ionospheric coherence patches than appropriate self-calibration sources, which drives us, for frequencies down to 300 MHz, to stations with a diameter larger than 20 m. Only when more advanced multi-patch self-calibration procedures have been proven to work satisfactorily, could we contemplate smaller stations.

To avoid side lobe confusion limiting the effective continuum sensitivity, complete U,V-sampling is required. This can already be provided by any array with order a hundred stations distributed along a "five-armed" exponential shell, provided that the fractional bandwidth is about 10%. Arrays with more telescopes need proportionally less fractional bandwidth, but the required correlation and post processing powers increase quadratically.

References

Boonstra, A. J. and van der Veen, A. J.: 2003, 'Gain Calibration Methods for Radio Telescope Arrays', *IEEE Trans. Signal Process.* **51**(1).
Bregman, J. D.: 1999, 'Design Concepts for a Sky Noise Limited Low Frequency Array', *Proceedings NFRA Symposium Technologies for Large Antenna Arrays*, Dwingeloo, The Netherlands.
Bregman, J. D.: 2000, 'Concept Design for a Low Frequency Array', *SPIE Proc. Astron. Telescopes Instrum.* **4015**.
Bunton, J. D.: 2003, 'Figure of Merit for SKA Survey Speed', SKA memo 40.
Bunton, J. D., Jackson, C. J. and Sadler, A. M.: 2002, 'Cylindrical Reflector SKA', White Paper, SKA Design Concept White Papers 1–7, Eng. Section on http://www.skatelescope.org.
Jones, D. L.: 2004, 'SKA Science Requirements: Version 2', SKA memo 45.
Kraus, J. D.: 1988, *Antennas*, 2nd Edn., ISBN 0-07-100482-3.
Noordam, J. E.: 2001a, 'LOFAR Calibration and Calibratability', LOFAR-ASTRON-MEM-001.
Noordam, J. E.: 2001b, 'Guidelines for the LOFAR Array Configuration', LOFAR-ASTRON-MEM-002.
Perley, R. A., Schwab, F. R. and Bridle, A. H.: 1994, 'Synthesis Imaging in Radio Astronomy', *Astronomical Society of the Pacific Conference Series* **6**, ISBN 0-937707-23-6.

SKA COST MODEL FOR WIDE FIELD-OF-VIEW OPTIONS

JOHN D. BUNTON* and STUART G. HAY
CSIRO, ICT Centre, Epping 1710, Australia
(*author for correspondence, e-mail: john.bunton@csiro.au*)

(Received 20 August 2004; accepted 28 February 2005)

Abstract. We describe a SKA hardware component model that includes various concepts for achieving a wide field-of-view (FoV) as desired for survey speed. Costs for each of the components in the model are derived and in some cases optimum parameters are estimated. The model identifies components of major cost and allows comparison of the relative costs of the wide FoV concepts.

1. Introduction

The SKA is a next generation radiotelescope to be built in the next decade. It will be an international project that will deliver an instrument with more than an order of magnitude more sensitivity than any existing radiotelescope in the frequency range 0.2–20 GHz. In this paper we propose a hardware model of the components that make up such a radiotelescope and develop a cost estimate for each of these components that is constrained by two key SKA specifications: point source sensitivity and survey speed requirement at low frequency. This last requirement dictates that the SKA will be a wide field-of-view instrument at frequencies below 1.4 GHz. The hardware component model presented here is largely based on the SKA design concept white papers (SKA, 2002, 2003). These component models have been refined by Horiuchi et al., 2004 to a general concept that can cover all SKA concepts. One of us, (Bunton), was a significant contributor to the general concept and the component model presented here broadly uses the same signal flow concept but groups the components differently. The major difference in the models is that here the station beamforming is done at the central site, not at the groups of antennas that make up a station.

Many concepts can meet the wide field-of-view requirements including the aperture arrays, Luneburg lenses and cylindrical reflectors already proposed for the SKA as well as parabolic dish reflectors with feed arrays, and small aperture concentrators. For these concepts relative cost estimates of the various hardware components in the model are derived. Three main hardware cost areas are identified: the cost of the concentrator, antenna electronics from the feed to RF downconverter and digital electronics. The results allow a direct comparison of the cost of the hardware components and, in the case of beamformers and correlators, an optimisation of the number of antenna stations. It would also be useful to derive the post correlation processing cost but this is beyond the scope of this paper.

2. Main global symbols

N_{fa} = no. feeds per antenna at the maximum frequency
N_{ba} = no. beams per antenna
N_{as} = no. antennas per station
N_{bs} = no. beams per station
N_s = no. stations
N_b = no. antenna beams = $N_a N_{ba}$
N_c = no. correlations
N_a = no. antennas
A_{eska} = Total effective area of SKA
A_{ea} = effective area per antenna
FoV = SKA Field-of-view (steradians)
FoVd = SKA Field-of-view (deg^2)
S_{ska} = Sensitivity of the SKA
T_{sys} = antenna system temperature
η_a = antenna aperture efficiency
α_{con} = cost per m^2 of concentrator
α_f = cost per feed including LNA
κ_a = antenna beam area divided by λ^2/A_{ea}
κ_s = antenna station aperture-filling factor
ν = frequency in GHz
f = maximum operating frequency in GHz
BW = Bandwidth processed per beam
k_{bands} = ratio of total no. of feed to the number at the maximum frequency
R_{fb} = number of feeds per antenna beam
R_{adb} = number of A/Ds per antenna beam
N_{rfi} = no. of inputs to RF beamformer
N_{adb} = no. of inputs to digital beamformer
foD = focal length to diameter ratio

3. SKA hardware component model

The component model presented here is largely based on the SKA design concept white papers (SKA, 2002, 2003) and, in particular, the white paper on Luneburg lenses. The major difference to the previous models is that here the station beamforming is done at the central site, not at the station. Figure 1 illustrates our model of the SKA antenna. Some concepts may use only a subset of the components

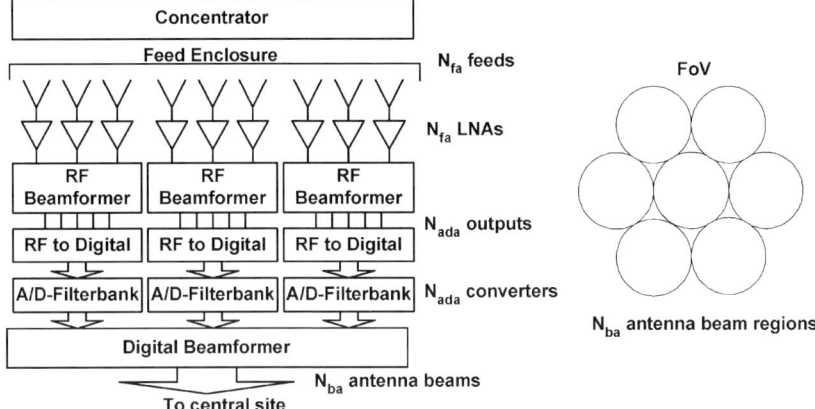

Figure 1. Model of the SKA antenna.

shown. The antenna is a multibeam antenna, producing N_{ba} simultaneous antenna beams, with the regions within 3 dB of the beam peaks being disjoint. The RF electronics is broken down into the separate LNA, RF beamformer and RF to digital converter, which includes an RF downconversion stage, A/D converter, and signal transport to the digital beamformer, which consists of filterbanks for each A/D converter and digital beamformer. The antenna has N_{fa} feeds and N_{ada} is the number of analogue-to-digital converters.

Figure 2 shows a block diagram of our model of the SKA array and associated signal transport and signal processing. The antennas are grouped into N_s stations, with N_{as} antennas per station. The signals from all antenna beams are transported to the central site. The reason for this is that if the full antenna FoV is beamformed then the station beamformer output data rate can be higher than the input. Thus putting the station beamformer at the central site will minimise the amount of

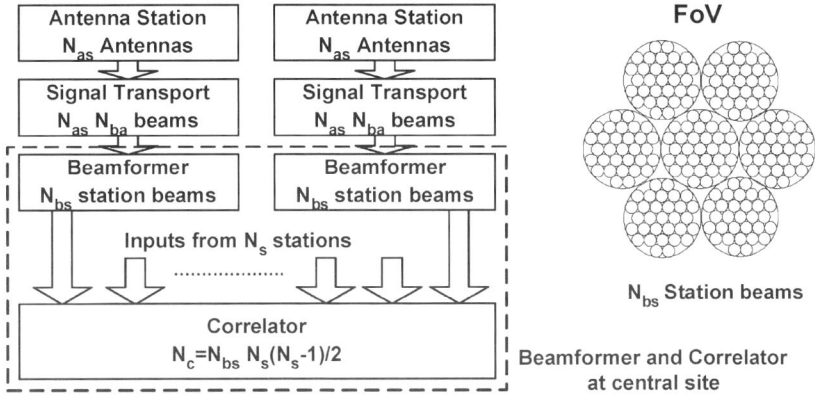

Figure 2. Model of SKA array and signal transport and processing.

data that must be transported on long haul links. At the central site the signals from each station are first combined by phased-array beamforming, producing N_{bs} station beams per station. Each beam is assumed to have a bandwidth BW. Station beamforming at the central site minimises signal transport cost in some concepts. All pairs of station beams with the same beam region are then cross-correlated, producing a total of $N_c = N_{bs} N_s (N_s - 1)/2$ correlations.

It will be seen in the following that N_s will have significant effects on the cost. The system temperature T_{sys} also has a major effect with the number of antennas needed approximately halving if cooling of the LNA is used. Note that in all cases the numbers of feeds will be of order 1 million. Cooling costs significantly less than $1000 per feed are needed to make cooling economic. This will probably make it impractical to implement cooled receivers on a wide FoV SKA. Thus, as a simplification, it will be assumed that T_{sys} is a constant. However, T_{sys} is included in all equations so that its effect can be determined for all parameters.

4. SKA specifications

The goals for the SKA now include a specification for survey speed. The specification takes the form (Jones, 2004)

$$\text{FoV} \cdot (A_{eska}/T_{sys})^2 \cdot \text{BW} = 4.57 \times 10^{15} \text{ steradians m}^4 \text{ K}^{-2} \text{ Hz at } 0.7 \text{ GHz}$$
$$= 9.14 \times 10^{13} \text{ steradians m}^4 \text{ K}^{-2} \text{ Hz at } 1.5 \text{ GHz} \quad (1)$$

where FoV is the field-of-view of the SKA in steradians (not deg^2 as in the original specification), A_{eska} is the total effective area of the SKA, T_{sys} is the system noise temperature and BW is the instantaneous bandwidth. The FoV is required to scale as λ^2, where λ is the wavelength (Jones, 2004). This scaling applies in two frequency ranges, namely 0.5–1 GHz and greater than 1.4 GHz. Here we are concerned with the 0.7 GHz specifications which is the more stringent of the two. Another goal is the point-source sensitivity S_{ska} specification

$$S_{ska} = A_{eska}/T_{sys} = 20000 \text{ m}^2 \text{ K}^{-1} \quad (2)$$

Together (1) and (2) imply a specification on the product of field-of-view and bandwidth:

$$\text{FoVd BW} = 37.5 \text{ deg}^2 \text{ GHz at } 0.7 \text{ GHz}$$
$$= 18.4/\nu^2 \text{ deg}^2 \text{ GHz at frequency } \nu \quad 0.5 \text{ GHz} < \nu < 1 \text{ GHz} \quad (3)$$

Therefore, for example a BW of 200–400 MHz corresponds to a FoVd in the range 200–100 square degrees at 0.7 GHz. A further requirement is that the bandwidth of the SKA is equal to that of two 25% fractional bandwidth bands. These two bands are equivalent to a total bandwidth of $0.4f$ where f is the maximum operating frequency. For simplicity a bandwidth of up to $0.5f$ is allowed.

5. Number of antennas, antenna beams, station beams and correlations

The two SKA specifications (1) and (2) impose limitations on the total number of antenna beams and correlations that are needed, which are calculated in this section. These will then be used to generate consistent cost measures for each of the hardware components in the SKA.

The number N_a of antennas in the SKA can be expressed in terms of the desired point source sensitivity $S_{ska} = A_{eska}/T_{sys}$. Since A_{eska} is equal to the number of antennas (N_a) multiplied by the effective area of each antenna (A_{ea}) after RF and digital beamforming it follows that

$$N_a = S_{ska} T_{sys}/A_{ea}. \tag{4}$$

The number N_{ba} of antenna beams per antenna can be expressed as

$$N_{ba} = \text{FoV}/\Omega_{-3dB} \tag{5}$$

where $\Omega_{-3\,dB}$ is the area of each antenna beam within 3 dB of its peak. The latter area may be expressed as

$$\Omega_{-3\,dB} = \kappa_a \lambda^2/A_{ea} \text{ steradians} \tag{6}$$

where κ_a is a factor dependent on the type of the antenna. For a uniformly-illuminated circular-aperture antenna $\kappa_a = 0.63$ (Silver, 1949). κ_a increases to ~ 0.7 for a parabolic dish with an illumination that tapers to $-18\,dB$ at the edge. For all antennas to be discussed κ_a will be in this range and, to a first approximation, κ_a can be considered to be a constant. Thus using (5) and (6) N_{ba} may be expressed as

$$N_{ba} = \text{SKA field-of-view/antenna field of view}$$
$$= (\text{FoV}/\lambda^2)(A_{ea}/\kappa_a) \tag{7}$$

Using (4) and (7) we can express the total number of antenna beams as (Bunton, 2000)

$$N_b = N_a N_{ba} = (\text{FoV}/\lambda^2)(S_{ska} T_{sys}/\kappa_a) \tag{8}$$

It is seen that, for a given FoV, the only concept-dependent parameters affecting the total number of antenna beams are T_{sys} and κ_a. As these are approximately constant for all wide FoV concepts it is seen that all concepts will generate the same number of beams, N_b, independent of antenna size or technology. Using the field-of-view bandwidth product (3) and sensitivity specifications (1),

$$N_b = K_1 T_{sys}/(\text{BW}\,\kappa_a) \qquad (9)$$

where $K_1 = 1.25 \times 10^{12}\,\text{Hz K}^{-1}$ for frequencies between 0.5 and 1.0 GHz. For example, if BW = 200 MHz, T_{sys} = 36 K and $\kappa_a = 0.7$ then the total number of antenna beams needed is 320,000.

For an antenna station consisting of N_{as} antennas the number N_{bs} of station beams can be expressed as

$$N_{bs} = N_{ba} N_{as}/\kappa_s \qquad (10)$$

where κ_s is the station aperture-filling factor. This has a value of about 1 if the station antennas can be arranged without shadowing so as to form a contiguous aperture and is about 1/10 for arrays of independently steered antennas such as parabolic dishes or Luneburg lenses (Bunton, 2003b). Using (8), (9), (10) and the fact that $N_a = N_{as} N_s$, the number of correlations can then be expressed as

$$\begin{aligned} N_c &= N_{bs} N_s (N_s - 1)/2 \\ &= N_b (N_s - 1)/2\kappa_s \\ &= (N_s - 1) K_1 T_{sys}/(2\,\text{BW}\,\kappa_a \kappa_s) \end{aligned} \qquad (11)$$

6. Cost summary

In the previous section the hardware components are broken down into concentrator, feeds and LNA, RF downconversion, A/D and filterbank including signal transport to the digital beamformer, antenna digital beamformer, signal transport to the central site, station beamformer, and correlator. The cost for all of these components are derived in Appendix 1 and the results summarised in Table I. The cost is broken up into two factors; one that is constant, or nearly so, and another that is a relative measure of the cost. The constant factors includes T_{sys}, which becomes a variable if cooling is possible, SKA dependent terms, S_{ska} and K_1, antenna dependent terms κ_a and η_a which are approximately the same for all concepts, k_{bands} which is approximately unity, and the cost factors for electronics α_{abf} to α_c which are concept independent.

For the concentrator the cost measure is simply α_{con} because all the concepts have the same total area if the aperture efficiency is assumed the same for all concepts.

TABLE I

Cost measures and factors for a wide field-of-view SKA

Cost component	Cost measure	Constant factors	Comments
Concentrator	α_{con}	$S_{ska}\ T_{sys}/\eta_a$	No. RF beamforming $R_{fb} = R_{adb}$
Feeds and LNAs	$\alpha_f\ R_{fb}/BW$	$K_1\ T_{sys}\ k_{bands}/\kappa_a$	
Antenna RF beamforming	$N_{rfi}\ \log_2(N_{rfi})\ R_{adb}/BW$	$\alpha_{abf}\ K_1\ T_{sys}\ k_{bands}/\kappa_a$	
RF downconvert	R_{adb}/BW	$\alpha_{dc}\ K_1\ T_{sys}\ k_{bands}/\kappa_a$	
A/D, signal transport and filterbanks	R_{adb}	$\alpha_{fb}\ K_1\ T_{sys}/\kappa_a$	
Antenna digital beamforming	N_{adb}	$\alpha_{dbf}\ K_1\ T_{sys}/\kappa_a$	$N_{adb} = 0$ if there is no antenna digital beamforming
Signal transport to central site	1	$\alpha_{st}\ K_1\ T_{sys}/\kappa_a$	
Station digital beamforming	N_{as}/κ_s	$\alpha_{dbf}\ K_1\ T_{sys}/\kappa_a$	$N_{as}/\kappa_s = 0$ if a station consists of a single antenna
Correlator	$(N_s-1)/2\kappa_s$	$\alpha_c\ K_1\ T_{sys}/\kappa_a$	

The cost of a feed α_f depends on the feed type (Vivaldi, log periodic etc.) and the total cost is proportional to the number of feeds per antenna beam, R_{fb}. The bandwidth term simply indicates that the number of beams N_b is bandwidth dependent, (Equation (9)) but the cost of the feed and LNA is approximately bandwidth independent. This applies to all RF electronics. Each RF beamformer has a downconverter and their cost is proportional to R_{adb} the ratio of the number of A/D converters to antenna beams. The cost of all other hardware components is bandwidth dependent and will be proportional to the total bandwidth transported to the central site, which is found to be a constant. Thus the cost measures for all the digital electronics R_{adb}, N_{adb}, N_{as}/κ_s, and $(N_s-1)/2\kappa_s$ are simply relative measures of the amount of computation that is needed for each data sample transported to the central site.

The cost measures break up naturally into three major areas, namely concentrator, antenna electronics from feed to RF downconverter, and digital electronics. Within the digital electronics the cost factors α_{dbf} for digital beamformers and α_c for correlators are very similar. The basic operation of both is a complex multiplication. The beamformer multiplication has a higher precision but the correlator needs an associated accumulator. Therefore, the total beamforming and correlator cost is approximately proportional to $N_{adb} + N_{as}/\kappa_s + (N_s-1)/2\kappa_s$. For each concept, N_{adb} or N_{as} is dependent on N_s and so it is possible to minimise the cost by adjusting N_s. For the cases discussed here, if there is station beamforming, N_{adb} is negligible and the optimum number of stations is $\sqrt{(2N_a)}$. This can be interpreted as the beamforming and correlator cost being proportional to the reciprocal of the concentrator diameter, and the larger the concentrator is the lower the cost.

For cases presented where there is antenna beamforming there is no station beamforming and $\kappa_s = 1$, $N_s = N_a$ and N_{adb} is equal to the number of signals available to the beamformer per antenna. This, in turn, is equal to the number of A/D converters and filterbanks, which is a constant for a given concept. For example, for the aperture array there are $44.4f^2$ feeds per square metre giving $44.4f^2/N_{rfi}$ A/D converters per square meter; whereas for a cylindrical reflector there are $6.66f$ feeds per metre of line feed and for a 15 m reflector there are $0.44f$ feeds per square meter giving $0.44f/N_{rfi}$ A/D converters per square metre of collecting area. Assuming η_a is approximately the same for all concepts then the total collecting area is the same and the total number of feeds is this area times the factors given above. The number of A/D converters is also equal to the number per antenna times the number of antennas. For cylindrical reflectors and aperture arrays N_{adb} is equal to the number of A/D converters per antenna. Thus $N_{adb}N_s$ is equal to the total number of A/D converters which is a fixed number for a given concept and N_{rfi}. As there is no station beamforming, the beamforming and correlator cost is approximately proportional to $N_{adb} + (N_s - 1)/2\kappa_s$ and the optimum number of antennas is $\sqrt{(2N_s N_{adb})}$.

7. Wide FoV concepts

The wide FoV concepts can be grouped into four categories. If small concentrators are used then the FoV of a single feed will be large and only a 'single feed and beam per antenna' is needed. Large antennas will need multiple feeds to reach the required FoV and there could either be a 'single feed per beam, multiple feeds per antenna' or a focal plane array (FPA) with 'full digital beamforming'. Lastly there are systems with both 'RF and digital beamforming' In the following, it will be assumed we have a collecting area of 1 million square metres, which meets the SKA specification (2) if T_{sys} is 35 K and η_a is 0.7.

7.1. Concentrator cost

For most concepts currently proposed the estimated concentrator costs can be obtained from the recent SKA conference (SKA, 2004). For the comparison of reflectors it is assumed cost scales as the cube root of frequency and all data brought to a frequency of 1 GHz.

7.1.1. Symmetrical reflector
One of the lowest cost symmetrical reflectors is the one being developed by Swarup and Shankar in India. A 12 m dish produced by this method has been estimated to cost $120000 giving a 1 GHz cost of $530/m^2 (Shankar, 2004).

7.1.2. Cylindrical reflector

For the cylinder the cost of a $\sim$4.5 GHz antenna is estimated to be $580/m^2 or $340/m^2 at 1 GHz. This was for a completely costed design that was far from optimal (Bunton, 2004). It is expected that the final cost will lie between this value and the original estimate of $133/m^2 (Bunton, 2002).

7.1.3. Dual reflector and 3 m dishes

The 12 m high frequency dishes were costed at $493 M for 2500 units or $197000 each. At 1 GHz this equals $605/m^2 for a front-fed design. An offset design is expected to cost 1.3 times as much (USSKA, 2003) or $786/m^2; this value will be used for the dual reflector concepts. For the cost of 3 m dishes it is assumed that the cost scales as diameter to the power 2.7 (Bunton, 2002, Appendix A). For a symmetric design the cost is $230/m^2.

7.1.4. Aperture array

For a 1.25 GHz aperture array a cost of $270 M was given for an area of 0.8 km^2 (van Ardenne et al., 2004). At 1 GHz this gives a cost of $216/m^2, assuming cost is proportional to frequency squared. This value relies on a continuous reduction in cost for the next 10 years. As a comparison the construction and development cost of aperture arrays for EMBRACE is about $80000/m^2 (Patel, 2004).

7.1.5. Luneburg lens

For a 7 m Luneburg lens the concentrator cost was estimated as $340/m^2 (Hall, 2002). The cost is approximately proportional to diameter so a 3 m lens would cost $145/m^2. For this concept the concentrator cost is independent of frequency. These cost estimates depend on the development of suitable manufacturing process to produce "extremely low density" dielectrics. However, SKA development on this process has stopped (Kot, 2004), in part because of a significant amount of up-front non-recurring engineering cost. Work continues on the Luneburg lens for commercial applications, which may see the cost goals being achieved.

7.2. SINGLE FEED PER BEAM, MULTIPLE FEEDS PER ANTENNA

The Luneburg lens (Hall, 2002) and wide FoV dual reflector antennas (Hay et al., 2002, 2004) are large diameter structures that can support multiple independent feeds and beams needed to achieve a large FoV. In general, if a focal plane array[1] is not used, the FoV of the various beams do not overlap and for surveys the operation is similar to that used on the Parkes Multibeam. As there is one feed per beam and no RF beamforming $R_{adb} = R_{fb} = 1$. The maximum bandwidth of $0.5f$

---

[1] Here we distinguish between multiple independent feeds each generating a beam in a focal plane and FPAs and arrays of feeds where groups of feeds need to be beamformed to generate a beam.

can always be achieved while satisfying Equation (3) by choosing an appropriate number of feeds. This strategy minimises feed costs. With this bandwidth then $R_{fb}/BW = 2/f$. There is no antenna digital beamforming, making N_{adb} zero for these concepts. These results also hold for concepts with a single feed and beam per antenna. In this case size of the antenna is adjusted to achieve the desired FoV.

The main difference between the two concepts is the configuration of the instantaneously accessible sky. For the wide FoV dual reflector antenna it is a single patch of up to 400 square degrees (Hay et al., 2002, 2004) and with uncooled receivers there are about 9000 12 m antennas. In the case of the Luneburg lens the beams can be steered independently over the sky but about 26000 7 m lenses are needed. With $\kappa_s = 0.1$ for both concepts, the cost of the dual reflector digital electronics is minimised if N_s is 134 and $(N_s - 1)/2\kappa_s = N_{as}/\kappa_s = 670$. For the Luneburg lens the corresponding quantities are $N_s = 228$ and $(N_s - 1)/2\kappa_s = N_{as}/\kappa_s = 1140$.

7.3. SINGLE FEED AND BEAM PER ANTENNA

If the aperture of an antenna is made small enough, then with a single feed the beam will have a wide FoV. Suitable small aperture antennas are small parabolic dishes and Luneburg lenses, log periodic antennas (Large and Thomas, 2000), small hemi-spherical dishes (Braun, 1996) or, horn antennas (Roddis et al., 1994). Of these, the hemi-spherical dishes have low aperture efficiency due to severe underillumination. Except for the small dishes and Luneburg lenses these antennas have about a 10:1 frequency range. For horns, dishes and Luneburg lenses, the aperture area is constant with frequency, for the log periodic antenna the effective aperture decreases with frequency.

Parabolic dishes and Luneburg lenses need to be about five wavelengths across before they operate efficiently. Thus at 0.5 GHz, a diameter of about 3 m is needed, which gives a FoV of about $30/f^2$ deg^2 and this approximately matches the FoV requirements derived from Equation (3). About 140000 3 m dishes or lenses are needed to give an area of 1 million square metres giving an optimum N_s of 531 and $(N_s - 1)/2\kappa_s = N_{as}/\kappa_s = 2650$. A high quality horn could operate over a frequency range of about 1:10. If the maximum frequency is 1.4 GHz then the diameter of the horn would be about 2 m and 318000 horns are needed for the SKA. In this case the optimum N_s is 798 and $(N_s - 1)/2\kappa_s = N_{as}/\kappa_s = 3940$.

The situation is a little different for the log-periodic antenna. At its lowest frequency of operation the effective aperture can be more than twice the physical area. But if the log-periodic antenna operates over a 10:1 frequency range the effective area has dropped by a factor of 100 at its highest frequency. The decrease in area with frequency makes it impractical at higher frequencies and greatly complicates the analysis of its cost. For this reason it is not considered further.

It is interesting to compare the small-dish option with a cooled 12 m design, such as the LNSD proposal (USSKA Consortium, 2003). The small dish is about one-fourth the width of the 12 m LNSD dish. Assuming cost scales as $D^{2.7}$ the small dishes are 2.6 times cheaper but as they are uncooled twice the area is needed. Therefore, small dishes are still cheaper by a factor of about 1.3.

7.4. FULL DIGITAL BEAMFORMING

A phased array provides a wide FoV when it is either:

- Used by itself as an aperture array (European SKA Consortium, 2002)
- As a linefeed at the focus of a cylindrical parabolic reflector (Bunton, 2003) or
- In the focal plane of a parabolic dish, a Luneburg lens (Hall, 2002), or a wide field-of-view dual reflector (Hay et al., 2002, 2004).

For the aperture array and cylindrical reflector each element of the phased array has a wide FoV and all feed elements within an antenna must be combined to generate beams. In both concepts an antenna station is a single antenna and there is no station beamforming.

For the parabolic dish, Luneburg lens or wide FoV dual reflectors, only a subset of elements need to be beamformed to generate beams as the concentrator has done much of the initial beamforming. However, an antenna station consists of many reflectors or lenses and a station beamformer is needed. Each concept has different antenna electronics cost as defined by R_{fb} and bandwidth BW. These are considered individually below.

7.4.1. Cylindrical reflector

Assuming the linefeed can be approximated by a continuous array of length m half wavelengths then the broadside beamwidth is $\sim 100/m$ degrees. As the beam scans an angle θ from transit the beam size increases to $\sim 100/(m \cdot \cos(\theta))$. If this beam size is averaged over the scan range of 0–60° then an average beam size is found to be $128/m$ degrees.

For a linefeed made of m feeds spaced $\lambda/2$ apart, beams can be formed over the 120° FoV. Thus for a cylindrical reflector there are approximately 1.07 feeds per beam if all beams are formed. The SKA requirement of an FoV of about 200 square degrees, can be met with a 15 m reflector. For beams formed over a limited scan range, sc degrees, the number of feeds per beam R_{fb} is $\sim 128/sc$. Unlike most other concepts this only holds true at the maximum frequency of operation, f. At lower frequencies the number of feeds per beam is proportional to f/ν and R_{fb} is equal to $\sim 128 f/\nu \cdot sc$. A 15 m antenna, operating at frequency ν GHz, has a total FoV of $\sim 1.4 sc/\nu$ degrees, which using (3) gives a bandwidth BW of $13/\nu \cdot sc$ GHz and the cost of feeds and LNA R_{fb}/BW is equal to $10f$. For the assumed total collecting area of 1 million square metres and line feeds that are 90% the length of the reflector there are $396000f$ feeds and the optimum number of antennas is $890\sqrt{f}$. This gives $(N_s - 1)/2\kappa_s = N_{adb} = 445\sqrt{f}$.

7.4.2. Aperture array

Consider a circular aperture array of m half wavelength feeds across the diameter. The array has $(\pi/4)m^2$ feeds and generates a beam width of $2.04/m$ radians. The beam area is $3.3/m^2$ steradians at zenith, increasing to double this at a zenith angle of $60°$. The averaged beam area over this area of sky is $4.6/m^2$ steradians. There are π steradians in the full FoV so the total number of beams needed to cover the FoV is $\pi m^2/4.6$. Dividing this into the number of feeds shows there are 1.15 feeds per beam if the full area of sky down to a zenith angle of $60°$ is imaged. In practice much less than this will be imaged at any one time. If the area imaged is FoVd square degrees then the number of feeds per beam R_{fb} is 12000/FoVd. From (3) the bandwidth needed for this FoVd is $18.4/(\mathrm{FoVd}.\nu^2)$ thus R_{fb}/BW is equal to $650\nu^2$. This concept also allows collecting area to be traded off for more bandwidth in survey only applications. The aperture array must work up to its maximum frequency f so R_{fb}/BW becomes $650f^2$. To meet SKA sensitivity the total area must be about 1 km^2 and the total number of feeds is $44000000f^2$. Thus the optimum number of antennas is $9400f$ and $(N_s - 1)/2\kappa_s = N_{\mathrm{adb}} = 4700\ f$.

7.4.3. Wide field-of-view dual reflector

With a dual reflector antenna it is possible to maintain an in-focus focal plane over a considerable FoV (Hay et al., 2002, 2004). The radius of the focal plane is equal to the effective focal length f_e times the sine of the half width of the FoV (θ). Therefore the area of the focal plane

$$= \pi(f_e \sin(\theta))^2$$

$$\sim \pi(D\theta)^2(f_e/D)^2$$

To calculate the number of feeds, the spacing of the feeds is needed. From Appendix 2 it is found that the element spacing δ is given by

$$\delta \leq \frac{1 + (4 f_e/D)^2}{[1 + (4 f_e/D)]^2}\lambda$$

The allowed maximum element spacing is shown in Figure 3.

The number of feeds in the focal plane is equal to $\pi(D\theta/\delta)^2(f_e/D)^2$.

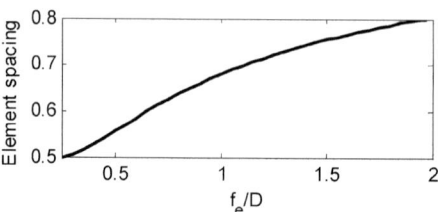

Figure 3. Allowed element spacing in wavelength as a function of focal length ratio.

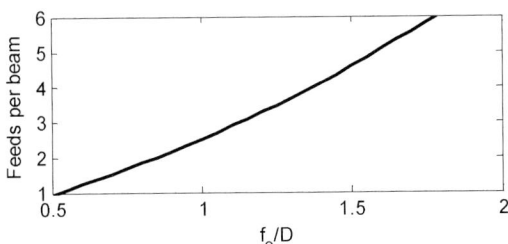

Figure 4. Number feeds per beam needed in the focal plane of a dual reflector antenna.

The beamwidth of each beam is approximately $1.2\lambda/D$ giving a beam area of $\pi(0.6\lambda/D)^2$ steradians and the total FoV of the antenna is $\pi\theta^2$. The number of beams needed to cover the FoV is $2.7(D\theta/\lambda)^2$. Dividing this into the number of feeds in the focal plane gives the number of feeds per beam $= 1.16\,(\lambda/\delta)^2\,(f_e/D)^2$

This is plotted in Figure 4.

For wide field-of-view dual reflectors an F_e/D of 1.0 is possible (Hay et al., 2004). For this foD R_{fb} is at least 2.5 at the maximum operating frequency. However the FoV is approximately constant. So at lower frequencies R_{fb} is equal to $2.5 f^2/\nu^2$. Physically this tells us that if the FoV is specified at some operating frequency ν then at the maximum frequency f the FoV is larger by a factor f^2/ν^2 than the SKA requirements, which have the FoV proportional to λ^2. Maximising the bandwidth minimises the cost and a bandwidth of $0.5f$ makes the feed and downconverter cost R_{fb}/BW equal to $5f/\nu^2$. Note, for this concept, the prime focus dish and Luneburg lenses the focusing of the concentrator illuminates a small section of the FPA and N_{adb} is of order 10 at the maximum frequency increasing to about 100 at half this frequency. The values of N_{as}/κ_s, $(N_s - 1)/2\kappa_s$ are the same as those for the dual reflector antenna with multiple feeds.

7.4.4. *Prime focus parabolic dish*

With a large parabolic dish a large FoV can also be obtained by placing a focal plane array (FPA) at the prime focus. At large foD ratios the focal plane is large because for a given angular offset the deflection of the focus in the FPA is proportional to foD. At small foD ratios the FPA is large due to poor focussing. Figure 5 shows the relative effective area for a number of values of foD and sizes of square FPAs as calculated by the approach described in Appendix 3.

In Figure 5 it is seen that the optimum foD depends on the allowed reduction in effective area for beams at the edge of the focal plane, with optimum foD increasing with increasing effective area. For edge beams limited to a 1 dB loss (horizontal line) the optimum foD is in the range 0.45–0.5. With this foD the side length of the focal plane is 0.18D, where D is the diameter of the dish. As the dish is 60 wavelengths across the rectangular FPA has an area of 116 wavelengths squared. From Figure 3 the allowed feed spacing is 0.53 wavelengths so there are 415 feeds

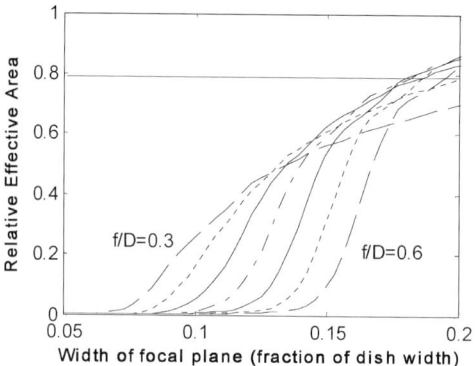

Figure 5. Effective area of a beam 7 degrees off axis relative to an on axis beam as a function of FPA side length for a parabolic dish 60 wavelengths in diameter. foD ratios range from 0.3 to 0.6 in steps of 0.05.

in the focal plane. The field-of-view of a single beam with an aperture efficiency of 0.8 is approximately $1.2\lambda/D$ radians across (Christiansen, 1987) or 1.15 degrees in this example. The FoV of the FPA is 196 square degrees so with hexagonal packing of the beams there are 172 beams and the number of feeds per beam, R_{fb}, is 2.4. As with the dual reflector concept this increases at lower frequencies. Thus for the single FPA R_{fb} is equal to $2.4 f^2/\nu^2$. For a bandwidth of $0.5f$ the RF downconverter cost measure R_{fb}/BW is equal to $4.8 f/\nu^2$. The values of N_{adb}, N_{as}/κ_s, $(N_s - 1)/2\kappa_s$ are the same as that for dual reflector antennas with FPAs.

The number of feeds per beams is also dependent on FoV, the diameter of the reflector and the allowed aperture efficiency loss for the edge beams. For example allowing a 2 dB aperture efficiency loss reduces R_{fb} at the maximum frequency to 1.8 but this improvement must be balanced against the loss of sensitivity. For the other factors it is estimated that the R_{fb} increases as the size of the dish decreases. Its behaviour with changes in FoV is complex due to the minimum size requirements which causes R_{fb} to increase as FoV approaches that of a single beam, but before that reduction in aberration as FoV decreases will cause R_{fb} to decrease.

7.4.5. Luneburg lens

When a FPA is used with a Luneburg lens the minimum number of feeds per beam occurs when the FPA is coincident with the surface of the lens. For a lens of diameter D the beam separation is about $1.2 \lambda/D$ radians and the separation of the beams on the focal plane is $1.2 f.\lambda/D$. For a focus on the surface of the lens foD is equal to 0.5 and the beam separation is 0.6λ. With feeds in the focal plane separated by $\lambda/2$ there are 1.44 feeds/beam for a square arrangement of beams and $R_{fb} = 1.1$ feeds/beam if the arrangement is hexagonal. And R_{fb}/BW is about $2.2/f$. The values of N_{as}/κ_s, $(N_s - 1)/2\kappa_s$ are the same as that for Luneburg lenses with multiple individual feeds.

7.5. RF AND DIGITAL BEAMFORMING

7.5.1. Cylindrical reflector

For scan ranges much less than 120 degrees, RF beamforming would be used to reduce the FoV, which reduces downconversion, A/D and filterbank costs. To avoid excessive loss imaging would be done for the FoV between the -1 dB points of the resulting beam. For RF beamforming over m feeds spaced $\lambda/2$ apart, the field-of-view between the -1 dB points is about $50/m$ degrees, or $100/m$ deg^2 for a 15 m wide reflector at 0.7 GHz. The maximum bandwidth limit and Equation (3) constrain $m = N_{\text{rfi}}$ to be at most 2. If k RF beamformed signal are then beamformed digitally the final beam width is $100/km$. Thus the number of beams in the RF beamformed field-of-view is $50k/100$ or $k/2$. Thus, for a linefeed with RF beamforming there are an average of two RF beamformed sections, that is $R_{\text{adb}} = 2$ and $R_{\text{adb}}/BW = 4/f$. With $m = 2$ the scan angle sc is 25 degrees at the maximum frequency f. At other frequencies it is $25f/v$ degrees. For a cylindrical reflector R_{fb} is equal to $128f/v.\text{sc.}$, substituting the scan angle gives $R_{\text{fb}} = 5$. Thus R_{fb} is approximately 5 and R_{fb}/BW is $10/f$. The RF beamforming reduces the total number of inputs into the digital beamformers to $396000f/m$. Thus the optimum number of antennas for $m = 2$ is $629\sqrt{f}$. This gives $(N_s - 1)/2\kappa_s = N_{\text{adb}} = 315\sqrt{f}$.

7.5.2. Aperture array

In aperture arrays, RF beamforming would greatly reduce the cost of downconverters, A/D converters, filterbanks and digital beamforming. The number of RF beamformed signals per beam can be estimated by beamforming first along rows of elements then along the columns of the resulting signal. At the first stage there are two signals per beam and after the second there are four, that is $R_{\text{adb}} = 4$. If the bandwidth is maximised then the downconverter cost is minimised and $R_{\text{adb}}/BW = 8/f$. If feeds contribute to only one beamformer and there are N_{rfi} inputs then FoVd is $\sim 1900/N_{\text{rfi}}$. The bandwidth needed to satisfy the survey speed specification (3) is 18.4 $N_{\text{rfi}}/1900f^2$ and for the maximum bandwidth, $0.5f$, N_{rfi} is equal to $50f^3$. Dividing the number of feeds, $44000000f^2$, by N_{rfi} gives the number of downconverters: $880000/f$. Thus the optimum number of antennas is $1330/\sqrt{f}$ This gives $(N_s - 1)/2\kappa_s = N_{\text{adb}} = 665/\sqrt{f}$.

8. Summary

In this section we summarise the results of the two previous sections and compare the cost measures of the various antenna options. The relative cost measures are summarised in the Table II.

Except for the aperture array concept, the feed, LNA and downconverter costs are within a factor of five at the maximum operating frequency. The most expensive of these is the cylinder where the cost is estimated to be $\sim\$25$M (Bunton, 2002,

TABLE II
Approximate relative cost measures of wide field-of-view options

System component	Aperture	Feed + LNA + downconverter and A/D + filterbank	Antenna & station beamforming, correlator	No. antenna stations
Cost measure	α_{con} \$/m^2	R_{fb}/BW, R_{fb}	N_{adb}, N_{as}/κ_s, $(N_s-1)/2\kappa_s$	N_s
Single feed per beam, multiple feeds per antenna				
Luneburg lens	340[a]	2/f, 1	0, 1140, 1140	228
Dual reflector	786	2/f, 1	0, 670, 670	134
Single feed and beam per antenna				
Small parabolic dish	230	2/f, 1	0, 2230, 2230	477
Small Luneburg lens	145[a]	2/f, 1	0, 2230, 2230	477
Small hemisphere	no data	2/f, 1	0, 2230, 2230	477
Horn	no data	2/f, 1	0, 3940, 3940	798
Full digital beamforming				
Luneburg lens	340[a]	2.2/f, 1.1	~10, 1140, 1140	228
Prime focus dish	540	4.8f/v^2, 2.4f^2/v^2	~10, 670, 670	134
Dual reflector	786	5f/v^2, 2.5f^2/v^2	~10, 670, 670	134
Aperture array	216	650f^2, 12000/FoVd	4700f, 0, 4700f	9400f
Cylindrical reflector	133–340	10f, 128f/v.sc	445$\sqrt{f}$, 0, 445$\sqrt{f}$	890$\sqrt{f}$
RF and digital beamforming		R_{fb}/BW, R_{adb}/BW, R_{adb}		
Aperture array[b]	216	650f^2, 8/f, 4	665/$\sqrt{f}$, 0, 665/$\sqrt{f}$	1300/$\sqrt{f}$
Cylindrical reflector[c]	133–340	10/f, 5/f, 2	315$\sqrt{f}$, 0, 315$\sqrt{f}$	629$\sqrt{f}$

[a]Depends on development of manufacturing process to produce extremely low density dielectrics.
[b]$N_{rfi} = 50f^3$ and RF beamforming cost $N_{rfi}\log_2(N_{rfi}) = 282f^3(1 + 0.5\log(f))$.
[c]$N_{rfi} = 2$ and RF beamforming cost $N_{rfi}\log_2(N_{rfi}) = 2$.

for a 0.5 GHz bandwidth). However, it is emphasised that this is small compared to the concentrator costs. For the aperture array these costs are much higher but these are included in the concentrator cost. Thus, in general, the feed, LNA and downconverter cost will not have a major effect compared to concentrator cost α_{con}. This leaves the digital backend costs: filterbanks, beamformers and correlator. These costs are currently high and depend largely on Moore's law to bring them down to an acceptable level. For this paper it assumed that the digital backend cost will be significant for the SKA. Because the correlator and beamforming costs have been optimised just the correlator cost is needed to compare both correlator and beamforming costs. This is high for the single feed and beam per antenna concepts and aperture arrays with full digital beamforming. These high costs may make these concepts uneconomic. But this must be balanced against the possibly low concentrator cost and the ability of the single feed concepts to cover a 10:1 frequency range. The correlator cost is also significant to the multibeam Luneburg lens concept, which reduces its competitiveness.

Many concepts have a correlator cost measure of about 600 including dual reflector antennas, prime focus dishes, cylinders with full digital beamforming and aperture arrays with RF beamforming. Of these, the prime focus dish and the aperture arrays only support a single frequency feed structure and require a new investment in collecting area for each new wide FoV frequency range. This could double the cost if a frequency coverage from 1.7 to 0.3 GHz is required and feed structure covers 2.5:1 frequency range. For the cylinders and dual reflector antennas there is sufficient space in the focal plane to accommodate multiple feed structures and these concepts are able to reuse the expensive collecting area at multiple frequencies. In terms of concentrator cost the aperture array and cylinder hold the greatest promise. It is hoped that a small investment in structural development might refine the mechanical cost of cylinders. For aperture arrays the current large investment in its development will need to continue and succeed over the next decade.

A final interesting concept is the cylinder with RF and digital beamforming. Its correlator and beamforming costs are about half as much as dish reflectors with digital beamforming and its downconverter, A/D and filterbank costs are similar. This together with the possibly low concentrator cost makes it an attractive option.

9. Conclusion

Major objectives of the SKA are high point source sensitivity and high survey speed at about 0.7 GHz. In this paper the relative costs of the hardware components needed to meet these specifications are derived. These results, for the numerous concepts that can meet the SKA specifications, are discussed at length in Section 8. However, two important results are emphasised:

1. Antenna station size can be optimised so that the costs of beamforming and correlation are minimised.
2. With station beamforming at the central site the cost of signal transport is concept independent. This result means that electronics costs at the antenna reflect the 'efficiency' with which data is generated and at the correlator and station beamformer costs are a measure of the 'efficiency' with which the data is converted into visibility data.

Appendix 1: Calculation of cost components

The SKA can be broken down into nine components: the concentrator, feed and LNA, RF beamformer, RF downconverters, antenna digital beamforming, signal transport to the central site, the correlator, and A/D converter, antenna signal transport and filterbanks. The last three are considered as one cost because for each an increase in one increases the others by the same amount. The cost measures for these components are derived in this appendix, and the reader is referred to Section 3 and Figures 1 and 2 for a description of the hardware model. Note, an important result, from the main text, that is used extensively is that the number of beams N_b is equal to $K_1 T_{sys}/(BW\, \kappa_a)$.

SIGNAL TRANSPORT AND CORRELATOR COSTS

The total data rate of the signal transport to the correlator is given by $D_{st} = N_b\, BW$ and using (9) can be expressed as

$$D_{st} = K_1 T_{sys}/\kappa_a \tag{A1}$$

This is same as the result given in Bunton (2003b) under the condition that FoV. BW is given by Equation (3). With beamforming at the central site the data rate is independent of the antenna technology and is largely dependent on T_{sys} only. The cost of the signal transport can then be written as

$$C_{st} = \alpha_{st} D_{st} = \alpha_{st} K_1 T_{sys}/\kappa_a \tag{A2}$$

where α_{st} is the cost per bit. Hz. With T_{sys} constant this is independent of antenna technology.

The total compute load of the correlator is given by $D_c = N_c\, BW$ and using (11) can be expressed as

$$\begin{aligned} D_c &= N_c BW \\ &= N_b(N_s - 1)BW/2\kappa_s \\ &= (N_s - 1)K_1 T_{sys}/(2\,\kappa_a\,\kappa_s) \end{aligned} \tag{A3}$$

It is seen that the correlator cost is independent of FoV and BW and can be written as

$$C_c = \alpha_c D_c = \alpha_c(N_s - 1)K_1 T_{sys}/(2\kappa_a\kappa_s). \tag{A4}$$

where α_c is the cost per Hz of cross correlation for one baseline. With T_{sys} a constant, C_c will depend mainly on the station filling factor and $(N_s - 1)$. Thus $(N_s - 1)/2\kappa_s$ is a measure of the relative correlator cost of a wide-FoV SKA. In this model, the correlator cost does not depend on bandwidth for a given survey speed as any change in bandwidth is balanced by an inverse change in the number of beams.

Concepts using small aperture concentrators minimise correlator costs by having antenna stations consisting of more than 10 concentrators. The reduction in N_s more than compensates for the increased cost due to the decrease in filling factor κ_s. The effect of the filling factor could be reduced if the area covered by the main sidelobes was imaged as well as the main beam. To do this it is preferable to use regular antenna arrays: rectangular or hexagonal. This puts most of the sidelobe energy into a small number of grating lobes. Imaging the grating lobes increases κ_s and hence reduces the correlator cost. However, there is no proposal to demonstrate the imaging of grating lobes. To maintain consistency with the current concepts it is assumed that only the main beam of an array is imaged.

CONCENTRATOR OR COLLECTING-AREA COSTS

The cost of the concentrators can be expressed as

$$C_{\text{con}} = \alpha_{\text{con}} A_{\text{eska}} / \eta_a = \alpha_{\text{con}} S_{\text{ska}} T_{\text{sys}} / \eta_a \tag{A5}$$

where α_{con} is the cost per square metre of concentrator and η_a is the aperture efficiency. Aperture efficiency is similar for all the concepts considered in this paper so it is sufficient to compare α_{con} for each antenna concept.

FEED AND LNA COSTS

Assuming one LNA per feed the cost of the feeds and LNAs can be expressed as

$$\begin{aligned} C_f &= \alpha_f N_f \\ &= \alpha_f N_b (N_{\text{fa}} / N_{\text{ba}}) k_{\text{bands}} \\ &= \alpha_f N_b R_{\text{fb}} k_{\text{bands}} \end{aligned} \tag{A6}$$

where α_f is the cost per feed plus LNA, N_f is the number of feeds, R_{fb} is the ratio of the number N_{fa} of feeds per antenna to the number N_{ba} of antenna beams per antenna at a given frequency and k_{bands} is a factor that accounts for the need for multiple feed systems due to the finite bandwidth of the feeds. With 3:1 bandwidth feeds, k_{bands} is 1.12 for aperture arrays and for parabolic dishes with a FPA, the number of feeds in each feed array is proportional to f^2 and each successively lower frequency array will have 1/9 as many feeds. For cylindrical reflectors the number of feeds is proportional to f and k_{bands} is about 1.5. However, for focal-plane arrays and aperture arrays it may not be possible to reuse the physical structure of the antenna for other frequency bands which negates any advantage they may have because k_{bands} is smaller. Using Equation (9) in (A6) the cost of feeds becomes

$$C_f = \alpha_f K_1 T_{\text{sys}} R_{\text{fb}} k_{\text{bands}} / (\text{BW} \, \kappa_a) \tag{A7}$$

In α_f we include the cost per feed of the structure enclosing and supporting the feeds. We assume the cost of a feed and LNA will be similar in all concepts that use focal or aperture plane technologies. It could be higher in concepts that have a single feed per beam or cooling. For example α_f for a cooled corrugated horn might be a hundred or a thousand times that of an uncooled Vivaldi feed. The values of K_1, k_{bands}, and κ_a are approximately constant for the SKA so for a given T_{sys}, the relative cost measure of feed and LNA costs is $\alpha_f R_{fb}/BW$. The main variables are α_f and R_{fb}/BW. In some concepts, the FoV is set by the technology and using Equation (1) and (2) defines BW.

RF DOWNCONVERSION DIGITAL CONVERTER, FILTERBANK AND ANTENNA BEAMFORMING COSTS

In the model each RF signal is down converted to baseband but then possibly only a subset of the signals from the multiple feeds needed to cover the full frequency range is converted to digital data. This data is then transported to a filterbank and beamformed. The size of the signal subset is large enough to process all signals from the highest frequency feed.

The number of down converters is equal to the number of analogue to digital converters A/D N_{ad} multiplied by k_{bands}. The cost of the down conversion using Equation (9) can be expressed as

$$\begin{aligned} C_{dc} &= \alpha_{dc} N_{ad} k_{bands} \\ &= \alpha_{dc} N_b R_{adb} k_{bands} \\ &= \alpha_{dc} (K_1 T_{sys}/\kappa_a)(R_{adb}/BW) k_{bands} \end{aligned} \tag{A8}$$

where R_{adb} is the ratio of the number of A/D converters to antenna beams, calculated on a per antenna basis, and α_{dc} is the cost of a single down converter, which is assumed to be approximately independent of bandwidth BW. For the A/D converters, digital transport to the filterbanks and the filterbanks, the number of units is $N_{ad} = N_b R_{adb}$. For all of these it is assumed that the cost per unit is proportional to bandwidth. Thus the cost of these can be expressed as

$$\begin{aligned} C_{fb} &= \alpha_{fb} N_b R_{adb} BW \\ &= \alpha_{fb} R_{adb} K_1 T_{sys}/\kappa_a \end{aligned} \tag{A9}$$

where α_{fb} is the cost of an A/D converter and filter bank plus the cost of signal transport from the feed to the digital processing unit. This cost will depend on whether analogue data is transported from the feed or the A/D is at the feed and the data is transported digitally, but for this analysis it is assumed to be the same for all concepts.

Comparing Equations (A8) and (A9) to Equations (A6) and (A7) it is seen that R_{adb}/BW is a relative measure of the cost of downconversion and R_{adb} is a measure of A/D, filterbank and signal transport costs. If there is no RF beamforming R_{adb} is equal to R_{fb} and the α_{dc} and α_f can be combined.

We assume that the cost of the antenna digital beamforming can be expressed as

$$\begin{aligned} C_{dbf} &= \alpha_{dbf}(\text{no. of inputs per beam})(\text{no. of beams generated})\,\text{BW} \\ &= \alpha_{dbf} N_b \, \text{BW} \, N_{adb} \\ &= \alpha_{dbf} N_{adb} K_1 T_{sys}/\kappa_a \end{aligned} \quad (A10)$$

where α_{dbf} is the cost per Hz per input to a beamformer and N_{adb} is the number of inputs used in forming each output beam. For a given T_{sys} a relative measure of the cost is given by N_{adb}. If there is no RF beamforming then the number of inputs to the antenna beamformer is equal to the number of feeds per antenna.

STATION BEAMFORMING COSTS

Where there are multiple concentrators in an antenna station it is necessary to beamform the signals from the separate concentrators. We assume that the cost of this station beamforming can be expressed as

$$\begin{aligned} C_{sdbf} &= \alpha_{dbf}(\text{no. of inputs per beam})(\text{no. of beams generated})\,\text{BW} \\ &= \alpha_{dbf}(N_{as})(N_{ds}N_s)\text{BW} \\ &= \alpha_{dbf}(N_{as})(N_{ba}N_{as}/\kappa_s \, N_s)\text{BW} \\ &= \alpha_{dbf} N_b \, \text{BW} \, N_{as}/\kappa_s \\ &= \alpha_{dbf} K_1 T_{sys} \, N_{as}/\kappa_s \kappa_a \end{aligned} \quad (A11)$$

Equations (A10) and (A11) have a similar form and the cost measure is N_{as}/κ_s.

ANTENNA RF BEAMFORMING COSTS

If there is RF beamforming then the number of beamformers is equal to the number of inputs to the RF downconverters $N_{ad} \cdot k_{bands}$. The cost of each beamformer is proportional to the number of RF inputs to each beamformer N_{rfi} and the number of delay stages in the beamformer approximately proportional to $\log_2(N_{rfi})$. The cost of RF beamforming can then be expressed as:

$$\begin{aligned} C_{rfb} &= \alpha_{abf} N_{ad} k_{bands} N_{rfi} \log_2(N_{rfi}) \\ &= \alpha_{abf} N_b R_{adb} N_{rfi} \log_2(N_{rfi}) k_{bands} \\ &= \alpha_{abf} K_1 T_{sys} N_{rfi} \log_2(N_{rfi}) R_{adb} k_{bands}/(\text{BW} \, \kappa_a) \end{aligned} \quad (A12)$$

where α_{abf} is the cost factor for analogue RF beamformers. The term N_{rfi} is technology and implementation dependent and ranges from 2 to 100, so $N_{rfi} \log_2(N_{rfi})$

ranges between about 2 and 700. Assuming α_{abf} is similar for all concepts then the relative cost is given by $N_{\text{rfi}} \log_2(N_{\text{rfi}}) R_{\text{adb}}/\text{BW}$, which is equal to the relative cost of RF-digital converters multiplied by $N_{\text{rfi}} \log_2(N_{\text{rfi}})$.

Appendix 2: Focal-plane-array element spacing

Consider any focus-linked sequence of quadrics of revolution (ellipsoids, hyperboloids, paraboloids, planes, spheres) where the last reflector is a paraboloid. The ray mapping between the input and output may be expressed in the fractional-linear form (Dragone, 1981)

$$w = \frac{2 f_e u}{1 - u \tan i_e}. \tag{A13}$$

where $w = x + jy$, x and y are Cartesian coordinates of the point where the output ray intersects a plane orthogonal to the axis of the output paraboloid, and $u = \Re\{u\} + j\Im\{u\}$ is the stereographic projection of the point where the input ray intersects the sphere of radius 1/2 and with centre at the input focal point. In terms of spherical-polar coordinates θ and ϕ, $u = \tan(\theta/2)\exp(j\phi)$. In Equation (A13), f_e and i_e are the effective or equivalent focal length and angle of incidence respectively of the equivalent paraboloid. Two or more offset-fed reflectors may be arranged such that $i_e=0$ and this eliminates cross-polarization and first-order astigmatism in scanned beams (Dragone, 1983).

The mapping (A13) is conformal and so maps circles into circles or straight lines. Thus if the aperture in the $w = x + jy$ plane is a circle of diameter D then the ray cone at the input is a circular cone of half-angle θ_c given by $\theta_c = (\theta_+ - \theta_-)/2$ where

$$\tan(\theta_\pm/2) = \frac{\pm D}{4 f_e \pm \tan i_e D}$$

and if $i_e = 0$ then

$$\tan(\theta_c/2) = \frac{D}{4 f_e}. \tag{A14}$$

The receive-mode focal-plane field may be represented in plane-wave-spectrum form (Clemmow, 1996)

$$E(x_0, y_0) = \iint_{\theta \leq \theta_c} dp\, dq\, \hat{e}(p, q) \exp(-jk(x_0 p + y_0 q)) \tag{A15}$$

where $p + jq = \sin\theta \exp j\phi = 2u/(1 + |u|^2)$ is the complex direction cosine and $k = 2\pi/\lambda$ where λ is the wavelength. We assume that the restriction of the integration domain in (A15) is implied by the ray representation (A13) and (A14). Thus the focal-plane field is a bandlimited function, with a circular bandpass region of diameter in the $p + jq$ plane of $2 \sin \theta_c$. The advantage of this representation is the sampling theory (Papoulis, 1968) which shows that sampling the function (A15) on a rectangular grid has the effect of replicating its spectrum also on a rectangular grid; some other sampling geometries such as hexagonal have the same property.

Now suppose we place an array of elements in the focal plane with the aperture centre of each element at a point of the sampling grid, referred to above. Suppose also that the transmit-mode radiation pattern of the array can be represented as the product of an array factor and an element pattern, and that the transmit-mode field at the centre of each element is the conjugate of the receive-mode field at the same point. The array factor is then just the replicated spectrum of the sampled receive-mode field.

For the array factor to have no replications or grating lobes in the propagating-region $|\sin \theta| \leq 1$ of the plane-wave spectrum or radiation pattern, the replica interval δ/λ (where δ is the focal-plane sampling interval) must be greater than 1 plus half the diameter of the original spectrum. Thus, in the case of rectangular sampling,

$$\frac{\delta}{\lambda} \leq \frac{1}{1 + \sin \theta_c} \tag{A16}$$

Using Equation (A13), condition (A16) can be expressed as

$$\frac{\delta}{\lambda} \leq \frac{1 + (4f_e/D)^2}{[1 + (4f_e/D)]^2} \tag{A17}$$

Appendix 3: Effective area of conjugate-matched focal-plane array

Using the reciprocity theorem the effective area of an antenna may be expressed as (Wood, 1980)

$$A_e = \frac{1}{16P_t} \left| \int_S (\mathbf{E}_r \times \mathbf{H}_t - \mathbf{E}_t \times \mathbf{H}_r) \, d\mathbf{S} \right|^2 \tag{A18}$$

where $\mathbf{E}_r$, $\mathbf{H}_r$ are the receive-mode electric and magnetic fields respectively produced by an incident plane of unit power density, $\mathbf{E}_t$, $\mathbf{H}_t$ are the transmit-mode electric and magnetic fields respectively, P_t is the power of the source for the transmit-mode field and S is any surface enclosing this source. In the case of a reflector antenna, S may be taken as the surface of the feed and, assuming the feed has no influence on the receive-mode field scattered by the reflector, in (1) the field $\mathbf{E}_r$, $\mathbf{H}_r$ may be replaced by its incident component that would exist on S in the absence of the feed. This replacement is possible because the component of $\mathbf{E}_r$, $\mathbf{H}_r$ due to scattering from the feed can be represented in terms of an equivalent source within S and the field coupling integral in (1) is identically zero for any pair of fields whose sources reside entirely on the same side of S. If on the feed aperture Σ $\mathbf{E}_t$ is the complex conjugate of $\mathbf{E}_r$ and $\mathbf{H}_t$ is the complex conjugate of $-\mathbf{H}_r$ whereas on the remainder of S the transmit-mode fields are assumed to be zero then (1) reduces to

$$A_e = \frac{1}{2}\left|\int_\Sigma \Re(\mathbf{E}_r \times \mathbf{H}_r^*)\, d\mathbf{S}\right| \quad (A19)$$

The field $\mathbf{E}_r$, $\mathbf{H}_r$ scattered by the reflector is calculated using the physical-optics method (Silver, 1949).

Acknowledgements

Thanks are extended to Trevor Bird, Graeme James and Peter Hall for their comments on this paper.

References

Braun, R.: 1996, 'Small Hemi-spherical Reflectors for SKAI?' NFRA note 642, March 1996, http://www.atnf.csiro.au/projects/ska/rbraun_960614.ps.

Bunton, J.D.: 2000, 'SKA Antenna Selection – Economics and Field of View', Technology Pathways to the Square Kilometer Array, Jodrell Bank, 3–5 August 2000, http://www.jb.man.ac.uk/ska/workshop/Bunton1.pdf.

Bunton, J.D. (ed): 2002, 'Panorama of the Universe: A Cylindrical Reflector SKA', SKA Design Concept White Paper 13, July 2002, http://www.skatelescope.org/documents/dcwp/Cylindrical_EMT_reply_4.pdf.

Bunton, J.D.: 2003a, 'SKA Station Cost Comparison ', SKA Memo 36, Aug 2003, http://www.skatelescope.org/documents/SKA_cost_comparison_5.pdf.

Bunton, J.D.: 2003b, 'SKA correlator input data rates', SKA memo 43, 14 Nov 2003 http://www.skatelescope.org/documents/Bunton_Correlator_input_data_rate_0402.pdf.

Bunton, J.D.: 2004, 'Cylindrical Reflector SKA', SKA 2004 – Penticton, 18–22 July, 2004.

Christiansen, W.N. and Hogbom, J.A.: 1987, *Radiotelescopes*, 2nd Edn., Cambridge University Press, Cambridge.

Clemmow, P.C.: 1996, *Plane-Wave Spectrum Representation of Electromagnetic Fields*, IEEE Press, Piscatawaty, NJ.

Dragone, C.: 1981, *BSTJ*, **60**(10), 2397–2421.

Dragone, C.: 1983, *IEEE Trans. Ant. Prop.* **AP-31**(5), 764–775.

European SKA Consortium: 2002, 'The European Concept for the SKA, Aperture Array Tiles', SKA Design Concept White Paper 3, July 2002, http://www.skatelescope.org/documents/SKA_EUR_CONCEPT_IntegratedApertureArrayPanels_17072002.pdf.

Hall, P.J. (ed): 2002, 'Eyes on the Sky: A Refracting Concentrator Approach to the SKA', SKA Design Concept White Paper, July 2002, http://www.skatelescope.org/documents/SKA_AUS_CONCEPT_Luneburg_17072002.pdf.

Hay, S., Bunton, J., Granet, C., Forsyth, R., Bird, T. and Sprey, M.: 2002, 'Multibeam Dual Reflector Antennas', International SKA Conference, Groningen, Netherlands, 13–15 August 2002. http://www.lofar.org/ska2002/pdfs/Hay_DualReflector2.pdf

Hay, S.G., Bird, T.S., Bunton, J.D. and Sprey M.A.: 2004, "Field-of-view and Beamforming Complexity in Array-Fed Single- and Dual-Reflector Antennas", SKA 2004 - Penticton, 18–22 July, 2004.

Horiuchi S., Chippendale A. and Hall, P.: 2004, 'SKA System Definition and Costing: A First Approach', SKA Memo 57, Oct 2004, http://www.skatelescope.org/pages/page_skares.htm

Jones, D.: 2004, 'SKA Science Requirements, Version 2', SKA memo 45, 26 Feb 2004, http://www.skatelescope.org/documents/Memo%2045_RTS_extra_sentence.pdf.

Kot, J.: 2004, 'Lenses', SKA 2004 – Penticton, 18–22 July, 2004.

Large, M.I. and Thomas, B.M.: 2000, 'Single or Multiple Antenna Array Stations?', Technology Pathways to the Square Kilometer Array, Jodrell Bank, 3–5 August 2000, http://www.jb.man.ac.uk/ska/workshop/Large1.pdf.

Papoulis, A.: 1968, *Systems and Transforms with Applications in Optics*, McGraw Hill, New York.

Patel, P.D.: 2004, 'EMBRACE', SKA 2004 – Penticton, 18–22 July, 2004.

Roddis, N., Baines, C. and Pedlar, A.: 1994, 'Horns for the 1 Square Kilometer Array?', Report to the URSI Large Telescope Working Group: – August 1994, http://www.atnf.csiro.au/projects/ska/roddis_940801.ps.

Shankar, N.U.: 2004, 'Pre-oaded Parabolic Dishes', SKA 2004 – Penticton, 18–22 July, 2004.

Silver, S. (ed): 1949, *Microwave Antenna Theory and Design*, McGraw-Hill, New York.

SKA: 2002 & 2003, 'SKA design concept white paper are found at' http://www.skatelescope.org/pages/documents_1.htm.

SKA: 2004 – Penticton, 18–22 July, 2004 at http://www.skatelescope.org/pages/conf%20_pentincton.htm.

USSKA Consortium: 2003, 'The Large-N-Small-D Concept for the SKA: Addendum to the 2002 Whitepaper', SKA Design Concept White Paper 9, May 2003, http://www.skatelescope.org/documents/dcwp/small_parabolas_May03.pdf.

van Ardenne, A., Lazio, J. and Patel, P.: 2004, 'Hybrid of Aperture Arrays and LargeNSmallDconcept for SKA; The aperture array part', SKA 2004 – Penticton, 18–22 July, 2004.

Wood, P.J.: 1980, *Reflector Antenna Analysis and Design*, Peter Peregrinus, London.

COST EFFECTIVE FREQUENCY RANGES FOR MULTI-BEAM DISHES, CYLINDERS, APERTURE ARRAYS, AND HYBRIDS

JAAP D. BREGMAN
ASTRON, P.O. Box 2, 7990 AA, Dwingeloo, The Netherlands
(e-mail: bregman@astron.nl)

(Received 14 September 2004; accepted 28 February 2005)

Abstract. Five out of six Square Kilometre Array (SKA) science programs need extensive surveys at frequencies below 1.4 GHz and only four need high-frequency observations. The latter ones drive to expensive high surface accuracy collecting area, while the former ask for multi-beam receiver systems and extensive post correlation processing. In this paper, we analyze the system cost of a SKA when the field-of-view (Fov) is extended from 1 deg^2 at 1.4 GHz to 200 deg^2 at 0.7 GHz for three different antenna concepts. We start our analysis by discussing the fundamental limitations and cost issues of wide-band focal plane arrays (FPA) in dishes and cylinders and of wide-band receptors in aperture arrays. We will show that a hybrid SKA in three different antenna technologies will give the highest effective sensitivity for all six key science programs.

Keywords: aperture efficiency, field-of-view, focal plane arrays, foreshortening efficiency, system optimization

1. Introduction

The Square Kilometre Array (SKA) consortium members have proposed various antenna concepts, where only two pay attention to very large field-of-view (Fov) and show decent cost if the observing frequencies are below 1.4 GHz. This has led to extension of the science requirements and 200 deg^2 should be provided at 0.7 GHz (Jones, 2004). In face of the discussion on hybrid designs we first need to compare optimized system designs based on each concept and show which part is responsible for which advantage and for which disadvantage. Then we can make an optimum distribution of the total SKA budget over different sub-arrays such that optimum performance is obtained for the total package of science programs.

Based on the sensitivity equation for an aperture synthesis array, Bregman (2004) has derived an optimum cost distribution over collecting area plus low-noise amplifier, receiver chains including digitisation with digital beam forming, and finally cross-correlation plus post processing. However, to maximize the sensitivity in a survey that covers a larger solid angle than the instantaneous Fov we need a fixed ratio between collecting area and the effective number of feeds that illuminate the aperture. In a multi-beam dish, every feed with its dual polarization receiver chain provides an independent Fov. Actual realisations of a multi-feed system with a focal plane array (FPA) will need slightly more receiver chains than independent feeds produced by the beam-former that recombines FPA element signals. Then the optimum system cost distribution requires that at least half of the available budget should be spent on collecting aperture and the rest on receiver chains, digital multi-beam signal processing electronics, correlation and post processing, which provide bandwidth and Fov that increases the effective integration time per deg^2 on the sky.

D'Addario (2002) estimated the cost of a digital cross-correlator with two independent beams of 0.8 GHz bandwidth for 2560 dual polarization antenna inputs and only coplanar polarization output at \$23 million, in 2003. The cost of antenna-based digitisation and tracking units is \$20 million. According to roadmaps of the semiconductor industry, we can expect Moore's law to continue till 2009 when current technology is hitting its limits. Price erosion will still to continue after that, which means that in the period from 2003 till 2015 we can expect seven more performance-doubling steps in digital processing technology at the same cost. The time for receiver chain performance doubling is higher (Walden, 1999) and we expect only two doublings between 2003 and 2015.

By 2015, the digital signal processing will be a mature technology without further performance doubling times less than 2 years. For all reasonable SKA designs, the cost of cross-correlation, calibration and image processing power will be smaller than 11% of the system budget that should maximally be allocated (Bregman, 2004). The LOFAR paradigm to make use of Moore's law and postpone purchase of processing resources till really needed is no longer valid in the SKA era.

So, the actual SKA design will be based on mature mechanical, electronic and digital processing technology, where the last one will no longer be a bottleneck.

2. Comparison of system concepts for frequencies below 1.4 GHz

There are five key science (Jones, 2004) programs, which require deep imaging of substantial fractions of the sky. Probing the Dark Ages investigates the epoch of reionization by looking for hydrogen line emission in the frequency range 0.1–0.18 GHz. The Evolution of Galaxies and Large Scale Structures program investigates the dark energy by looking for hydrogen line emission in galaxies in the frequency range 0.47–0.71 GHz. Pulsar timing for the Strong Field Test of Gravity key program could be done best in the frequency range 0.7–1.4 GHz and the Magnetic Universe needs frequency coverage from 0.3 to 1.4 GHz. Observations above 1.4 GHz are needed by the Cradle of Life key science program, for Probing the Dark Ages with red-shifted carbon monoxide, and by the Magnetic Universe key program to determine high rotation measures with additional observations up to ~5 GHz. In summary, out of five key science programs, one needs deep imaging of substantial fractions of the sky at frequencies below 0.2 GHz and three programs need the frequency range from 0.3 to 1.4 GHz. This indicates that the SKA should be designed for effective surveying with high sensitivity, especially for frequencies below 1.4 GHz.

The survey sensitivity reached in a given time by an aperture synthesis array depends on the effective aperture, the system temperature and the processed bandwidth. The latter includes the Fov, which is to first order only determined by the effective number of dual polarization receiver chains between the digital cross-correlator and the low noise amplifiers, each of which illuminates a fraction of the total physical aperture (Bregman, 2004). Dishes and cylinders with a FPA need a dual polarization receiver chain for each feed formed by a few focal plane receptors; aperture arrays need such a chain for each beam formed by RF combining a set of many receptors. Since the electronics cost for Fov is first order independent of the system choice, cost comparison of systems with equal Fov should only involve comparing the cost of pointed effective area per receiver chain. This cost has components related to the physical aperture area, to the system temperature and to the steering. The first is mechanical cost, which can be established on technologies known in 2004. The cost for system temperature can be well extrapolated from current transistor and cooling technology. Pointability is a mechanical aspect for dishes and cylinders, and is well known. For aperture arrays, we need RF beam forming electronics, for which the cost extrapolation has considerable uncertainty.

From the required Fov for the surveying applications follows the required number of receiver chains with a digital output, as well as the aperture size for the RF beam former. It has to be realised that the Fov is first order independent of frequency for a dish with FPA, scales linearly with wavelength for a cylinder and quadratically for an aperture array. When we cover a certain frequency range, this will lead to a different effective Fov for a given number of receiver chains.

We will discuss the pros and cons of different antenna concepts each with a feed and an IF system optimized for large-scale surveying at frequencies below 1.4 GHz. We do not compare the actual white paper proposals with their updates, but use their key cost and define hypothetical systems based on the proposed antenna concepts but without the digital cross-correlator system. For proper comparison, we will compare basic elements that all have the same effective collecting area.

2.1. LOW-FREQUENCY LIMITATIONS OF THE SMALL DISH ARRAY

For a 12 m dish with a 1.5 m FPA the maximum aperture efficiency at low frequencies is limited by diffraction scattering at the rim of the dish and reaches at 0.1 GHz a value of 30%, and 55% at 0.175 GHz (Ivashina, 2004). Practical efficiencies will be even lower, leading to a total effective area comparable with the currently planned LOFAR. Sky noise temperatures of 1000 K and 260 K, respectively, dominate the 30 K of the uncooled receivers, so for these low frequencies sensitivity can only be bought by large aperture and not by receiver technology. At 0.3 GHz the sky noise temperature is about 60 K, and 30 K at 0.4 GHz, indicating a clear transition regime from sky noise to receiver noise-dominated system temperature, demarcating where sub-optimum use of expensive high frequency collecting area turns over in a waste of observing time.

We discuss a FPA using Vivaldi elements with an element separation equal to the Nyquist spacing of 0.21 m, which is half a wavelength at a frequency of 0.7 GHz. The lowest frequency with sufficient impedance to match to a LNA is a factor 2.3 lower, i.e. 0.3 GHz. In such an array, we need at least 3×3 elements to form a high-efficiency feed at the highest frequency and 5×5 elements at the lowest frequency. With more elements per feed we could improve the beam quality (Ivashina et al., 2004). An 8×8 FPA without connected corner elements could then provide 32 independent feeds at 0.7 GHz, giving 130 deg^2 effective sky coverage. The aggregate Fov is not determined by the diffraction limit of the 12 m dish, but by the extent of the feed array in the focal plane and the focal distance of 6 m. Almost the same aggregate Fov would be covered at 0.47 GHz or even at 0.35 GHz, however with fewer effective feeds. In all situations we could reach 70% aperture efficiency and 40 K system temperature, typical for uncooled receiver systems. This leads to an effective aperture of 80 m^2 and requires 60 dual polarization receiver chains. A dual polarization receptor element costs $69 and a digital receiver chain with fibre optical output and 0.4 GHz bandwidth could cost $42 in 2010 (Bunton, 2004). The cost of the focal plane system would become of the order $10,000 which then includes 60 digital receivers as well as a digital beam former to create 9–32 output beams for the correlator system. The mechanical structure of the telescope has to be enhanced to handle the additional weight, which we estimate at $5,000. These costs are marginal compared with the $150,000 telescope cost providing 113 m^2 physical aperture area and the $39,000 for the cooled receiver in the secondary focus (USSKA Consortium, 2002). Alternatively, the Indian Preloaded Parabolic Dish concept demonstration program has built one 12 m telescope with a surface accuracy that supports observing to 8 GHz for about $500/m^2 and it is expected that production in large quantities will reduce the cost considerably (Udaya Shankar and Ananthakrishnan, 2004). Making the dishes larger and accepting the reduced surface accuracy will lead to a further cost reduction for a dish that operates up to 2 GHz.

For the frequency range 0.7–1.6 GHz we would need a separate FPA also with 8×8 elements that will then cover 25 deg^2 with 32 beams at the highest frequency and the same aggregate Fov with fewer beams at lower frequencies. The cost for an additional 0.8 GHz bandwidth system would be order $16,000.

The conclusion is that for frequencies between 0.3 and 1.6 GHz two focal plane arrays would be needed. The additional cost of the FPA the receiver chains and the feed former of $88 m^2 meter physical telescope aperture is small compared to the investment in the high-frequency telescope cost of $1717/m^2 including the mechanical enhancement. This results for the low-frequency survey programs into an investment distribution far from the optimum one to one ratio (Bregman, 2004). To avoid leaving the high-frequency surface idle while low frequency surveys occupy the dishes and the correlator, it is attractive to replace a part of the dishes by a larger cylindrical surface, or by an aperture array dedicated to the low frequencies where high sensitivity surveying is imperative. Both systems should have their own correlator, and additional fibres in the trenches to the correlator form only a marginal cost.

2.2. SKY SURVEYING WITH THE CYLINDER SYSTEM

We discuss the situation of Vivaldi or tapered slot line receptor elements in a line feed. We have basically the same situation as with Vivaldi elements in a 2-D crate structure. Diffraction scattering at the reflector rim causes a section of about half a wavelength wide to scatter the incoming radiation around instead of focussing it on the feed, reducing the effective aperture at long wavelengths. However, the effect is less severe than for a circular dish, which is smaller and has more rim per unit area. A feed is typically five Vivaldi elements wide (Ivashina et al., 2004), and needs a RF combiner to provide a proper illumination efficiency that would become about 0.77. Such a five-element feed can provide a proper reflector beam if not located too far from the focal line. This allows for three-feed systems, each with a factor 2.3 frequency range, and we choose the frequency range 0.3–0.7–1.6–3.7 GHz. The highest frequency is centred at the focal line and the other two feeds are located each at a side. This particular choice gives easy comparison with the dishes with FPA that cover the same sub-ranges. Alternatively the high-frequency range could be replaced by a low frequency one covering 0.13–0.3 GHz, which should be placed at a side. The fan beams have over the −60 to +60° scan range an effective foreshortening factor of 0.83, leading to a combined aperture efficiency of 0.64 for surveying.

A cylinder with 15 m diameter then needs a length of 8.3 m to give 80 m^2 effective area. With Nyquist spacing of 0.21 m we need 40 feeds each with their RF combiner followed by a digital receiver. The digital beam former only makes 32 beams that cover the full 120° scan range at 0.7 GHz and needs proportionally less beams at lower frequency. The cylindrical reflector gives a beam width in the elevation direction of 2.2°, leading to a total FoV of 260 deg^2. The cost for such a piece of feed with digital receivers of 0.4 GHz bandwidth is about \$7,000 including the digital beam former, which leads to \$56/m^2 physical telescope aperture. A feed system for the 0.7–1.6 GHz range provides the same FoV at 0.7 GHz but needs 2.3 times as many elements, produces 102 output beams at 1.6 GHz and with 0.9 GHz bandwidth the cost is estimated at ∼\$200/m^2 physical telescope aperture.

According to Bunton (2004), the cost of a 15 m cylindrical antenna with 1.6 GHz maximum frequency is \$153/m^2 for the cylinder telescope without line feeds. Comparing this cost with the cost of the 0.7–1.6 GHz feed system shows that the FoV component exceeds the 50% for an optimum cost distribution, indicating that a too large FoV is provided with too little collecting area. Given the total budget a larger fraction should be spent on collecting area providing increased survey sensitivity.

Compared to the dish solution we need the same amount of beam signals per unit collecting area with the cylinder solution but we get the double effective

Fov. However, we require three times the amount of Vivaldi elements that need to be combined at RF level with a frequency-dependent weight feeding a digital receiver chain. In view of the large relative bandwidth, it might be contemplated to go for a digital solution that will require then 200 digital receiver chains.

An advantage of the cylinder is its increase in Fov at lower frequencies proportional with wavelength, due to the increased beam width in elevation, while for the small dishes the Fov is determined by the FPA size and independent of frequency.

A disadvantage of the cylinder system is that we cannot track the sky for all beams simultaneously since they have a common elevation, which makes surveying of less than the full sky less efficient.

2.3. ALL SKY IMAGING WITH AN APERTURE ARRAY

A much more attractive scheme for the 0.1–0.3 GHz range is an aperture array based on vertical bowtie type elements. Using a Nyquist spacing of half a wavelength at 0.17 GHz we are in the dense array mode for lower frequencies with strong mutual coupling between the elements. This results effectively in a receptor with constant effective area and an antenna pattern with cosine zenith angle shape. For frequencies above 0.2 GHz the array is sparse, which means that we get more a cosine squared zenith angle shaped antenna pattern for each element with less coupling and an effective area given by $\sim \lambda^2/3$. At 0.25 GHz the effective area is then reduced to 64%. Due to impedance mismatching the effective receiver noise temperature increases at the lower frequencies, but remains below the fast increasing sky noise temperature. Since the system temperature is in the considered frequency range dominated by the sky noise, the noise power at the output of an array beam is reduced by the effective element beam pattern. The same applies to the signal of a point source at the peak of the array beam when pointed at low elevation, resulting in an elevation independent signal-to-noise ratio provided that the array main beam efficiency is close to unity, which is the case for an appropriately tapered dense array (Bregman, 2000).

As an example we take a large array of abutted 2×2 tiles, which have a simple reflector plate of 1.8 m $\times$ 1.8 m with four dual polarization elements and a RF delay beam former. Ignoring mutual coupling we find a Fov of about $0.9/\cos(\theta)$ steradian at 170 MHz, which is proportional with wavelength squared. Based on the LOFAR costing we estimate a price of \$60 per dual polarization receptor pair including beam former, and including reflector we find \$90/m^2. Bunton (2004) estimates \$60 for a dual polarization digital receiver chain with 0.2 GHz bandwidth and more than 8 bits by 2010, which adds \$19/m^2.

A big advantage of any aperture array is that we can make stations of arbitrary size and shape to match observing system requirements of the array. Whatever we do, there is always complete coverage of the 0.9-sr tile beam by the set of station beams that can be produced by the number of tiles in a station. However, when we fill the tile beam up to half power width with digital station beams, their sensitivity is on average decreased to 75%.

For our 80 m² effective reference aperture we need 33 tiles and form 33 beams to fill the Fov. A seven-steradian sky survey could be completed in a day and a half, and reach the thermal noise sensitivity after 5-h integration per field. Here we assumed an array configuration that provides full U,V-coverage in about 5 h of tracking (Bregman, 2004).

Above 0.26 GHz we need antenna elements that give good matching to low noise receivers. Vivaldi elements separated at Nyquist spacing of $\lambda_N/2$ at frequency f_N have a fairly constant impedance in the dense regime from 0.43 f_N to f_N with a $\cos(\theta)$ pattern and area $\lambda_N^2/4$. This $\cos(\theta)$ pattern leads for the receiver noise dominated regime to an average 75% foreshortening efficiency for elevations larger than 30°. Up to 1.15 f_N the area remains constant but the element pattern gradually changes to a $\sim\cos^2(\theta)$ pattern and stays so in the sparse regime above 1.15 f_N with and effective area $\sim\lambda^2/3$. For the focal plane array we need to stay in the dense regime to get full field sampling, limiting the frequency range to a factor 2.3 instead of 2.7 for constant aperture use.

An array with spatial Nyquist sampling at 0.6 GHz has a constant effective area from 0.26 to 0.7 GHz and with element separation of 0.25 m we get 6 × 6 tiles of 1.5m × 1.5m. van Ardenne and Lazio (2004) estimate for such a tile \$140/m² using phase control and 0.2 GHz bandwidth by chip receivers, we adopt \$200/m² for tiles with delay control covering about 0.4 GHz bandwidth. Each tile needs a dual polarization digital receiver chain, which adds \$20/m² plus \$10/m² for the beam former. Ignoring mutual coupling for the abutted array of tiles we calculate a nominal Fov by the tile beam of about 415/cos(θ) deg² at 0.7 GHz and proportional with wavelength squared. For comparison with the cylinder system we fill only 260/cos(θ) with digital station beams or 63%. Then the average sensitivity of the digital station beams is reduced by a factor 0.84 leading to an average efficiency of 0.63. For the standard 80 m² effective area we then need 56 tiles and receiver chains and need only 35 beams to be formed. A seven-steradian sky survey at full sensitivity (of 5 h integration per field) could be completed in about 2 weeks, taking into account the increased Fov at low elevation.

A constant aperture array for the frequency range 0.53–1.4 GHz will need an element separation of 0.125 m and we estimate the cost of a 12 × 12 element tile with delay control at \$400/m². To provide at 0.7 GHz the same Fov we need four digital receiver chains per tile, which at 0.9 GHz bandwidth cost \$180/m² plus \$90/m² for the beam former. In principle, we could observe at 1.6 GHz with an additional sensitivity decrease of 77% but require a larger output bandwidth.

Given the fast electronic pointing flexibility we could even complete a hemispheric snapshot with 35 fields in 1 s and get 7-sr sky coverage in 24 h. It should be realized that the correlator output data rate is then also increased by a factor 35 and that standard calibration of this data within 24 h will require processing resources that are an order of magnitude costlier than the correlator.

3. Optimum SKA hybrids

Let us assume a total SKA antenna budget of $1.1 billion and see what would be the best distribution over available sub-arrays that are optimized for their specific frequency range.

As mentioned in the section on comparison of system concepts, there are five key science programs all needing extensive surveys. A sixth program, Exploration of the Unknown, was introduced at the SKA2004 workshop, and needs observations in all frequency bands. In summary, out of six science programs, two use frequencies below 0.2 GHz, four use the range from 0.3 till 1.4 GHz and four use the higher frequencies. This justifies an allocation in observing time roughly proportional in the same ratio.

For the lowest frequencies only a separate aperture array in LOFAR technology can provide adequate sensitivity with a 1 km^2 array for ~$100 million.

A Small Dish array of 0.5 km^2 could be equipped with two focal plane arrays covering the 0.3–1.6 GHz range with appropriate sensitivity and FoV. Then the high surface accuracy parabolas costing $1 billion are used less than half the time for their intended high frequency range.

However, if we spend only $0.7 billion on 0.35 km^2 of a small dish system with an optimized secondary focus system, then the high frequency programs could get twice as much observing time and obtain the same sensitivity as with the factor 1.4 larger array that is only available half the time.

For the remaining $0.3 billion a separate survey array for the 0.3–1.6 GHz could be built in three different technologies, that which all three offering more than ~1 km^2 collecting area. Since this array is used full time, the effective sensitivity is about three times that which would be obtained with the small dishes in half the time.

We could argue that four programs that need the low frequencies deserve at least the same amount of antenna money allocated to them as the four programs that need high frequencies. Then we could afford ~1.6 km^2 low-frequency array but have only 0.25 km^2 for the high-frequencies.

We have shown that a separate array covering the 0.3 to 1.6 GHz range is the most adequate solution for large-scale surveying around 0.7 GHz. Such an array could be built with aperture array stations, large cylinders or even parabolas larger than 12 m in pre-loaded dish technology.

4. Conclusions

For all practical types of aperture synthesis arrays the digital correlator cost is a minor issue after 2015. Digital receiver chains will then provide sufficient FoV and bandwidth for large-scale surveying below 1.4 GHz and are no cost driver either.

We have shown that for a given effective collecting area multi-beam dishes, cylinders and aperture arrays need equal amounts of beam inputs into the correlator and then provide, to first order, the same FoV. The FoV of a dish is determined by the size of the FPA, while in a cylindrical array the FoV is the product of effective scan range in azimuth and the diffraction limit of the reflector in elevation. In practice the FoV of a cylinder could be up to a factor of two larger than of a dish, but depends on how actual designs handle the frequency range, the beam forming and the edge issues.

Optimum system designs for survey applications could spend up to half the budget on multi-beam receiver chains, digital signal and subsequent data processing, provided that all these elements are based on cost-effective state-of-the-art technology. Since the electronics cost for a given FoV is to first order independent of the particular system choice, be it parabolic dishes or cylinders with focal plane arrays or multi-beam aperture arrays, cheap primary collecting aperture is key in systems for large-scale surveying.

In the following section we summarize our systems comparison where all systems have the same effective aperture of $80\,m^2$ and $0.4\,GHz$ bandwidth, where the cost of equal functionality is referred to a single reference and where different functionality such as collecting aperture has its cost referred to the appropriate sections of the various white papers. All cost figures are expressed per square meter of physical primary aperture.

A 12 m diameter dish with a FPA could have 70% aperture efficiency.

- Sixty Vivaldi elements with digital receiver chains provide only 32 beams.
- FoV of 130 deg^2 almost independent of frequency for the 0.3–0.7 GHz range.
- The FPA with beam former adds $\$88/m^2$ to the telescope cost of $\$1717/m^2$.
- A second 0.9 GHz FPA for 0.7–1.6 GHz would add $\$142/m^2$ for 25 deg^2 FoV.
- Low frequency prime focus use seems however a waste of high frequency dish surface.
- A preloaded dish larger than 12 m and operating up to 2 GHz would cost considerable less than $\$500/m^2$.

A 15 m cylinder with a line feed could have 64% average aperture efficiency.

- Forty feed segments each five Vivaldi elements wide with RF beam forming.
- Forty digital receiver chains provide 14–32 beams for 0.3–0.7 GHz.
- FoV of 260 deg^2 at 0.7 GHz, proportionally with wavelength.
- The FPA with beam former adds $\$56/m^2$ to the cylinder cost of $\$153/m^2$.
- A second 0.9 GHz FPA for 0.7–1.6 GHz would add $\$200/m^2$ for the same FoV.
- Elongated FoV is only practical for very large surveys

An aperture array could have 63% average aperture efficiency.

- Fifty six tiles of 1.5m × 1.5m with 6 × 6 Vivaldi elements spaced at 0.25 m.
- Fifty six digital receiver chains provide 35 beams for 0.26–0.7 GHz, which cover
- FoV of 260 deg^2 at 0.7 GHz, proportionally with wavelength squared.

- The digital receivers with beam former contribute $30/m^2$ to the tile cost of $200/m^2$.
- A 12×12 tile with 0.125 m separation for 0.53–1.4 GHz would cost $400/m^2$ and needs four receiver chains, that including the beam former for 0.9 GHz bandwidth, would add $270/m^2$ for the same FoV. At 1.6 GHz the sensitivity is reduced by an additional factor 0.77.
- In fast pointing mode a 7-sr sky survey could be completed in 24 h.

A sky noise limited aperture array could have 75% average aperture efficiency.
- Thirty three tiles of 1.8 m × 1.8 m with 2 × 2 vertical bowtie elements spaced at 0.9 m.
- Thirty three digital receiver chains provide 33 beams for 0.1–0.3 GHz.
- FoV of $2900/\cos(\theta)$ deg^2 at 0.17 GHz, proportionally with wavelength squared.
- Above 0.2 GHz the effective aperture is proportional with wavelength squared.
- A 7-sr sky survey could be completed in a day and a half with 5 h integration per field.

A hybrid SKA using three different sub-arrays for the three frequency ranges 0.1–0.3 GHz, 0.3–1.6 GHz and above 1.6 GHz, each with their own correlator and processing pipeline, would give the highest sensitivity for all key science programs in a given total budget.

References

van Ardenne, A. and Lazio, J.: 2004, 'SKA Hybrid, Three Aperture Arrays+Small Dishes', Hybrid White Papers, Eng. Section on http://www.skatelescope.org.

Bunton, J.D.: 2004, 'A Cylinder plus 12-m Dish Hybrid', November 2003, Section in SKA Memo 48, Contributions & Summary of First SKA Convergence Workshop Held in Cape Town, Eng. Section on http://www.skatelescope.org.

Bregman, J.D.: 2000, 'Concept Design For a Low Frequency Array', *SPIE Conf. Proc.* 4015.

Bregman, J.D.: 2004, 'Survey Performance of Multi-Beam Aperture Synthesis Arrays', in *Exp. Astrono*, this issue.

D'Addario, L.R.: 2002, 'Digital Processing for SKA', SKA Memo 25.

Ivashina, M.V.: 2004, 'Efficiency Limitations of a 12 meter Dish System', ASTRON Report-SKA-00916.

Ivashina, M.V., Simons, J. and de bij Vaate, J.G.: 2004, 'Efficiency Analysis of Focal Plane Arrays in Deep Dishes', *Exp. Astrono.*, this issue.

Jones, D.L.: 2004, 'SKA Science Requirements: Version 2', SKA Memo 45.

Udaya Shankar, N. and Ananthakrishnan, S.: 2004, 'Present Indian Efforts Relating to SKA Demonstrator', SKA demonstrator plans, Eng. Section on http://www.skatelescope.org.

USSKA Consortium: 2002, 'The Square Kilometer Array Preliminary Strawman Design LargeN – Small-D', White Paper 2, SKA Design Concept White Papers 1–7, Eng. Section on http://www.skatelescope.org.

Walden, R.H.: 1999, 'Analog-to-Digital Converter Survey and Analysis', *IEEE J. Selected Areas Commun.* **17**(4).

CYLINDER – SMALL DISH HYBRID FOR THE SKA

JOHN D. BUNTON[1,*] and T. JOSEPH W. LAZIO[2]

[1]*CSIRO ICT Centre, PO Box 76 Epping NSW 1710 Australia;* [2]*Naval Research Laboratory, Washington, U.S.A.*
(*author for correspondence, e-mail: john.bunton@csiro.au)

(Received 20 August 2004; accepted 30 December 2004)

Abstract. No single technology can meet the current science specifications for the Square Kilometre Array (SKA). In this paper a hybrid option is explored that uses cylindrical reflectors to satisfy low frequency sensitivity and field-of-view requirements and small parabolic dishes to work at frequencies above 0.5 GHz.

Keywords: cylindrical reflector, hybrid, parabolic reflector, SKA

Introduction

None of the current design concepts for the Square Kilometre Array (SKA) can meet all the science requirements in terms of frequency coverage, sensitivity at high and low frequencies, and field-of-view. In this paper a hybrid consisting of cylindrical reflectors, to cover the lower frequencies, and small parabolic reflectors, to cover the higher frequencies, is considered. This hybrid maintains full frequency coverage and meets field-of-view and low-frequency sensitivity requirements. It does so at the expense of sensitivity at high frequencies.

1. Selection of hybrid components

One of the major challenges posed by the SKA Science Requirements (Jones, 2004) is the sensitivity requirements at high and low frequencies. At high frequencies fully steerable parabolic dishes have been the technology of choice for all existing radiotelescopes. For the SKA a small dish about 12 m in diameter represents an optimum between the costs of the reflector itself and the associated electronics (USSKA Consortium, 2002, 2003). Maximum performance is achieved if cooled receivers are used. In order to meet the SKA sensitivity specifications above approximately 1 GHz, a total effective area of 0.36 km^2 is needed, equivalent to 4500 dishes.

At low frequencies where sky noise dominates receiver noise larger effective areas (about 1 km^2) are needed to meet the science requirements. At 0.5 GHz this could be reduced to 0.7 km^2, if cooled receivers were used, but remains 1 km^2 for uncooled systems. Thus, additional collecting area is needed at low frequencies. Moreover, an additional SKA science requirement is a high survey speed specified at 700 MHz, but in practice this requirement may extend to frequencies as high as 1 GHz. For the ~100 deg^2 field-of-view needed to meet the survey speed specification, at least one million feed elements (Bunton, 2000) are needed, each with a separate receiver. The number of feeds will probably dictate the use of uncooled receivers.

A low cost solution for providing the large collecting area and feed numbers is a cylindrical reflector. The cylindrical reflector has only one axis of rotation reducing the cost of the drive system. The antenna mounts are also cheap as they can be short, they are distributed along the length of cylinder, and they only experience torque around an axis aligned along the cylinder. The cylinder also provides a solution to the minimisation of feeds numbers for a field-of-view (FoV) that matches the SKA science requirements (Bunton and Hay, 2005). The resulting FoV is highly elongated, which does not impair the cylindrical reflector's survey capabilities but limits its ability to target multiple long track observation. The converse of this is that the area of sky accessible for short observations is significantly increased.

2. Hybrid components

The small dish for this proposal is based on the SKA Design Concept white papers 2 and 9 (USSKA Consortium, 2002, 2003). For the hybrid, the antenna does not have to operate at low frequencies. This allows a cheaper symmetric design to be used with a comparatively small subreflector as shown in Figure 1. The reflector is 12 m in diameter with a 1.6 m Gregorian subreflector allowing the dish to operate down to 0.47 GHz. With a shaped surface the aperture efficiency is 72%. Full efficiency is available up to 25 GHz. Above this frequency it is proposed to under illuminate the dish to avoid pointing problems. This allows the small dish to operate at 36 GHz with an effective aperture that is half its 25 GHz value.

The cylindrical reflector is an offset feed design 110 m × 15 m illustrated in Figure 1 (Bunton et al., 2002; Bunton, 2003). At low frequencies it will be a light

Figure 1. Illustrations of the proposed cylindrical reflector and 12-m symmetric parabolic dish.

weight structure with a mesh reflecting surface. In order to reduce end effects, the linefeed at the focus of the cilinder is 100 m long and contains 1000 feed elements at 1.4 GHz. It is proposed that all feed elements will be digitised with data being fed to a digital beamformer over optical fibre. At higher frequencies RF analogue beamforming will be used to reduce the total data rate. As with the small dishes the cylinders will operate at half efficiency at a frequency 40% above their nominal maximum. This reduction is due to limitations in the performance of the focal plane array, or line feed, at the focus of the reflector.

2.1. THE HYBRID MIX

The target cost for the SKA is one billion dollars (US), but all current proposals exceed this target. In this paper the total antenna related costs are constrained to be less than one billion dollars. If approximately $ 490M is allocated to a high-frequency sub-array consisting of 12-m small dishes, then 2500 small dishes with 15 K cryogenic receivers can be built (USSKA consortium, 2003). This high-frequency sub-array would provide a sensitivity of 11,000 m^2/K at 5 GHz, or 55% of the SKA specification.

To just satisfy the low-frequency sensitivity requirement, a low-frequency sub-array with a minimum collecting area of 1.5 km^2 is needed. This can be satisfied by 900 cylindrical reflectors operating up to 0.5 GHz. A second stringent requirement is the SKA survey speed specification at 0.7 GHz (Jones, 2004). Increasing the maximum frequency for this low-frequency sub-array to this frequency reduces the requirements of field-of-view and bandwidth. The antenna related cost for a low-frequency sub-array consisting of cylindrical reflectors is estimated to be $200M (Bunton, 2003). Within the one billion dollar constraint, this leaves $300M available to increase the sensitivity for one of these sub-arrays. Options include increasing the number of small dishes or increasing the maximum frequency of the cylinders.

Four of the five SKA Key Science Projects have frequency requirements in the range 0.1–2 GHz: The Evolution of Galaxies and Cosmic Large Scale Structure, Probing the Dark Ages, Strong Field Tests of Gravity Using Pulsars and Black Holes, and The Origin and Evolution of Cosmic Magnetism (Gaensler, 2004). Much of this work involves surveys which greatly benefit from the wide field-of-view of cylindrical reflectors. This suggests that cylindrical reflectors with full SKA sensitivity would be useful up to 2 GHz. Searching for the Epoch of Reionization signature, part of the Probing the Dark Ages Key Project, would also benefit from extra collecting area. One possible means of obtaining this additional collecting area is to use 600 high-frequency (HF) cylindrical reflectors covering 0.1 to 2 GHz, augmented with 600 low-frequency (LF) cylindrical reflectors covering 0.1 to 0.7 GHz. The HF cylinders are estimated to cost $226M and the LF ones $140M.

In this second option the 1200 cylinders are cheaper than the small dishes sub-array. At additional cost (roughly equal to that of the small dishes), the maximum frequency of the HF cylinder could be increased to 5 GHz. This would provide additional sensitivity for portions of the key science projects involving the Strong Field Tests of Gravity Using Pulsars and Black Holes, The Origin and Evolution of Cosmic Magnetism, and the Cradle of Life. However there is the unresolved question of the polarisation performance of cylindrical reflectors. Theoretical work on this problem has been promising but as yet there has not been a practical demonstration. If the polarisation performance were poor this would limit the usefulness of cylindrical reflectors for pulsar timing and studies of magnetism, leading to this science being done by small dishes.

In this case a better option for improving high frequency sensitivity is to use the minimum of cylindrical reflectors to meet low-frequency sensitivity and survey requirements and build more small dishes. For $789M, 4000 small dishes can be built increasing their sensitivity to 88% of the full SKA specifications. The sensitivity of 2500 and 4000 small dishes, the three cylinder options, and the SKA specifications are shown in Figure 2.

The 0.7 GHz cylinders meet the SKA specification between 0.2 and 0.9 GHz. Coupled with 4000 small dishes (88% of a full SKA), the proposed hybrid nearly matches the SKA specifications. The option of having the cylinders operate to 2 GHz meets SKA specification up to 3 GHz if the cylinders and 2500 small dishes (55% of a full SKA) are operated together, and the option of having the cylinders operate to 5 GHz can extend full SKA functionality to 7 GHz. The costs of all components in the 2 GHz option are given by Bunton and Lazio (2004). The 0.7 GHz option with 2500 small dishes is $156 M cheaper and the 5 GHz option $113M more expensive. These costs are summarised in the Table I. The 4000 small dish plus 0.7 GHz option is similar in cost to the 5 GHz cylindrical reflectors and 2500 small dishes.

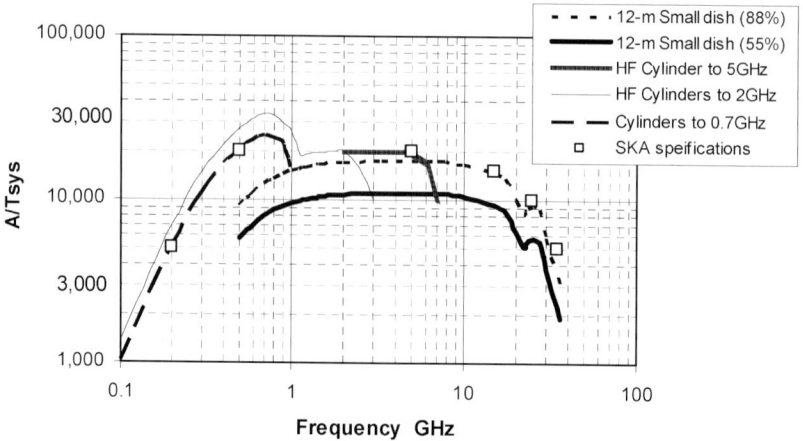

Figure 2. Sensitivity of the various components of a cylinder – small dish SKA hybrid.

TABLE I
Cost of SKA hybrid options (millions of dollars)

Cylinder Maximum frequency GHz/No. small dishes	Small dishes	LF cylinders	HF cylinders	Infrastructure, Signal processing, Computing & NRE	Total cost $ M
0.7, 4000	789	210	–	414	1412
0.7, 2500	493	210	–	414	1117
2, 2500	493	140	226	414	1273
5, 2500	493	140	339	414	1386

It is seen that there are a range of performance and cost options for the hybrid. In the following only the 2 GHz option is explored.

3. Observing with multiple components

The signal from the line feed of the cylindrical reflector can be configured in a number of ways. For operation in conjunction with the small dishes it is proposed that the number of feed elements beamformed be dynamically adjusted so that the beam generated matches that of the small dishes. The field-of-view for the two components would then be approximately the same. Processing the combined data should be no more difficult than existing arrays with slightly mismatched antennas such as the DRAO (Willis, 1999). The high-frequency cylindrical reflectors are broken up into 5000 antennas sections in this mode which together with the 2500 small dishes increases the number of baselines by a factor of 9 compared to small dishes alone. However, the maximum frequency of operation is 2 GHz and the bandwidth needed is 0.5 GHz or one eighth of the 4 GHz needed above 16 GHz. It is seen that the small dish correlator alone has sufficient compute capacity to form the extra baselines. The instantaneous sensitivity achieved with the two components operating together is 30,000 m^2/K.

Below 700 MHz, the feeds and electronics of the LF and HF cylindrical reflectors are identical. Operating these together doubles the sensitivity.

The components of the hybrid can also be operated separately giving in effect two or three independent radiotelescopes, increasing the total available observing time. With the three components operating independently the sensitivity for non-transient observing is increased by a factor of 1.7. For example, the effective sensitivity of the 2500 small dishes would be equal to 95% of full SKA specification. As the LF cylindrical reflectors and small dishes will usually operate in mutually exclusive frequency bands the effective sensitivity of the small dishes alone is at least 77% of the full SKA specification. The HF cylinders could operate alone or co-operatively with either the LF cylinders or, the small dishes. An example of stand alone operation is surveying at frequencies above ~1 GHz. The three components could also be used for multi-frequency observing; for example simultaneous observing at 0.25, 0.5, 1, 2, 4, and 8 GHz.

4. Conclusion

Small dishes are a proven method of meeting the SKA specification at high frequencies. However, they have difficulty meeting the sensitivity and field-of-view requirements at low frequencies. Cylindrical reflectors can meet the latter requirement at low cost while still allowing a significant fraction of the small dish array to be built. If a minimum 0.7 GHz cylindrical reflector array was built, then 88% of the small dish array could be built. If half the small dishes are built, then the added cylinders could have enhanced sensitivity below 1 GHz and allow full sensitivity up to 7 GHz. Scaling back to the maximum frequency of the cylinders reduces cost by 10% with full SKA sensitivity maintained up to 3 GHz.

Acknowledgement

Basic research in radio astronomy at the NRL is supported by the Office of Naval Research.

References

Bunton, J. D.: 2000, 'SKA antenna Selection, Economics and the Field of View', *SKA workshop, Technology Pathways to the Square Kilometre Array*, 3rd–5th August 2000, Jodrell Bank Observatory, UK, http://www.jb.man.ac.uk/ska/workshop/

Bunton, J. D., Jackson, C. A. and Sadler, E. M.: 2002, 'Cylindrical Reflector SKA', *SKA Design Concept White Paper*, 7, July 2002 http://www.skatelescope.org/documents/SKA_AUS_CONCEPT_Cylindrical_17072002.pdf

Bunton, J. D. (ed.): 2003, 'Panorama of the Universe A Cylindrical Reflector SKA', Ed. *SKA Design Concept White Paper*, 13, May 2003 http://www.skatelescope.org/documents/dcwp/Cylindrical_EMT_reply_4.pdf

Bunton, J. D. and Lazio, J. W.: 2004, 'CR-small dishes, SKA Hybrid design concept', 2004, http://www. skatelescope.org/PDF/CR_LNSD.pdf

Bunton, J. D. and Hay, S. G.: 2005, *Exp. Astron.* this issue.

Jones, D. L.: 2004, 'SKA Science Requirements', *SKA memo 45*, Feb 2004, http://www.skatelescope.org/pages/page_skares.htm

Gaensler, B.: 2004, 'Key Science Projects for the SKA', *SKA Memo 44*, Feb 2004, http://www.skatelescope.org/pages/page_skares.htm

USSKA Consortium.: 2002, 'The Square Kilometer Array, Preliminary Strawman Design, Large N – Small D', *SKA Design Concept White Paper 2*, July 2002, http://www.skatelescope.org/documents/SKA_USA_CONCEPT_Large-N_Small-D_17072002.pdf

USSKA Consortium.: 2003, 'The Large-N-Small-D Concept for the SKA: Addendum to the 2002 Whitepaper', *SKA Design Concept White Paper 2*, May 2003, http://www.skatelescope.org/documents/dcwp/small_parabolas_May03.pdf

Willis, A. G.: 1999, *A& AS* **136**, 603–614.

RFI TEST OBSERVATIONS AT A CANDIDATE SKA SITE IN CHINA

B. PENG[1,*], J. M. SUN[2], H. Y. ZHANG[1], T. Y. PIAO[1], J. Q. LI[2], L. LEI[2], T. LUO[3], D. H. LI[3], Y. J. ZHENG[1] and R. NAN[1]

[1]*National Astronomical Observatories, Chinese Academy of Sciences, Beijing 100012, P.R. China;*
[2]*Radio Monitoring Station of Guizhou, Guizhou 550001, P.R. China;* [3]*Radio Monitoring Station of Qiannan, Guizhou 558000, P.R. China;*
(*author for correspondence, e-mail: pb@bao.ac.cn)

(Received 15 July 2004; accepted 31 December 2004)

Abstract. Radio frequency interference (RFI) test observations were carried out at one of the candidate Square Kilometre Array (SKA) sites in Guizhou province, following the "RFI Measurement Protocol for Candidate SKA Sites" (hereafter RFI protocol). All data (raw and calibrated) are preserved in some suitable format, such as that set by the international RFI working group of the Site Evaluation and Selection Committee (SESC). An RFI test in December 2003 was performed according to Mode 1 of the RFI Protocol, in order to identify technical difficulties which might arise during a co-ordinated RFI measurement campaign over a period of 1 year. In this paper we describe the current equipment, observational technique and data presentation. The preliminary results demonstrate that the RFI situation at Dawodang depression in Guizhou province makes it quite a promising location for the proposed SKA. Furthermore, the first session of the RFI monitoring program, which was made in May 2004, showed that a complete RFI measurement including both modes 1 and 2 of the RFI Protocol would take about 2 weeks. The possible ways to minimize some limitations of the current equipment are also discussed, which will enable us to meet the RFI protocol.

Keywords: antenna, measurement, radio frequency interference, radio spectrum, telescope site

1. Introduction

In 1993, a large radio telescope with a collecting area of about 1 km^2 (namely Square Kilometre Array, SKA), was proposed by astronomers from 10 countries at the 24th General Assembly of URSI. There are various concepts to realize the SKA (see details at www.skatelescope.org), including the one Chinese astronomers propose to build a set of large (Arecibo-style) spherical reflectors by making use of the extensively existing karst landforms, which are bowl-shaped limestone sinkholes. The Chinese SKA, KARST (Kilometer-square Area Radio Synthesis Telescope), consists of about 30 individual elements, each roughly 200 m in diameter (Peng and Nan, 1997).

There are many factors affecting the choice of site for the construction of a radio telescope, and the ambient RFI (radio frequency interference) environment is one that is particularly critical. For siting the SKA, we have carried out a series of RFI measurements since 1994 at some potential locations in Guizhou province in southwestern China, and an additional one at the Urumqi astronomical observatory in the Xinjiang autonomous region of northwestern China. These RFI monitoring sessions consisted of, firstly, measurements at eight karst depressions for 10 days in November 1994, and, secondly, additional ones at some of the same locations for comparison in March 1995. These included monitoring on one site in Puding county for a month to obtain more complete statistics, and additional sites along the direction between Puding and Pingtang counties in an attempt to understand distance effects. The results of these measurements provide information about the frequency, strength and characteristics of the radio interference signals in Guizhou province (Peng et al., 2001).

In November 2003, an RFI survey group was established within the Radio Monitoring Station of Guizhou (RMSG), in order to conduct an RFI measurement campaign for a period of at least 1 year in Guizhou according to the RFI protocol (Ambrosini et al., 2003). It was not intended to push to the ITU radio astronomy thresholds, but to meet the RFI protocol for the initial screening of SKA sites. In this paper we report on a new RFI testing observation made in December 2003, and the first session of the RFI monitoring program conducted in May 2004. Current equipment, the observational technique and data presentation are described briefly. Some discussion and concluding remarks are made about meeting the RFI protocol.

2. Method and measurements

The observational purpose is to obtain data on background levels at potential SKA site, focusing on transmissions from horizontal directions. The RFI protocol divides RFI observations into two modes: a survey of strong RFI potentially causing saturation and thus linearity, which is referred to as a mode 1 measurement, and that of weak RFI which obscures signals of interest, which is designated mode 2 (Ambrosini et al., 2003). An instrument setup to execute the campaign is presented schematically in Figure 1. It mainly consists of an antenna, low-noise amplifier (LNA), spectrum analyzer, antenna controller and control computer.

Two different antennas are used for the broad frequency range from 70 MHz to 22 GHz by the RFI survey team of the RMSG. A log periodic antenna, HL023A2, covers the frequency range 70–1300 MHz with mean gain (over frequency) 6.5 dBi and 65° (at 3 dB power) beam width (BW) in both horizontal and vertical polarizations. A parabolic antenna of 0.8 m in diameter covers the frequency band from 800 MHz up to 30 GHz with mean gain 13.7 dBi and 33.9° BW at 800 MHz, and 43.2 dBi gain and 1.2° BW at 22 GHz in both polarizations. A spectrum analyzer, FSP 30, made by the Rhode & Schwarz Company, is employed for the spectrum 9 kHz–30 GHz. The noise figure of the spectrum analyser is between about 24 dB at 70 MHz and 32 dB at 22 GHz, and that of the cable attenuation between antenna and LNA is 0.2 dB. The low-noise amplifiers (LNAs) have mean gains of 25 dB and noise figures (NF) of 3.0 dB in the frequency band 0.07–1 GHz, a mean gain

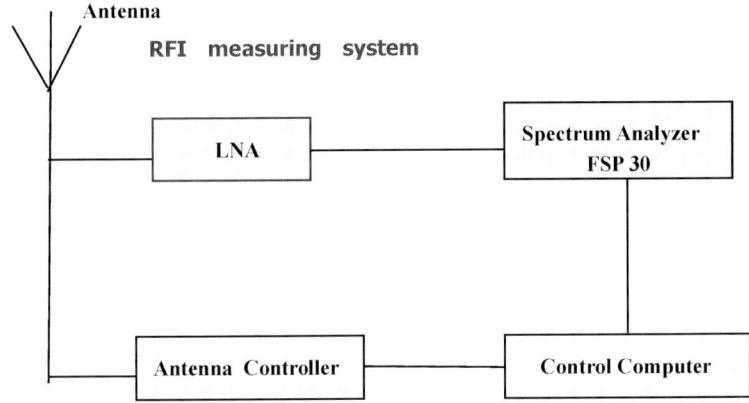

Figure 1. Schematic configuration of the RFI measurement system set up at the Guizhou sites in China.

of 40 dB and a noise figure of 2.5 dB in the frequency bands 1–4 GHz, 4–9 GHz, 9–14 GHz, 14–18 GHz, and 18–30 GHz, which agree with the specifications of the LNA manufacturer and were confirmed before our observations by using a signal generator, but only below 3 GHz due to limited resources. We will check all these values by making actual measurements using calibration metrology before the next monitoring session in August 2004 or later, although field measurements cannot be made daily due to lack of suitable noise sources in lieu of the antenna. The statements above are outlined in Table I. The measurements have limited accuracy due to errors such as the lack of appropriate calibration, multi-path interference from the ground and surrounding terrain, unknown antenna gain in the direction(s) of arrival, and changes to the antenna gain due to interaction with the mount structures.

During December 16–23 in 2003, an RFI test observation was performed about 24 h a day at one of the best candidate SKA sites (from the point of view of geometric

TABLE I

Location of RFI instruments for test observations made at Dawodang in December 2003

Measurement period	16–23 December, 2003
Location	Longitude: 106°51′20.20″E;
	Latitude: 25°38′59.55″N;
	Altitude: 1002.3-m above the sea level
Antennas	Below 1 GHz: HL023A2 log periodic antenna
	Above 1 GHz: Parabolic antenna (Diameter = 0.8 m)
Antenna height	4-m above the ground
Low noise amplifier (LNA)	0.07–1 GHz: gain 25 dB, NF 3.0 dB;
	1–30 GHz: gain 40 dB, NF 2.5 dB
Spectrum analyzer	FSP30

shape for the KARST concept), Dawodang in Guizhou. It was made only in mode 1 of the RFI protocol to identify any technical difficulties, and to prepare for a long-term campaign. The monitoring system was located at a longitude of 106°51′20.20″ east, a latitude of 25°38′59.55″ north, and a height of 1002.3 m above sea level. The measurements were realized with antennas at a height of 4 m above the ground. An effort was made to ensure that the coverage of frequencies with respect to time-of-day was as uniform as possible for multiple cycles during our observations. It should be noted that the spectrum analyzer was set up with a zero span mode in the time domain at L band. The parameters were set with a RBW of 1 MHz and a sweep time of 2 s, to meet the suggested settings of the RFI protocol, RBW = 1 MHz, dwell = 0.002 ms, Reps. = 10^6. This is needed to detect radar bursts e.g. from airplanes. This gives 440 pointings for the frequency band 960–1400 MHz. Where RBW is defined by the RFI protocol as "resolution bandwidth", and dwell is the length of time that a channel (one slice of the spectrum having width equal to the specified RBW) is examined. Reps. is the number of times the experiment should be repeated per iteration of the measurement cycle (Ambrosini et al., 2003).

It transpired that the criteria in SESC's RFI Protocol can be mostly met with current equipment from the RMSG, suggesting that a session of RFI measurements including both modes 1 and 2 could be achieved in about 2 weeks (about 24 h a day). The main instrumental settings and results are briefly summarized in Table II, where columns 1–6 are the observing frequency bands in MHz, the bandwidth in

TABLE II

Mode 1 measurement: RFI settings and brief results of test observations made at Dawodang in December 2003

Frequency band (MHz)	RBW (kHz)	S_{0h} (dB Wm^{-2} Hz^{-1})	Dwell (ms)	Reps.	Antenna mean Gain (dBi)	Min., Max. Signal (dB Wm^{-2} Hz^{-1})
70–150	3	−201.67	10	5	6.5	−182.84, −149.94
150–300	3	−200.04	10	1	6.5	−186.24, −159.84
300–800	30	−194.02	10	1	6.5	−196.04, −168.54
800–960	30	−185.50	10	20	6.5	−196.24, −174.74
960–1000	1000	−183.75	0.002	10^6	6.5 or 17.4	−184.77, −176.57
1000–1400					(>1325 MHz)	−202.47, −192.27
1400–2000	30	−160.47	10	1	18.9	−210.33, −192.02
2000–3000					23.1	−214.18, −195.18
3000–4000	1000	−172.68	10	1	24.8	−202.03, −187.03
4000–8000					30.6	−205.53, −184.47
8000–9000					33.0	−184.31, −184.01
9000–14000					36.1	−195.79, −176.04
14000–18000					38.1	−203.45, −191.55
18000–22000					41.0	−210.06, −191.55

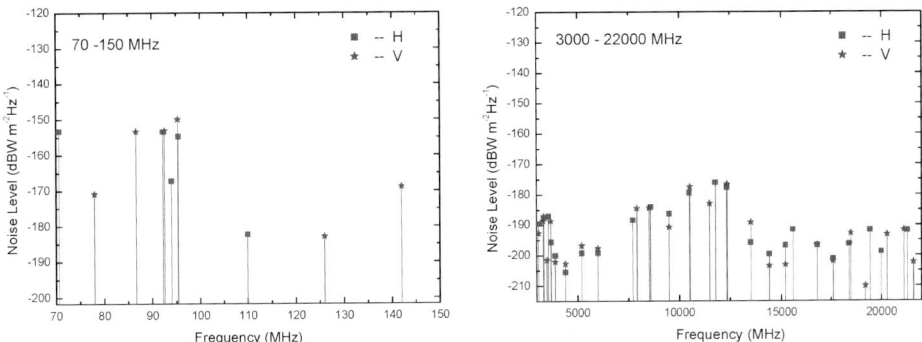

Figure 2. Mode 1 measurement cycle: Partial RFI test results recorded at Dawodang depression in December 2003. The frequency is along the *x*-axis in units of MHz, the power is along the *y*-axis in dB Wm^{-2} Hz^{-1}; horizontal polarization is shown by squares, and vertical polarization is by stars. The noise floor is the expected system sensitivity, i.e., -201.7 and -215 dB Wm^{-2} Hz^{-1}, respectively.

kHz, (calculated) system sensitivity in dB Wm^{-2} Hz^{-1}, length of dwell time in seconds, number of times the experiment was repeated, and the antenna mean gain in dBi respectively; the last column indicates the minimum and maximum flux densities measured in each frequency band. The results for two frequency bands, 70–150 MHz, and 3000–22000 MHz, are illustrated as an example in Figure 2, where the frequency is along the *x*-axis in unit of MHz, and the power is along the *y*-axis in dB Wm^{-2} Hz^{-1}; horizontal polarization is shown by squares, and vertical polarization is shown by stars.

3. Results

The output of the spectrum analyzer is referenced by combining information on the polarization (H or V), the lowest frequency in each swept band in MHz, the measurement cycles, and the azimuth angle with respect to the north in degrees. For example, the output file (data or map) H70–1–0 refers to the first cycle measurement covering frequencies of 70–150 MHz in horizontal polarization, pointing to the north horizon.

The power level of a signal is calculated by

$$P_s \text{ (dBm)} = P_r - G_{\text{LNA}} + L_{\text{C}}, \tag{1}$$

where P_r is the readout of the spectrometer in dBm, G_{LNA} the gain of the LNA in dB, and L_C the cable loss in dB. The field strength is given by

$$E \text{ (dB}\mu\text{v/m)} = P_s + A + 107, \tag{2}$$

where A is the antenna gain factor in dB, and 107 is a constant due to unit transformation.

The spectral flux density is then

$$S\,(\text{dBm Wm}^{-2}\,\text{Hz}^{-1}) = [E - 28.77 - 90]/B_f, \qquad (3)$$

where B_f is the frequency bandwidth of the observation in Hz, the value 28.77 corresponds to the free space impedance (120 pi) in one polarization, and 90 is a constant due to unit transformation.

For a signal-to-noise ratio of 1, a signal occupying bandwidth B_f incident on the antenna with the spectral flux density,

$$S_0 = 2kT_{\text{sys}}/A_{\text{eff}} \qquad (4)$$

where $k = 1.38 \times 10^{-23}$ J/K, T_{sys} is the system temperature (K), A_{eff} is the antenna collecting area (m^2). For any given direction of incidence ψ, A_{eff} is a function of ψ and related to the directivity $D(\psi)$ by

$$A_{\text{eff}}(\psi) = \eta \lambda^2 D(\psi)/4\pi \qquad (5)$$

where η is the antenna efficiency and λ is wavelength. Vertically polarized dipoles and discones have $D \approx 2$ along the horizon (Ambrosini et al., 2003). We have that S_0 along the horizon (i.e., the expected flux at zero-elevation) is approximately

$$S_{0h} = 4\pi k T_{\text{sys}} f^2/(c^2 \eta)/(B_f \tau)^{0.5} \qquad (6)$$

Here $\eta = 0.9$ is assumed, T_{sys} is taken to be the measured noise temperatures, τ refers to integration time(s).

It is noticeable that some of the measured interference signal levels, as listed in the last column of Table II, are lower than that of the instrument sensitivity S_{0h}. This is not surprising, because S_{0h} in the third column in Table II is calculated from Eq. (6) and is approximately consistent with results listed in Table II for the frequency range below 1 GHz where a log-periodic antenna was employed. In our case we took a parabolic antenna for measurements above 1 GHz; it produces a D from 63.2 to 30541.1, or 18.0 to 44.8 in dB. This can explain reasonably well why the measured interference level appears lower.

The calibrated data in our case is tabulated in Excel in a sequence of the frame name, observing frequency, readout P_r of the spectrum analyzer, gain of antenna and LNA, cable loss L_C, antenna gain factor A, field strength E, signal power P_s, RBW, spectral density S in dBm Wm^{-2} Hz^{-1} (and in Jy and dBJy), in order to facilitate examination later. It will be available upon request by the SESC at a IP controlled ftp site. A graduate student from the Guizhou Institute of Technology has been assigned to work on data presentation following the instructions of the RFI protocol.

4. Discussion

Based on our experience in recent years, and considering the limited resources, RFI measurements for both modes 1 and 2 at the candidate Dawodang site are being performed once every 3 months, roughly in May, August, November (2004) and early 2005 by the regional RFI survey group of the RMSG. It should be noted that the last measurements could be simultaneous with a visit of the international RFI calibration group from ASTRON in the Netherlands.

Recently, the 2004 May session was completed by May 30, though a problem remained. The RFI measurements using the parabolic antenna will be too time consuming for mode 2 observations at the higher frequency bands, which are divided into more sub-bands than for mode 1, although it is acceptable to make measurements there for mode 1. Therefore the antenna was replaced by a horn ETS-3115, made by the Lindgren company, with a mean gain (over frequency) of 10 dBi and a beam width of 10–60° in both polarizations working over a frequency range of 1–18 GHz. The antennas we used are shown in Figure 3. Now the question is how to conduct measurements at above 18 GHz for mode 2. One option based on current instruments, to cover the frequency range 18–22 GHz, is to use the parabolic antenna but with reduced pointings, for example, pointing randomly to do all cycles in a day or so. This will have the disadvantage of missing some RFI detections in directions not covered, but might be reasonable with limited resources. Alternatively, a costly way is to order another log-periodic antenna, like the HL 050 proposed by the international RFI calibration group, to cover the wider frequency band of 850 MHz–26.5 GHz. This has to be settled for forthcoming RFI sessions.

It would be helpful and beneficial to individual RFI information collecting groups to have a uniform format for data (raw and calibrated) preservation, and to work out a standard flow chart for data reduction.

Figure 3. Photos of the log periodic antenna HL023A2 and the ETS-3115 horn antenna used by the RMSG survey group in May 2004 at Dawodang.

5. Concluding remarks

There was interference detected at low levels, possibly from radio broadcasting at 70–110 MHz, from beeper call services at 150–160 MHz, and the GSM mobile service around 942 MHz. At the L band there were signals detected at low levels from airplane radar during our test observations. But radio signals at frequencies of 211.8, 942.72 and 965 MHz were identified as radiation from the control computer. In the frequency band 14–18 GHz, three strong interference signals at 14.016, 15.592 and 17.144 GHz were found in all directions, and were tentatively identified as radiation from the LNAs. But further work about the calibration is to be done.

A site measurement report including information requested by the RFI protocol, such as the "spectral occupancy" statistics, will be provided when the whole campaign is finished. All in all we conclude that, due to the remoteness of the proposed region and terrain shielding, the RFI situation in Guizhou province looks quite promising, with relatively little interference found in the region 70 MHz–22 GHz.

Acknowledgments

We would like to thank the technical support provided by the Office of the Guizhou Radio Regulatory Committee. We are very grateful to financial support from the Chinese Academy of Sciences, the Department of Science and Technology of Guizhou province and the Office of the Guizhou Radio Regulatory Committee. We acknowledge grant NKBRSF2003CB716703. Particular thanks should be given to Prof Shengyin Wu of the National astronomical observatories, who translated the "RFI measurement protocol for candidate SKA sites" into Chinese, which aided the observations locally.

References

Ambrosini, R., Beresford, R., Boonstra, A.-J., Ellingson, S. and Tapping, K.: May 23, 2003, RFI measurement protocol for candidate SKA sites, internal report from the SKA International Working Group on RFI Measurements.
Peng, B. and Nan, R.: 1997, IAU Symposium, Kluwer Academic Publishers, p. 93, Vol. 179.
Peng, B., Strom, R. G. and Nan, R.: 2001, in: Cohen R. J. and Sullivan W. T. (eds.), IAU Symposium, p. 276, Vol. 196.